Handbook of Exploration and Environmental Geochemistry

VOLUME 11

Geochemical Anomaly and Mineral Prospectivity Mapping in GIS

HANDBOOK OF EXPLORATION GEOCHEMISTRY

G.J.S. GOVETT (Editor)

1. ANALYTICAL METHODS IN GEOCHEMICAL PROSPECTING
2. STATISTICS AND DATA ANALYSIS IN GEOCHEMICAL PROSPECTING
3. ROCK GEOCHEMISTRY IN MINERAL EXPLORATION
4. REGOLITH EXPLORATION GEOCHEMISTRY IN TROPICAL AND SUB-TROPICAL TERRAINS
5. REGOLITH EXPLORATION GEOCHEMISTRY IN ARCTIC AND TEMPERATE TERRAINS
6. DRAINAGE GEOCHEMISTRY
7. GEOCHEMICAL REMOTE SENSING OF THE SUB-SURFACE

HANDBOOK OF EXPLORATION AND ENVIRONMENTAL GEOCHEMISTRY

M. HALE (Editor)

8. LIFE CYCLE OF THE PHOSPHORIA FORMATION – FROM DEPOSITION TO THE POST-MINING ENVIRONMENT
9. BIOGEOCHEMISTRY IN MINERAL EXPLORATION
10. THE INDIAN OCEAN NODULE FIELD
11. GEOCHEMICAL ANOMALY AND MINERAL PROSPECTIVITY MAPPING IN GIS

Handbook of Exploration and Environmental Geochemistry

VOLUME 11

Geochemical Anomaly and Mineral Prospectivity Mapping in GIS

EMMANUEL JOHN M. CARRANZA

Department of Earth Systems Analysis
International Institute for Geo-Information Science and Earth Observation (ITC)
Enschede, The Netherlands

AMSTERDAM • BOSTON • HEIDELBERG • LONDON
NEW YORK • OXFORD • PARIS • SAN DIEGO
SAN FRANCISCO • SINGAPORE • SYDNEY • TOKYO

ELSEVIER

Elsevier
Radarweg 29, PO Box 211, 1000 AE Amsterdam, The Netherlands
Linacre House, Jordan Hill, Oxford OX2 8DP, UK

First edition 2009

British Library Cataloguing in Publication Data
A catalogue record for this book is available from the British Library

Library of Congress Cataloging-in-Publication Data
Carranza, Emmanuel John M. (Emmanuel John Muico, 1962–)
Geochemical anomaly and mineral prospectivity mapping in GIS / Emmanuel John M. Carranza. — 1st ed.
p. cm. — (Handbook of exploration and environmental geochemistry ; vol. 11)
ISBN 978-0-444-51325-0
1. Geochemistry—Remote sensing. 2. Mines and mineral resources—Remote sensing. 3. Geological mapping—Remote sensing. 4. Geographic information systems. I. Title.
QE515.5.R45C37 2009
622′.13—dc22

2008039925

ISBN: 978-0-444-51325-0

For information on all Elsevier publications
visit our website at books.elsevier.com

Printed and bound in Hungary

08 09 10 11 10 9 8 7 6 5 4 3 2 1

EDITOR'S FOREWORD

In this volume John Carranza not only offers a comprehensive review of the current state-of-the-art of processing geochemical data, their integration with complementary geodata sets and multivariate data analysis using spatial statistics to create maps enhanced for mineral exploration, but also brings the Handbook series to something of a milestone. This marks the first volume in which the topic of an earlier volume (Vol. 2) is in effect revisited and updated; though the approach and format are – appropriately – entirely fresh.

Part I of the volume (chapters 1-2) introduces the concepts and methods of handling spatial data in a geographical information system for the purpose of predictive modeling for mineral exploration. Part II (chapters 3-5) looks in detail at geochemical data and how they are analyzed, classified, synthesized and attributed to catchment basins prior to their application in predictive modeling. Part III (chapters 6-8) begins by emphasising the importance of additional relevant spatial information and culminates in predictive modeling of mineral prospectivity by means of a range of knowledge-driven and data-driven methods. Throughout the volume there is a wealth of well-illustrated real-world examples. The author admirably demonstrates modern approaches to data analysis and interpretation in mineral exploration in ways which exploration professionals can appreciate and adapt to their exploration programmes.

This volume is the first in the series to go to press after the death in 2007 of John S Webb, whose achievements and influence in exploration and environmental geochemistry did much to lay the foundations for the series. In the 1950s he established the Geochemical Prospecting Research Centre at Imperial College, London, where his pioneering work in exploration geochemistry was soon extended to regional geochemical mapping and environmental geochemistry (leading to the centre being renamed the Applied Geochemistry Research Group). Many destined later to be closely associated with the Handbook series were Webb's PhD students or colleagues at Imperial College: K Fletcher (Vol. 1); Richard Howarth (Vol. 2); Gerry Govett (Vol. 3, series founder and series editor Vols. 1-7); Charles Butt (Vol. 4); Martin Hale (Vols. 6-7 and series editor Vols. 8-11); and Colin Dunn (Vol. 9). In authoring Volume 11, John Carranza, being a former PhD student (and now professional colleague) of Martin Hale, has extended this tradition into the third generation.

Having in some ways brought the series full circle, Volume 11 also shows that geochemical data now constitute just one of several types of data brought together and analyzed together using geographical information systems to yield information to guide mineral exploration. The Handbook series has amply fulfilled Gerry Govett's original vision of thoroughly documenting the value of exploration and environmental geochemistry. Whilst the value of geochemical data remain undiminished in mineral exploration, John Carranza most eloquently shows here that the time has come when this value is best realized when geochemical data are part of wider armoury of complementary geodata sets and modeling techniques.

MARTIN HALE
The Netherlands
May 2008

PREFACE

Twenty-five years ago, when the second volume of the Handbook of Exploration Geochemistry was published (Howarth, 1983), computers were just becoming useful tools in the analysis of mineral exploration data sets but mapping of geochemical anomalies and prospective areas still usually involved overlaying transparent geochemical map(s) and a geological map on a light table. The late 1980s through the 1990s saw rapid and far-reaching developments in quantitative techniques for mapping geochemical anomalies and mineral prospectivity due to the substantial improvements in the efficiency and availability of computer hardware and software (Agterberg, 1989) including geographic information system (GIS) technology (Burrough, 1987; Bonham-Carter and Agterberg, 1990; Maguire et al., 1991). Two textbooks and several papers published in exploration-related literature have explained and documented various GIS-aided and/or GIS-based methods for analysis of multiple geoscience spatial data sets in order to derive and synthesise pieces of geo-information that are pertinent to the decision-making process at every scale of target generation in mineral exploration. In "*Geographic Information Systems for Geoscientists: Modelling with GIS*", Bonham-Carter (1994) introduced ideas and methods of spatial analysis and modeling in GIS, especially those that are useful for characterising spatial associations between a set of geo-objects of interest (e.g., deposit-type locations) and individual sets of (indicative) spatial features (e.g., geochemical anomalies) in order to develop predictive models of the former set. In "*Information Synthesis for Mineral Exploration*", Pan and Harris (2000) introduced various methods for optimal assimilation of specific pieces of geo-information extracted from various spatial data sets in order to derive optimised geo-information for decision-making in mineral exploration. Nowadays, mapping of geochemical anomalies and/or prospective areas involves stacking digital geochemical and geological maps on top of each other on an electronic light table (i.e., in a GIS).

The objective of this book is to document, survey and demonstrate various GIS-aided and/or GIS-based techniques for mapping of geochemical anomalies and prospective areas during the target generation phase of mineral exploration. This volume consists of three parts, all centred on the theme *predictive modeling* or *mapping* and built upon particular notions and/or methods presented in the aforementioned textbooks and in various papers in exploration-related literature. Built upon the natural link between mapping of exploration targets and GIS, the chapters in Part I review and couple the concepts of (1) mapping geochemical anomalies and mineral prospectivity and (2) spatial data models, management and operations in a GIS. Built upon the remarks of Reimann (2005, pp. 369) that "*Although GIS techniques appear to have simplified geochemical mapping tremendously, most systems do not allow for fast and correct class selection for mapping…*", the chapters in Part II demonstrate GIS-aided and GIS-based methods for analysis of robust thresholds in mapping of geochemical anomalies. Built upon the notion that locations of mineral deposits of the type sought are intrinsic samples of mineralised landscapes, which are results of interactions of geological processes, the chapters in Part III explain GIS-aided and GIS-based techniques for spatial data analysis and geo-information synthesis for conceptual modeling and predictive modeling of mineral prospectivity. The essence of this book is, therefore, the prudent (thus, not black box) utilisation of GIS in mapping of geochemical anomalies and prospective areas through the application of understanding of relevant earth systems or processes that led to the formation (and/or alteration) of these geo-objects.

Each chapter in this volume is meant to be self-contained. The chapters in Parts II and III are, however, coherently linked by a common case study. The concepts and methods described here are demonstrated with real exploration data sets. Although the geochemical data used here represent Earth materials most commonly sampled in reconnaissance exploration surveys (i.e., stream sediments) and the geological data sets used here represent 'data-poor' situations of mapping exploration targets for epithermal Au deposits, the concepts and methods described here apply equally to geochemical data from different sampling media and to 'data-rich' situations of mapping exploration targets for various types of mineral deposits. In addition, whilst there is neither reference to nor endorsement of any GIS software throughout this volume, the concepts and methods described in every chapter are generic such that they are readily implemented with or in any GIS software. This volume is thus intended to be an instructional textbook and general reference manual for exploration geochemists and/or exploration geologists, who are enthusiastic and already possess skills in applying GIS or who are interested in applying GIS. It is also hoped that geoscience academics and graduate students not only in the knowledge fields of geo-resource exploration but also in the knowledge fields of geo-hazard mapping and/or geo-environmental characterisation would find the concepts and methods described in this volume useful in their work.

I thank the International Institute for Geo-Information Science and Earth Observation (ITC) for resources and a pleasant environment for working (both teaching and doing research in geological predictive modeling) that allowed me to write this book. I thank also my graduate (PhD/MSc) students, from 2002 to the present, with whom I have developed some of the ideas presented in this volume. Most of all, I thank Professor Martin Hale, for coaching me during the years I was a graduate (MSc to PhD) student of mineral exploration at ITC and TU Delft, for continuing to work with me thereafter, for inviting me to write a volume for the series of Handbook of Exploration and Environmental Geochemistry and for editing this volume. The errors in this volume remain mine.

E.J.M. CARRANZA
Enschede, The Netherlands
May 2008

CONTENTS

Editor's Foreword ... V
Preface ... VII

PART I. MODELS IN MINERAL EXPLORATION AND GIS

Chapter 1. Predictive Modeling of Mineral Exploration Targets ... 3
  Introduction ... 3
  What is Predictive Modeling? ... 4
    Approaches to predictive modeling ... 5
    Types of predictive modeling ... 6
  Predictive Modeling of Significant Geochemical Anomalies ... 9
  Predictive Modeling of Mineral Prospectivity ... 12
  Predictive Modeling with a GIS ... 18
    Data analysis ... 18
    Data integration ... 19
    Visualisation of spatial data or geo-information ... 20
  Summary ... 21

Chapter 2. Spatial Data Models, Management and Operations ... 23
  Introduction ... 23
  Models of Spatial Data ... 23
    Geo-objects ... 23
    Vector Model ... 24
    Raster Model ... 27
    Attribute Data Models ... 28
  Management of Spatial Data ... 29
    Spatial Data Capture ... 29
    Spatial Database Creation ... 31
  Operations on Spatial Data ... 33
    Spatial query and selection ... 33
    Classification and re-classification ... 35
    Measurements ... 36
    Transformations ... 37
    Neighbourhood operations ... 42
    Map overlay operations ... 44
  Summary ... 47

PART II. GEOCHEMICAL ANOMALY MAPPING

Chapter 3. Exploratory Analysis of Geochemical Anomalies 51
Introduction 51
Exploratory Data Analysis 53
Graphical tools in EDA 53
Boxplot and 5-number statistics of univariate data 54
Classification of uni-element geochemical data 57
Standardisation of classified uni-element geochemical data 57
Mapping of classified uni-element geochemical data 59
Applications of GIS in EDA 62
Management of spatial data 62
Operations on spatial data 62
Visualisation of spatial data or geo-information 62
Case study 63
Study area and spatial data 63
Analysis of uni-element data distributions 65
Analysis of uni-element threshold values and anomalies 72
Analysis of inter-element relationships 75
Analysis and mapping of multi-element associations 78
Summary 83

Chapter 4. Fractal Analysis of Geochemical Anomalies 85
Introduction 85
Geochemical Landscapes as Fractals 86
What are fractals? 86
Fractal geometry 86
Perimeter-area relationship in geochemical patterns 88
Multifractality of geochemical landscapes 92
The Concentration-Area Method for Threshold Recognition 92
Application of GIS in the Concentration-Area Fractal Method 94
Generation and discretisation of geochemical surfaces 95
Attribute table operation for area calculations 96
Case Study 98
Creation and classification of uni-element geochemical surface maps 98
Analysis of uni-element threshold values and anomalies 100
Analysis and mapping of anomalous multi-element signature 107
Synthesis of multi-element associations 110
Conclusions 113

Chapter 5. Catchment Basin Analysis of Stream Sediment Anomalies 115
Introduction 115
Estimation of Local Uni-element Background due to Lithology 117
Multiple regression analysis 117
Analysis of weighted mean uni-element concentrations due to lithology 118
Dilution Correction of Uni-element Residuals 120
Analysis of Anomalous Multi-element Signatures 122
Application of GIS in Catchment Basin Analysis 125
Creation of sample catchment basins 125
Measuring areas and areal proportions of lithologic units in catchment basins 127
Estimation of local background uni-element concentrations 127
Correction for downstream dilution 127
Analysis and classification of geochemical anomalies 128
Case Study 128
Estimation of local uni-element background 129
Derivation and standardisation of uni-element residuals 133
Correction for downstream dilution and classification of anomalies 136
Modeling of multi-element anomalies via principal components analysis 136
Screening of multi-element anomalies with fault/fracture density 141
Discussion and Conclusions 142

PART III. MINERAL PROSPECTIVITY MAPPING

Chapter 6. Analysis of Geologic Controls on Mineral Occurrence 147
Introduction 147
Spatial Distribution of Mineral Deposits 148
Point pattern analysis 148
Fractal analysis 152
Fry analysis 155
Knowledge synthesis and results of spatial analysis 158
Spatial Association of Mineral Deposits and Geologic Features 162
Distance distribution method 162
Distance correlation method 173
Synthesis and discussion of results 182
Conclusions 186

Chapter 7. Knowledge-Driven Modeling of Mineral Prospectivity 189
Introduction 189
General Purpose Applications of GIS 191
Modeling with Binary Evidential Maps 194
Boolean logic modeling 196
Binary index overlay modeling 199

Calibration of predictive modeling with binary evidential maps 204
Modeling with Multi-class Evidential Maps 205
Multi-class index overlay modeling 207
Fuzzy logic modeling 210
Evidential belief modeling 227
Calibration of predictive modeling with multi-class evidential maps 235
Wildcat Modeling of Mineral Prospectivity 236
Conclusions 246

Chapter 8. Data-Driven Modeling of Mineral Prospectivity 249
Introduction 249
Selection of Suitable Unit Cell Size for Modeling 253
Selection of Coherent Deposit-type Locations for Modeling 260
Analysis of mineral occurrence favourability scores at deposit-type locations 261
Analysis of coherent deposit-type locations 263
Cross-validation of Data-driven Models of Prospectivity 270
N–n strategies 271
Deposit-type classification strategies 271
Spatial subdivision strategies 272
Other strategies of cross-validation 272
Evidential Belief Modeling of Mineral Prospectivity 273
GIS-based data-driven estimation of EBFs 275
Calibration of data-driven estimation of EBFs 279
Integration of data-driven EBFs 285
Case study application of data-driven EBFs 286
Discriminant Analysis of Mineral Prospectivity 293
Basic principles and assumptions of linear discriminant analysis 293
GIS-based spatial evidence representation for discriminant analysis 295
Case study application of discriminant analysis 295
Discussion and Conclusions 308

References 311
Online Sources 337
Author Index 339
Subject Index 347

PART I. MODELS IN MINERAL EXPLORATION AND GIS

Geochemical Anomaly and Mineral Prospectivity Mapping in GIS
by E.J.M. Carranza
Handbook of Exploration and Environmental Geochemistry, Vol. 11 (M. Hale, Editor)

Chapter 1

PREDICTIVE MODELING OF MINERAL EXPLORATION TARGETS

INTRODUCTION

Mineral exploration endeavours to find mineral deposits, especially those with commercially viable concentrations of minerals or metals, for mining purposes. It has four phases, namely (1) area selection, (2) target generation, (3) resource evaluation and (4) reserve definition. Area selection defines *permissive* regions where mineral deposits of the type sought plausibly exist based on knowledge of environments at or near the surface of the Earth's crust where the geological processes (e.g., plate tectonics) are or were favourable for mineral deposit formation (Singer, 1993). Target generation demarcates, within permissive regions, *prospective* areas for further investigations until mineral deposits of interest are discovered based on exploration models for the deposit-type sought and on relevant thematic geoscience (geological, geochemical and geophysical) data sets. Resource evaluation estimates grade and tonnage of specific minerals or metals in discovered mineral deposits based largely on systematic drilling. Reserve definition classifies the various parts of mineral deposits as ore reserves (proved, probable) or mineral resources (measured, indicated, inferred) based on economic and technical feasibility analysis. This volume is concerned with only the target generation phase in mineral exploration.

Target generation is a multi-stage mapping activity from regional-scale to local-scale. Every scale of target generation involves collection, analysis and integration of various thematic geoscience data sets in order to extract pieces of spatial geo-information, namely (a) geological, geochemical and/or geophysical anomalies associated with mineral deposits of the type sought and (b) prospective areas defined by intersections of such anomalies. An example of a *geological anomaly* is hydrothermal alteration, although it may not necessarily be accompanied by mineral deposits. A *geophysical anomaly* is a variation from normal background patterns of measured physical properties of the Earth's upper crust (e.g., magnetism), which can be attributed to localised near-surface or subsurface materials such as metallic mineral deposits. A *geochemical anomaly* is a departure from the geochemical patterns that are normal for a given area. It can represent either geogenic (i.e., natural) or anthropogenic (i.e., industry-induced) enrichment in one or more elements in Earth materials. In mineral exploration, geochemical anomalies associated with mineral deposits are called *significant anomalies*, whereas geochemical anomalies associated with other natural processes or anthropogenic processes are called non-significant anomalies. Because not every anomaly is associated, genetically and/or spatially, with mineral deposits of the type sought, intersecting or integrated anomalies of various types are of interest in target generation. The process of analysing and integrating such pieces of spatial geo-information is called *predictive modeling*. This volume is further concerned with predictive modeling of only geochemical anomalies and prospective areas.

This chapter explains the concepts of (a) predictive modeling, (b) predictive modeling of geochemical anomalies and prospective areas and (c) application of a geographic information system (GIS) in predictive modeling of geochemical anomalies and prospective areas. A GIS consists of computer hardware, computer software, geographically-referenced or spatial data sets and personnel.

WHAT IS PREDICTIVE MODELING?

To understand the concepts of predictive modeling of geochemical anomalies and prospective areas via applications of GIS, it is imperative to define and understand what *model* means. The Wiktionary (Wikimedia Foundation, 2007) defines model as "*a simplified representation (usually mathematical) used to explain the workings of a real world system or event*". The Oxford English Dictionary (Oxford University Press, 2007) defines model as "*a simplified or idealized description or conception of a particular system, situation, or process, often in mathematical terms, that is put forward as a basis for theoretical or empirical understanding, or for calculations, predictions, etc.; a conceptual or mental representation of something*". The Glossary of Geology (American Geological Institute, 2007) defines model as "*a working hypothesis or precise simulation, by means of description, statistical data, or analogy, of a phenomenon or process that cannot be observed directly or that is difficult to observe directly. Models may be derived by various methods, e.g. by computer, from stereoscopic photographs, or by scaled experiments*".

Based on the definitions of a model, predictive modeling can be defined as "making descriptions, representations or predictions about an indirectly observable and complex real-world system via (quantitative) analysis of relevant data". It involves a target variable of interest, which is usually the behaviour (e.g., presence or absence) of an indirectly observable and complex real-world system (e.g., mineralisation), and a number of explanatory or predictor variables or properties that are directly observable or measurable as well as considered to be inter-related with each other and related to that system. Predictive modeling is therefore based on (a) inter-relationships amongst predictor variables, which may reveal patterns related to the target variable and (b) relationships between the target and predictor variables. The latter means that some quantity of data associated directly with the target variable must be available in order to create and to validate a predictive model. A predictive model is a temporal snap-shot of the system of interest, meaning that it embodies the knowledge and/or data sets used at the time of its creation and thus it must be updated whenever new knowledge and/or relevant data sets become available. Therefore, any cartographic representation of an indirectly observable and complex real-world system is a predictive model.

Predictive modeling is relevant to mapping of geochemical anomalies and prospective areas because these real-world systems are complex and indirectly observable. The objective of predictive modeling of geochemical anomalies and/or prospective areas is to represent or map them as discrete spatial entities or *geo-objects*, i.e., with conceivable boundaries. This entails various forms of data analysis in order to derive and integrate pieces of information that allow description, representation or prediction of geochemical anomalies and/or prospective areas. The term *representation* or *prediction* is equivalent to *mapping*, which embodies results of analyses and interpretations of various relevant geoscience spatial data sets according to hypotheses or propositions about geochemical anomalies and/or prospective areas. Therefore, a map of geochemical anomalies and/or prospective areas is a predictive model of where mineral deposits of interest are likely to exist.

Approaches to predictive modeling

There are two approaches to predictive modeling – induction and deduction. Induction is the process of making generalisations about particular instances in a set of observations or data. A generalisation is derived by studying patterns in a data set. Deduction – the inverse of induction – is the process of confirming particular instances based on a generalisation about patterns in a set of observations or data. The confirmation of particular instances is made by testing a generalisation against every observation or datum. The distinction between induction and deduction can be illustrated via a hypothetical example in Fig. 1-1 (cf. Bonham-Carter, 1994, pp. 180). An initial visual analysis of the distributions of Fe contents in soil vis-à-vis the distributions of lithologic units (Fig. 1-1A), may lead one to hypothesise that soils in areas underlain by basalts have higher Fe contents than soils in areas underlain by other lithologic units. The hypothesis may be supported by a simple linear model of Fe contents generally decaying with distance from the basalt unit (Fig. 1-1B). The map of spatial distributions of Fe contents and the simple linear model may further lead one to make a generalisation that basalts influence Fe contents in soils and that Fe data in soils are useful aids to lithologic mapping. Up to this point, one has performed an induction because a generalisation was made based on particular instances in a set of observations or data. To test the generalisation, because there is always an 'exception to the rule', one has to perform deduction. To do so, one may use the simple linear model to predict Fe contents in soil as function of distance to basalt and then map the spatial distributions of residual Fe contents (Fig. 1-1C). The presence of enriched Fe in soils (i.e., positive residuals) in certain parts of the area may lead one to hypothesise that there are unmapped basalt units. Confirmation of this hypothesis requires re-visiting the sample sites (i.e., every particular instance), which could result in updating of the lithologic map (Fig. 1-1C).

Induction and deduction are therefore complementary to each other, such that switching from induction to deduction or vice versa at intermediate steps in predictive modeling could provide better description, understanding and discovery of the system of interest. Thus, despite the approach in the preceding hypothetical example, it is not necessary to initiate predictive modeling with induction. The evolution of scientific

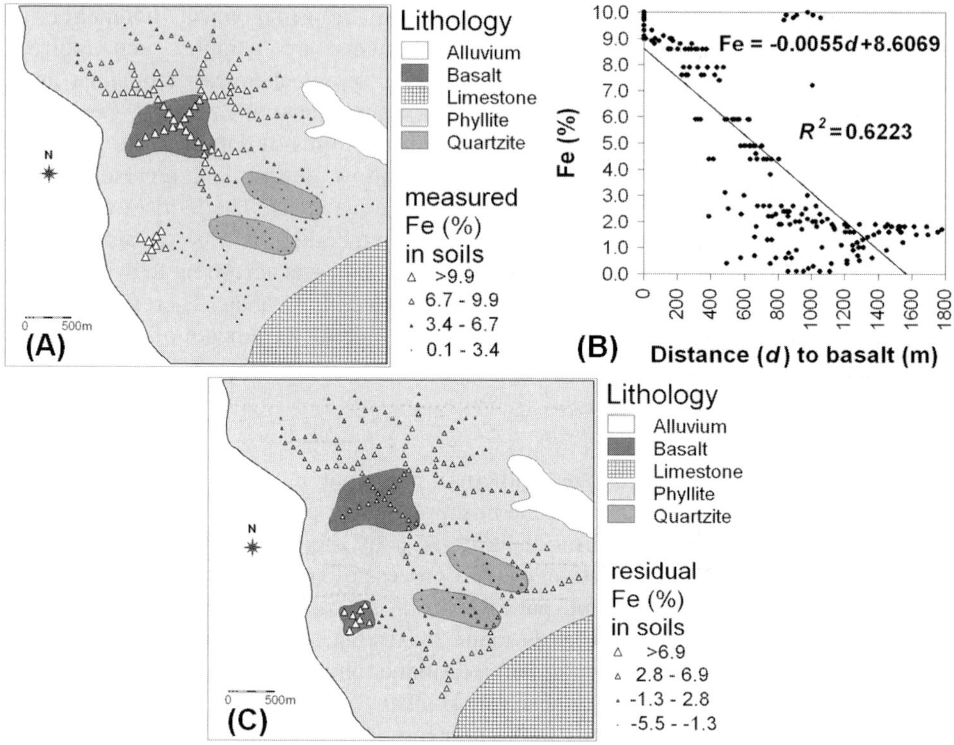

Fig. 1-1. (**A**) Existing lithologic map and measured Fe (%) contents in ridge-and-spur soil samples. The measured Fe contents in soil generally decrease with distance from the basalt. (**B**) Best-fit line model for measured Fe contents in soil and distance to the basalt. (**C**) Updated lithologic map after field investigation of residuals (measured-predicted) of Fe in soil.

knowledge has led to theories or generalisations about dispersion of elements and surface geochemical expressions of mineralisation (Bradshaw, 1975; Kauranne, 1975; Lovering and McCarthy, 1978; Butt and Smith, 1980; Smith, 1987) and genesis of mineral deposits (e.g., Lindgren, 1933; Pirajno, 1992; Evans, 1993; Richards and Tosdal, 2001; Robb, 2004). Therefore, many cases of predictive modeling involved in target generation commence with deduction, although switching to induction may be necessary at intermediate steps until a final predictive model is obtained.

Types of predictive modeling

This section reviews the types of predictive modeling that are relevant to mineral exploration, especially in the target generation phase. There is no generally accepted classification of types of predictive modeling of Earth systems such as geochemical anomalies and prospective areas. However, based on the way inter-predictor relationships and target-predictor relationships are described or represented, two types of predictive modeling – *mechanistic* and *empirical* – and hybrids of these two types can be distinguished (cf. Harbaugh and Bonham-Carter, 1970).

Mechanistic modeling applies fundamental or theoretical knowledge of individual predictor variables (i.e., processes) and their interactions in order to predict or understand the target variable of interest. Mechanistic modeling is therefore equivalent to *theoretical* modeling. Mechanistic modeling relies on mathematical equations to describe the interactions of processes that control the behaviour of system of interest. It applies relevant physical laws and is often based on laboratory studies, field experiments and physical models. Solving the theoretical equations in mechanistic modeling can be complex and require application of generalising or simplifying assumptions (e.g., simplified geometry, homogeneity, idealised initial conditions and boundary conditions). Mechanistic modeling therefore invariably follows a deductive approach. The predictive capability of a mechanistic model can be determined and then improved via probabilistic uncertainty analyses to investigate sensitivity of prediction to one or more predictor variables or assumptions.

There are two sub-types of mechanistic modeling – *deterministic* and *stochastic*. Deterministic modeling applies mathematical representations (e.g., differential equations) of the processes that control the behaviour of system of interest. It makes definite predictions of quantities (e.g., metal concentrations) without considering any randomness in the distributions of the variables in the mathematical equations. Stochastic modeling also applies mathematical representations of the processes that control the behaviour of system of interest, but it considers the presence of some random distribution in one or more predictor variables and in the target variable. Stochastic modeling therefore does not result in single estimates of the target variable but a probability distribution of estimates, which is derived from a large number of simulations (stochastic projections), reflecting random distributions in the predictor and target variables. Purely deterministic modeling has been rarely, if not never, used in mineral exploration, except in laboratory studies of mineral deposit formation (e.g., L'Heureux and Katsev, 2006). Purely stochastic modeling is seldom used in the target generation phase of mineral exploration, but it has been applied, however, in the resource estimation and reserve definition phases of mineral exploration (e.g., Sahu, 1982; Harris, 1984; Sahu and Raiker, 1985).

An interesting application of stochastic modeling is where the target variable sought represents fractal geo-objects as a result of stochastic rather than deterministic processes. A fractal geo-object is one which can be fragmented into various parts, and each fragment has similar geometry as the whole geo-object (Mandelbrot, 1983). Geochemical dispersion patterns and spatial distributions of mineral deposits are postulated to be fractals (Bölviken et al., 1992; Agterberg et al., 1993b). Agterberg (2001) and Rantitsch (2001) have demonstrated the utility of stochastic modeling to examine the fractal geometry of geochemical landscapes, as conventional geostatistical methods are not able to do so when the spatial variability of geochemical anomalies exceeds the spatial resolution (i.e., sampling density) of geochemical data sets. Hybrids of stochastic modeling (not based on assumption of fractals) and quantitative empirical modeling (see below) have been applied, however, in mapping of significant geochemical anomalies (e.g., Singer and Kouda, 2001; Agterberg, 2007) and prospective areas (e.g., Agterberg, 1974; Pan et al., 1992; Grunsky et al., 1994).

In contrast to mechanistic modeling, empirical modeling is appropriate when the underlying geochemical and/or physical processes that control the behaviour of the system of interest are insufficiently or indirectly known. Methods for empirical modeling do not take into account interactions of such processes in a mathematical sense as in mechanistic modeling. Instead, they characterise or quantify the influence of one or more of such processes on the behaviour of the system of interest via empirical model equations. Empirical modeling is therefore equivalent to *symbolic* modeling and generally follows an inductive approach. The equations in empirical modeling are constructed to define relationships between the target variable and a number of predictor variables representing processes in order to describe or symbolise the observed or predicted behaviour of the system of interest. Empirical modeling requires substantial amounts of data of both the target and predictor variables in order to quantify accurately their relationships. In terms of sufficiency of data of the target variable, there are two sub-types of empirical modeling – *quantitative* and *qualitative*.

Quantitative empirical modeling is appropriate when data of the target variable are sufficient to obtain, say, statistically significant results. Data of the target variable are usually divided into a training set and a testing set. Based on a training set, relationships between the target and predictor variables are quantified and then used for prediction. Methods for quantitative empirical modeling can be *statistical*, *probabilistic* or *mathematical*. The quality of a quantitative empirical model is described by its goodness-of-fit to data in a training set and its predictive ability against data in a testing set. In mapping of prospective areas, quantitative empirical modeling is also known as *data-driven* modeling. In contrast, qualitative or *heuristic* modeling is appropriate when data of the target variable are insufficient or absent. In qualitative modeling, relationships between the target and predictor variables are defined based on expert opinion. Qualitative modeling thus seems to follow a deductive approach. The quality of a qualitative empirical model can be described by its predictive ability against available (albeit insufficient) data of the target variable. In mapping of prospective areas, qualitative empirical modeling is also known as *knowledge-driven* modeling.

Based on the preceding discussion, further distinctions between mechanistic modeling and empirical modeling can be made as follows. Whereas mechanistic modeling attempts to characterise and understand the fundamental or theoretical processes that control the behaviour of the system of interest, empirical modeling attempts to depict quantitatively the influence of well-understood processes on the behaviour of the system of interest. Therefore, mechanistic modeling strives to derive realistic predictive models, whereas empirical modeling endeavours to derive approximate yet plausible predictive models. Furthermore, mechanistic modeling is *dynamic*, because it can contain the time variable in the mathematical equations especially in deterministic modeling; whereas empirical modeling usually ignores the time variable and is therefore *static*. Predictive modeling of geochemical anomalies, however, can be dynamic, especially modeling of drainage and soil geochemical anomalies (e.g., Tardy et al., 2004). Predictive modeling of prospective areas, however, is usually static. Finally, mechanistic and empirical modeling can be one-dimensional (1-D), two-dimensional (2-D), three-dimensional (3-D) or four-dimensional (4-D). The following sections review the concepts and practises of only 2-D mapping or predictive modeling of significant geochemical anomalies and prospective areas, which comprise the scope of this volume.

PREDICTIVE MODELING OF SIGNIFICANT GEOCHEMICAL ANOMALIES

Exploration geochemical data are generated in mineral exploration programmes through systematic measurements of one or more chemical properties of samples of certain Earth materials (sediment, water, soil, rock, vegetation, gas, etc.). Detailed explanations of concepts and practises of sampling different Earth materials for mineral exploration can be found in Levinson (1974), Rose et al. (1979), Govett (1983), Butt and Zeegers (1992), Kauranne et al. (1992), Hale and Plant (1994) and Hale (2000). In most cases, the chemical property determined from each sample is the concentration of one or more elements. Such element content determinations aim to find areas with enriched concentrations of one or more *indicator* or *pathfinder* elements, which could betray the presence of mineral deposits of the type sought.

The normal concentration of an element in non-mineralised Earth materials is referred to as *background*. It is more realistically viewed as a range of values rather than an absolute value because the distribution of any element in any particular Earth material is rarely uniform and varies considerably from one type of Earth material to another and from one location to another. Therefore, uni-element background is determined by stochastic, empirical or hybrid stochastic-empirical modeling techniques whenever a new area is explored for certain types of mineral deposits. The upper limit of the range of background values is called the *threshold*. Uni-element concentrations greater than the threshold are collectively called *anomaly*. Anomalous uni-element concentrations that indicate presence of mineral deposits are called *significant anomalies*.

The spatial variations of concentrations of any element can be called geochemical landscape; therefore, background and anomaly both occupy space. Areas characterised by normal concentrations of indicator or pathfinder elements are called background, whereas areas characterised by concentrations of at least one indicator or pathfinder element greater than the threshold are called anomalies. In an exploration area, anomalies can be delineated once threshold values in individual uni-element data sets are determined. The different methods used traditionally in modeling of geochemical thresholds can be classified broadly into two categories of analysis, namely (Levinson, 1974; Rose et al., 1979; Howarth, 1983b): (1) analysis of frequency distributions of uni-element concentrations; and (2) analysis of multi-element associations. In the latter, the concept of threshold in uni-element concentration data is extended to data values derived from analysis and synthesis of multi-element data sets by application of multivariate statistical methods.

The traditional methods applied in modeling of uni-element geochemical anomalies include (Levinson, 1974; Govett et al., 1975; Rose et al., 1979; Sinclair, 1983; Howarth and Sinding-Larsen, 1983): (a) comparison of data from the literature; (b) comparison of data with results of an orientation geochemical survey; (c) graphical discrimination from a data histogram; (d) calculation of threshold as the sum of the mean and some multiples of the standard deviation of data; (e) plotting cumulative frequency distributions of the data and then partitioning the data into background and anomalous populations; (f) (e) constructing normal probability plots of the data and then partitioning the data into background and anomalous populations and (f) recognition of clusters of anomalous samples when data are plotted on a map.

The traditional methods applied in modeling of multi-element geochemical anomalies include (Rose et al., 1979; Howarth and Sinding-Larsen, 1983): (a) stacking uni-element anomaly maps of the same scale on top of each other on a light table and then outlining areas with multiple intersections of uni-element anomalies; (b) determining inter-element correlations, usually by calculating Pearson correlation coefficient; (c) recognising and quantifying multi-element associations based on their correlations; and (d) mapping of quantified multi-element association scores. Applications of certain methods to model multi-element geochemical anomalies invariably require computer processing because such techniques are prohibitively difficult and time-consuming to perform manually. In addition, conventional methods for recognising and mapping multi-element associations vary depending on whether a-priori knowledge is available or not about certain controls on geochemical variations.

Principal components analysis and cluster analysis are conventional methods in identifying and mapping multi-element associations without a-priori information about controls on geochemical variations or which samples were collected at or near mineralised zones. In more recent times, fuzzy cluster analysis based on the fuzzy set theory (Zadeh, 1965) has been demonstrated to be more advantageous than the conventional cluster analysis in identifying boundaries of anomalous multi-element clusters and quantifying degrees of membership of samples to every fuzzy cluster (Yu and Xie, 1985; Vriend et al., 1988; Kramar, 1995; Rantitsch, 2000).

In situations where some controls on geochemical variations are known a-priori, a simple conventional approach to recognise and map significant geochemical anomalies is to apply element ratios (Plimer and Elliot, 1979; Brand, 1999; Garrett and Lalor, 2005). When scavenging effects of Fe-Mn oxides on variations of certain elements are known a-priori or recognised during the analysis of inter-element correlations, a usual approach is to apply regression analysis to estimate concentrations of certain elements as a function of Fe and Mn contents in order to interpret if geochemical residuals depict anomalous patterns or not. When it is known a-priori that some samples were collected at or near mineralised zones, such samples can be used as a training set in discriminant analysis to determine if the other samples are associated with mineralisation or not. Alternatively, logistic regression analysis can be performed by using binary data [0,1] of mineral deposit occurrence as target variable and the multi-element data as predictor variables to derive predicted mineral deposit occurrence scores for each sample.

Traditionally, modeling and mapping of significant geochemical anomalies were based mostly on geochemical data sets, probably because in the past (say, before the 1970s when the development of GIS was at its infancy) many outcropping mineral deposits were still undiscovered such that their associated geochemical anomalies are obvious in the data sets. However, not all geochemical anomalies indicate the presence of mineral deposits. Mineral deposits are themselves significant geochemical anomalies. In more recent times, it can be surmised that most, if not all, outcropping mineral deposits have already been discovered. In addition, mining and other industries have already altered the geochemical landscapes. Therefore, an obvious geochemical anomaly may be geogenic and significant (i.e., associated with mineral deposits), geogenic but non-significant (i.e., related to certain high background non-mineralised lithologies), or anthropogenic and non-significant (e.g., due to industrial contamination). In contrast, a subtle geochemical anomaly or absence of a geochemical anomaly does not necessarily indicate absence of mineral deposits, but may suggest either that weathering and erosion rates were insufficient to mobilise and disperse metals from mineralised sources or that mineral deposits are 'blind' or 'buried' (i.e., non-outcropping or unexposed to the surface). It is clear, therefore, that effective modeling of significant geochemical anomalies requires that all available relevant exploration data sets or pieces of geo-information are analysed and integrated in the light of fundamental or theoretical principles of exploration and environmental geochemistry.

Many modern methods for modeling of significant geochemical anomalies subsequently consider not only geochemical data frequency distributions but also other types of geo-information from uni-element or multi-element geochemical data sets (e.g., Grunsky and Agterberg, 1992; Bellehumeur et al., 1994; Cheng et al., 1996, 1997, 2000; Cheng, 1999b; Harris et al., 2001a; Karger and Sandomirsky, 2001), namely: spatial variability and correlations; geometry (shape and orientation, as well as fractal dimensions) of anomalies; and scale independence of anomalies. Consideration of the scale independence of anomalies aims to reduce the effects of sampling density and geo-analytical techniques on the spatial distributions of geochemical anomalies. Considerations of the spatial variability and correlations and the geochemical properties of geochemical anomalies aim to enhance anomaly patterns that reflect controls by geological processes and thus facilitate recognition of significant geochemical anomalies. Enhancement of geochemical anomaly patterns that reflect controls by geological processes can also be addressed by integrating geochemical data with other types of relevant spatial data or pieces of spatial geo-information that explicitly represent individual processes. For example, significant drainage geochemical anomalies can be enhanced and thus recognised by integrating geochemical data with area of drainage catchment basins (Polikarpochkin, 1971; Hawkes, 1976; Moon, 1999), lithologic units (Rose et al., 1970; Bonham-Carter and Goodfellow, 1986; Bonham-Carter et al., 1987), proximity to faults (Carranza and Hale, 1997); drainage sinuosity (Seoane and De Barros Silva, 1999), and stream order (Carranza, 2004a). Significant soil geochemical anomalies can be enhanced and thus recognised by integrating soil geochemical data chiefly with lithology (e.g., Lombard et al., 1999; Garrett and Lalor, 2005; Jordan et al.,

2007) and sometimes with topographic elevation (Grunsky and Smee, 1999; Grunsky, 2006). Subtle but significant geochemical anomalies can also be enhanced and recognised by applications of selective or partial extraction geo-analytical techniques to measure element concentrations (e.g., Smee, 1998); however, geo-analysis of element contents in geochemical samples is a topic that is beyond the scope of this volume. We now turn to modeling of mineral prospectivity by integrating mapped evidential features such as significant geochemical anomalies.

PREDICTIVE MODELING OF MINERAL PROSPECTIVITY

The term *mineral prospectivity* refers to the chance or likelihood that mineral deposits of the type sought can be found in a piece of land. It is similar to the terms *mineral potential* and *mineral favourability*, which refer to the chance or likelihood that mineral deposits of the type sought are contained in a piece of land. The terms mineral prospectivity, mineral potential and mineral favourability are therefore synonymous and can be used interchangeably. For consistency, the term mineral prospectivity is used in this volume. Because the presence of mineral deposits of the type sought is betrayed by the presence of certain evidential features (e.g., significant geochemical anomalies), mineral prospectivity is thus related to the degree of presence of evidential features. Thus, modeling mineral prospectivity intrinsically assumes, as illustrated in Fig. 1-2, that (a) a specific location is prospective if it is characterised by the same or similar evidential features as known locations of mineral deposits of the type sought and (b) if more important evidential features are present in one location than in another location in a mineralised landscape, then the former has higher mineral prospectivity than the latter..

Modeling of mineral prospectivity is involved at every scale, from regional-scale to local-scale, of exploration target generation (cf. Hronsky and Groves, 2008). It is concerned with the analysis and integration of evidential features derived from multi-source geoscience spatial data sets in order to delineate and rank prospective areas for further exploration of undiscovered mineral deposits of the type sought. In regional-scale target generation, modeling of mineral prospectivity aims to delineate the most prospective areas within large permissive regions. In district- to local-scale target generation, modeling of mineral prospectivity aims to define the most prospective zones or sites within regional-scale prospective areas. This means that from regional-scale to local-scale mineral exploration, geoscience data sets used in modeling of mineral prospectivity should have increasing detail and accuracy both in terms of spatial resolution and information content.

A predictive model of mineral prospectivity must pertain to just one type of mineral deposit. Thus, a mineral prospectivity model for epithermal Au deposits is not applicable to guide exploration for porphyry Cu deposits, and vice versa. In any scale of target generation, however, modeling of mineral prospectivity follows specific steps starting with the definition of a *conceptual model of mineral prospectivity* for mineral deposits of the type sought (Fig. 1-3). Such a conceptual model is *prescriptive* rather than predictive, as it specifies in words and/or diagrams the theoretical relationships between various

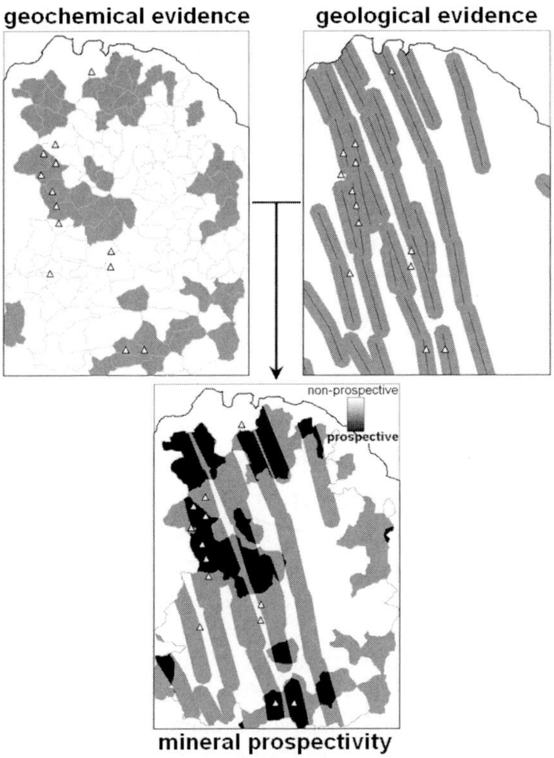

Fig. 1-2. Concept of modeling mineral prospectivity by integration of evidential features (in grey) like stream sediment catchment basin geochemical anomalies, proximity to faults/fractures, etc. The degree of mineral prospectivity is related to the degree of spatial coincidence of evidential features. Locations with the same or similar degree of prospectivity as most known locations (triangles) of mineral deposits of the type sought are considered new exploration targets.

geologic processes or controls in terms of how and especially *where* mineral deposits of the type sought are likely to occur. Defining a conceptual model of prospectivity for mineral deposits of the sought in a study area requires support of published knowledge of the various geological processes relevant to the formation of mineral deposits of the type sought. It is important to review mineral deposit models (e.g., Cox and Singer, 1986; Roberts et al., 1988; Berger and Drew, 2002), which describe the geological characteristics of specific types of mineral deposits, especially the type of mineral deposits sought in a study area, and their regional geological environments. Because the occurrence of several, if not all, types of mineral deposits is related to plate tectonics (e.g., Mitchell and Garson, 1981; Sawkins, 1989; Pirajno, 1992; Robb, 2004), it is imperative to review the geotectonic setting of a study area. Moreover, it is useful to review knowledge about certain geological systems (e.g., faults/fractures) that could act as controls for mineralisation at geological scales relevant to the stage of target

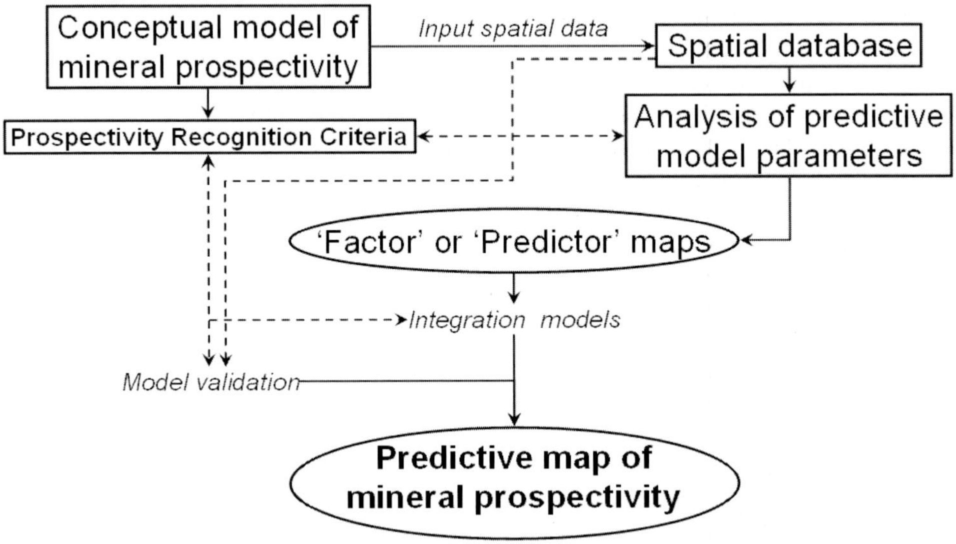

Fig. 1-3. Elements of predictive modeling of mineral prospectivity.

generation and to the size of the study area. Furthermore, analysis of spatial distributions of mineral deposits of the type sought (e.g., Carlson, 1991; Vearncombe and Vearncombe, 1999) and analysis of spatial associations between mineral deposits of the type sought and certain geological features (e.g., Bonham-Carter, 1985; Carranza and Hale, 2002b) are useful in defining, testing and, if necessary, re-defining a conceptual model of mineral prospectivity and the prospectivity recognition criteria. This aspect of mineral prospectivity modeling is demonstrated in Chapter 6.

The geological, geophysical and geochemical characteristics of areas that contain mineral deposits of the type sought constitute the *prospectivity recognition criteria*. The conceptual model of mineral prospectivity and the prospectivity recognition criteria provide the framework for mineral prospectivity modeling in terms of determining the suitable (a) geoscience spatial data sets to be used, (b) evidential features to enhance and extract from individual geoscience spatial data sets, (c) method of transforming mapped evidential features into maps of individual prospectivity recognition criteria, (d) method of weighting classes of individual prospectivity recognition criteria to create predictor maps and (e) method of integrating predictor maps to create a predictive model or map of mineral prospectivity. The preceding items (b), (c) and (d) constitute the analysis of *predictive model parameters*.

Methods of enhancing and extracting evidential features representing a prospectivity recognition criterion are specific to evidential themes (i.e., geochemical, geological, geophysical) and types of geoscience spatial data. The concepts of mapping significant geochemical anomalies are discussed briefly in the previous section and are treated further in Chapters 3 to 5. Mapping of evidential geological features of certain mineral deposits, such as hydrothermal alterations, could be performed in the field and/or by using remotely-sensed data sets (e.g., Spatz, 1997; Sabins, 1999; Carranza and Hale, 2002a). Mapping of geological features such as faults and intrusive rocks representing, respectively, structural and heat-source controls of certain mineral deposits, could be performed in the field and/or by analysis and interpretation of appropriate geophysical data sets (Telford et al., 1990; Parasnis, 1997; Kearey et al., 2002). Procedures for enhancement and extraction of evidential features from geological and geophysical data sets, however, are beyond the scope of this volume.

Certain geoscience spatial data or certain mapped evidential features require manipulation or transformation in order to represent a prospectivity recognition criterion. Data manipulation or transformation involves one or more types of map operations (Chapter 2), the choice of which depends on the prospectivity recognition criterion to be represented. For example, a prospectivity recognition criterion of presence of or proximity to strike-slip faults first requires selection of such faults from the database, followed by creation of a map of distances to such faults and then discretization of distances into proximity classes. Likewise, a prospectivity criterion of presence of geochemical, say Cu, anomalies first requires suitable interpolation of Cu data measured at discrete locations and then discretization of the interpolated Cu data (Fig. 1-4). The purpose of manipulating or transforming spatial data or a map of evidential features to represent certain prospectivity recognition criteria is to model and discretise (or classify in order to create geo-objects representing) the degree of presence of evidential features at every location. Methods of weighting of classes of individual prospectivity recognition criteria in order to create predictor maps involve either knowledge-driven or data-driven modeling of their spatial associations to mineral deposits of the type sought (Bonham-Carter, 1994). Data-driven methods of quantifying spatial associations between classes of prospectivity recognition criteria and mineral deposits of the type sought are explained further in Chapters 6 and 8. Knowledge-driven methods of weighting classes of prospectivity recognition criteria with respect to the mineral deposits of the type sought are explained further in Chapter 7.

Not every data-driven method of quantifying spatial associations between classes of prospectivity recognition criteria leads directly to creation and then integration of predictor maps to obtain a predictive map of mineral prospectivity (see Chapter 6). In addition, not every data-driven method that leads directly to creation of predictor maps applies knowledge or a conceptual model of the inter-play of geologic controls of mineral deposits of the type sought in integrating the predictor maps to obtain a predictive map of mineral prospectivity (see Chapter 8). In contrast, every knowledge-driven method of creating predictor maps can be used directly in integrating such predictor maps, although not all such methods apply knowledge or a conceptual model of the inter-play of geologic controls of mineral deposits of the type sought in integrating the predictor maps to obtain a predictive map of mineral prospectivity (see Chapter 7). Therefore, every method for creating and/or integrating predictor maps has inherent *systemic* (or procedural) *errors* with respect to interactions of geological processes involved in mineral deposit formation. In addition, every input geoscience spatial data

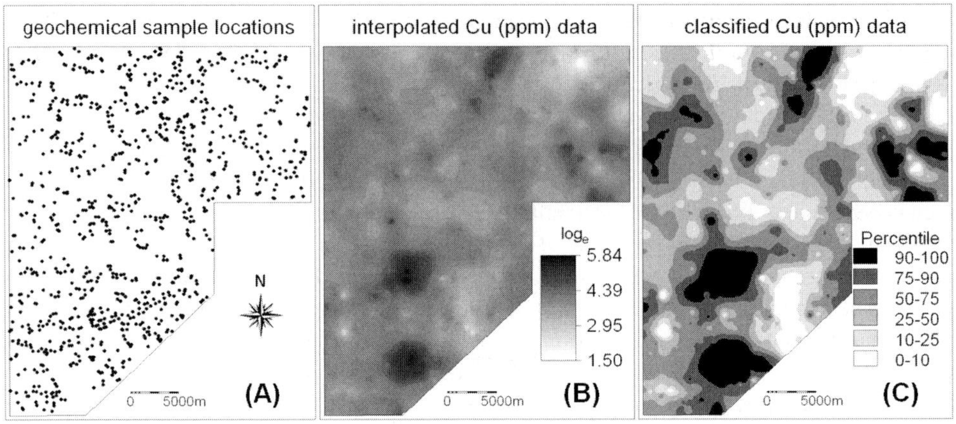

Fig. 1-4. (**A**) Geochemical sample locations where Cu is measured. In order to model or map discrete spatial entities or geo-objects representing degree of Cu anomaly at every location, the point data of Cu values are subjected first to point-to-surface transformation and then surface-to-area transformation. (**B**) Surface or continuous field of Cu values derived by via inverse-distance interpolation method. (**C**) Percentile classification of the interpolated Cu values.

set or mapped evidential features used in modeling of mineral prospectivity invariably contains *parametric* (or data-related) *errors* with respect to distribution of discovered and undiscovered mineral deposits. These errors are inevitable in empirical modeling. However, because such errors are propagated finally into the predictive map of mineral prospectivity, it is imperative to apply measures for *model cross-validation* and, if necessary, reduction of errors in every step of mineral prospectivity modeling.

The objective of model validation is to provide an answer to one or both of the following two basic questions:

Question 1: *Given at least two predictive models of mineral prospectivity, which one has more high prediction values corresponding spatially with known occurrences of mineral deposits of the type sought?*

This question pertains to either knowledge-driven or data-driven predictive modeling of mineral prospectivity. This question also pertains to predictive models of mineral prospectivity derived by one type or different types of either knowledge-driven or data-driven techniques for predictive modeling of mineral prospectivity (see Chapters 7 and 8). If one opts to apply only one type of either knowledge-driven or data-driven technique for predictive modeling of mineral prospectivity, then he/she must strive to derive at least two predictive models of mineral prospectivity in order to answer Question 1. The best possible predictive model of mineral prospectivity is, generally, the one which has the highest number of high prediction values corresponding spatially with known occurrences of mineral deposits of the type sought.

Question 2: *Suppose that, in a study area, we divide the set of known occurrences of mineral deposits of the type sought into two subsets, and we use the first subset to create a predictive model of mineral prospectivity. How much of the second subset of known occurrences of mineral deposits of the type correspond spatially to the high prediction values resulting from the first subset? Suppose further that we use the second subset to create a predictive model of mineral prospectivity. How much of the first subset of known occurrences of mineral deposits of the type correspond spatially to the high prediction values resulting from the first subset?*

This question pertains to and is the essence of the so-called blind testing of data-driven predictive models of mineral prospectivity (Fabbri and Chung, 2008; Chapter 8 of this volume). This question also pertains to predictive models of mineral prospectivity derived by one type or different types of data-driven techniques for predictive modeling of mineral prospectivity (see Chapter 8). The best possible predictive model of mineral prospectivity is, generally, the one which has high prediction values corresponding spatially with the highest proportion or percentage of the known occurrences of mineral deposits of the type sought.

Model validation thus aims at deriving the best possible predictive model of mineral prospectivity. Deriving the best possible prediction model of mineral prospectivity entails *model calibration*. Procedures for model calibration vary in every step of mineral prospectivity modeling. Analysis of spatial distributions of mineral deposits of the type sought (e.g., Carlson, 1991; Vearncombe and Vearncombe, 1999) and analysis of spatial associations between mineral deposits of the type sought and certain geological features (e.g., Bonham-Carter, 1985; Carranza and Hale, 2002b; Chapter 6 of this volume) can be useful in testing and, if necessary, re-defining (thus, calibrating) a conceptual model of mineral prospectivity and the prospectivity recognition criteria. Prior to the analysis of predictive model parameters, training deposit-type locations to be used in data-driven methods of creating predictor maps must be selected (thus, calibrated) systematically instead of randomly (Stensgaard et al., 2006; Carranza et al., 2008b; Chapter 8 of this volume). Every data-driven method of creating predictor maps has intrinsic ways of analyzing and representing (thus, calibrating to reduce) parametric errors of uncertainties in predictor maps, whereas knowledge-driven methods rely on expert opinion in judging (thus, calibrating to reduce) parametric uncertainties in predictor maps. Finally, one must quantify (thus, calibrate) *fitting-rate* and *prediction-rate* to characterise the performance of a mineral prospectivity map (Agterberg and Bonham-Carter, 2005; Chung and Fabbri, 2005). The fitting-rate quantifies the goodness-of-fit between a predictive map of mineral prospectivity and the training deposit-type locations. The prediction-rate quantifies how well a predictive map of mineral prospectivity delineates the testing deposit-type locations. The prediction-rate suggests the ability of a mineral prospectivity map to direct further exploration activities toward undiscovered mineral deposits of the type sought. The fitting-rate is pertinent only to data-driven mineral prospectivity maps, whilst the prediction-rate is pertinent to either data- or knowledge-driven mineral prospectivity maps. The fitting- and prediction-rates also quantify Type I (false-positive) and Type II (false-negative) errors in a predictive model. These errors in a predictive model, if not remedied, could render failure in mineral deposit discovery and thus investment loss in the succeeding scales of target generation or phase of mineral exploration. The various procedures for model calibration in every step of mineral prospectivity modeling are therefore necessary in order to derive reliable mineral prospectivity maps, upon which mineral exploration decisions or plans can be based.

PREDICTIVE MODELING WITH A GIS

A GIS aims to provide pertinent and reliable pieces of spatial geo-information to support decision-making in many fields of endeavour, including mineral exploration. At every scale, from region-scale to local-scale, of exploration target generation, GIS has become a decision-making tool since about the late 1980s (Bonham-Carter, 1994). A major basis for a 'go' or a 'no-go' decision to proceed into the next higher scale of exploration target generation or phase of mineral exploration is a set of geochemical anomaly models or a mineral prospectivity model derived in the previous lower scale(s) of target generation. To derive and visualise such pieces of spatial geo-information, a GIS can be used for efficient capture, storage, organisation, query, manipulation, transformation, analysis and integration of substantial multi-source geoscience data sets collected in the different scales of exploration target generation. The tasks involved in modeling of geochemical anomalies and/or prospective areas at any scale of target generation are therefore numerous, tedious and complex. A GIS does not reduce but facilitates those tasks to allow rapid yet efficient accomplishment of the pieces of spatial geo-information of interest. Of the different GIS functionalities mentioned, this volume is mainly concerned with analysis and integration of data sets to derive and visualise predictive models of geochemical anomalies and prospective areas. Such functionalities are briefly discussed below and demonstrated further in the succeeding chapters.

Data analysis

Predictive modeling of geochemical anomalies and/or mineral prospectivity involves characterisations of the statistical and spatial properties of variables and the spatial relationships among variables. Such operations are supported by query, manipulation and transformation of certain data. In a GIS, these operations can be performed using data attribute tables or data attribute maps. On the one hand, data attribute tables are useful for summarising the statistical properties of univariate data and for characterising statistical relationships among univariate data sets. For example, a data attribute table would be useful in predictive modeling of significant geochemical anomalies via logistic regression by using binary (i.e., presence or absence of) mineral deposit occurrence as the target variable and concentrations of various elements as predictor variables. The predicted values of mineral deposit occurrence ranging from [0] to [1] represent multi-element geochemical anomalies, which can then be visualised by creating the corresponding attribute map. On the other hand, data attribute maps are useful for characterising the spatial properties of univariate data and for characterising spatial relationships among univariate data sets. For example, an overlay or cross operation using a map of classified interpolated point-data Cu values and a map of mineral deposit occurrences would result in a cross-table indicating which class/classes of Cu values represent significant geochemical anomalies (Fig. 1-5).

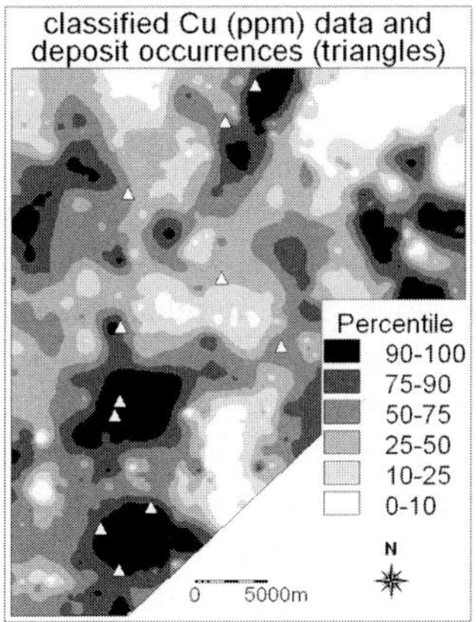

	Deposit occurrence (unit cells)
90-100%le Cu	6
75-90%le Cu	3
50-75%le Cu	1
25-50%le Cu	1
10-25%le Cu	0
0-10%le Cu	0

Fig. 1-5. Deposit occurrences overlaid on (or crossed with) classified Cu data. Cross-table output showing number of deposit occurrences in every class of Cu data. The analysis indicates that >75 percentile Cu values are significant Cu anomalies.

Specific tools for certain data analysis may not be available in some GIS software packages. In such cases, data must be exported or converted to formats supported by other computer software packages that provide the specific tools of interest. The examples given here and several other forms of data analysis demonstrated in the succeeding chapters support the creation of maps of evidential features (e.g., significant geochemical anomalies), which are eventually integrated to model mineral prospectivity. Note that the example of data analysis via two-map overlay of Cu data and mineral deposit occurrence data (Fig. 1-5) is already a form of data integration.

Data integration

The behaviour of indirectly observable and complex real-world system of interest, such as a geochemical anomaly or mineralisation, is controlled by several interacting processes. In order to predict the behaviour of such systems, it is instructive to combine or integrate sets of data, pieces of geo-information or models representing the individual processes involved. In a GIS, a predictive model or map is usually derived by combining predictor maps (Fig. 1-2) via a computational function that aptly characterises the interactions or relationships among the processes that control the behaviour of a system of interest:

$$\textit{predictive model} = f(\textit{predictor maps}).$$

There are different forms of the computational function *f*. The choice of a computational function depends on whether the predictive model is stochastic, empirical or hybrid stochastic-empirical. Unlike predictive modeling of geochemical anomalies, which can be stochastic, predictive modeling of mineral prospectivity usually makes use of computational functions that express empirical relationships of the predictor variables with the target variable. Knowledge-driven empirical methods, on the one hand, usually employ logical functions (e.g., AND and/or OR operators; see Chapter 7) for sequential integration of predictor maps through so-called *inference networks* (see Chapter 7). An inference network depicts knowledge about the inter-play of processes represented by individual predictor maps. Data-driven empirical methods, on the other hand, usually employ mathematical functions for simultaneous integration of predictor maps regardless of knowledge about the interactions of processes depicted by each predictor map (see Chapter 8). Some data-driven methods employ functions representing logical operations (e.g., Dempster's (1968) rule of combining of evidential belief functions) for sequential integration of predictor maps through an inference engine. Likewise, some knowledge-driven methods apply mathematical functions for simultaneous integration of predictor maps.

The way by which data are integrated in a GIS is controlled precisely by the spatial topology and linkage of data at every location to their map coordinates (see Chapter 2), although the topology and map coordinates do not directly play a role in a computational function applied to integrate data. Note that topology and map coordinates also provide precise control in data analysis involving at least one map. The choice of a spatial data model (vector or raster; see Chapter 2) could affect, however, computation during data integration. Although a vector model represents geometry of geo-objects better than a raster model, data integration using raster maps is faster and more precise than using vector maps (Brown et al., 2005). That is because raster maps represent continuous variables, such as element concentrations, better than vector maps. In addition, because of the one-to-one coordination of pixels referring to the same location in every raster map, building of topology of so-called *unique conditions* geo-objects (see Chapter 2) during data integration with raster maps is simpler than during data integration with vector maps (Mineter, 2003). GIS technology is still advancing, however, toward achieving routine capability to integrate data in both vector and raster maps (Winter, 1998; Winter and Frank, 2000). Vector maps are nevertheless preferable to raster maps in visualisation of many types of spatial data or geo-information.

Visualisation of spatial data or geo-information

Displaying spatial data or geo-information on-screen is perhaps the most exploited functionality of a GIS. Exploration geochemists usually 'eye-ball' the data for patterns of interest before actually performing quantitative analysis of the data. Graphical, especially interactive or dynamic, display of spatial data or geo-information is especially useful in the early stages of predictive modeling of geochemical anomalies (Haslett et al., 1991). Most GIS software packages contain, however, only a few dynamic graphical display functionalities. Visualisation of spatial data in a GIS is also useful in selective query, retrieval and analysis of certain data in a database (e.g., Harris et al., 1999). Finally, a GIS provides capability for mapping (i.e., preparing analogue maps in contrast to modeling). More than two decades ago, Howarth (1983a) has explained the types of useful geochemical maps and the techniques for preparing such maps. Recently,

Reimann (2005) reviewed and recommended various techniques for producing fully informative geochemical maps and concluded that such specialised techniques are not provided in most GIS. Data sets and predictive models of geochemical anomalies and/or prospective areas have to be exported to specialised software packages to create fully informative geochemical maps suitable for decision-making in mineral exploration.

SUMMARY

Predictive models of either geochemical anomalies or prospective areas are generally empirical models, which depict locations where mineral deposits of the type sought plausibly exist. Such pieces of spatial geo-information are important for decision-making in mineral exploration programmes. The tasks involved in predictive modeling of significant geochemical anomalies and/or prospective areas, however, are tedious and complex. A GIS provides various functionalities that facilitate rapid yet efficient accomplishment of predictive models of geochemical anomalies and/or prospective areas. This chapter has introduced the relevance of a GIS to predictive modeling of geochemical anomalies and/or prospective areas in terms of data analysis, integration and visualisation. The next chapter provides introductions to the types of data models by which geoscience data sets used in target generation can be represented in a GIS, the ways by which such data sets can be organised in a database, and the various types of operations by which such data sets can be manipulated in the predictive modeling process.

Geochemical Anomaly and Mineral Prospectivity Mapping in GIS
by E.J.M. Carranza
Handbook of Exploration and Environmental Geochemistry, Vol. 11 (M. Hale, Editor)

Chapter 2

SPATIAL DATA MODELS, MANAGEMENT AND OPERATIONS

INTRODUCTION

Geochemical and other types of data sets for target generation in mineral exploration are spatial (or geographically-referenced) data that come from either primary or secondary sources and are stored in either digital or non-digital (analogue) formats. The diversity in storage formats of such data sets calls for proper data management in order to achieve efficiency in *modeling* of geochemical anomalies and prospective zones via various forms of spatial data analysis. Target generation in mineral exploration thus requires a computerised system such as a GIS so that the pieces of spatial geo-information of interest are mapped as discrete spatial entities or *geo-objects* (i.e., with perceivable boundaries, sizes and shapes). In a GIS, geo-objects are represented either as vector or raster spatial data models. The range of operations for spatial data analysis supported by a GIS depends on (a) geometric model of geo-objects (point, line or polygon), (b) spatial data models (vector or raster), (c) type of attribute data (quantitative or qualitative), (d) objectives of analysis and (e) GIS software package used. The last factor is least but must be considered important because many GIS software packages that are available at present support certain types of spatial data analyses using either only vector or only raster spatial data models.

This chapter explains briefly the concepts of spatial data models, especially which model is appropriate for representation of certain types of geoscience spatial data in a GIS, and the concepts for capturing and organising spatial data in a GIS database. The various types of GIS operations for spatial data analysis are also discussed briefly, because these will be the topics in the succeeding chapters.

MODELS OF SPATIAL DATA

The definition of model in this context is different from the definitions given in Chapter 1. In the present context, a *data model* refers to (a) the schema or ways of organising data about real-world systems or (b) the symbolic representation of relationships between geo-objects and their data attributes.

Geo-objects

Many types of geological features with distinct boundaries, such as lithologic units, are clearly geo-objects. There are several types of geological features with no distinct boundaries, such as geochemical anomalies, which require modeling of pertinent spatial data to represent them as geo-objects. Modeling, therefore, involves various forms of analysis to partition or discretise pertinent spatial data in order to represent certain geological features of interest as geo-objects. For example, a threshold for background element concentrations must be determined in order to map geochemical anomalies.

The geometry of geo-objects can be represented based on their spatial dimensions. Point geo-objects are without length or area and thus 0-dimensional (0-D). Geochemical sample locations, although strictly not dimensionless, are usually depicted as points because they are usually too small to be represented in most map scales. Linear geo-objects (e.g., faults) are one-dimensional (1-D) and only have length as spatial measure. Polygonal geo-objects (e.g., geochemical anomalies) are two-dimensional (2-D) and have area and perimeter as spatial measures. Some geo-objects (e.g., geochemical landscape) require so-called 2.5-dimensional (2.5-D) representation, because they cannot be strictly described in two or three dimensions. Geo-objects characterised by their volume (e.g., orebody) require three-dimensional (3-D) representation. In addition, many geo-objects require fractal modeling to describe their geometry (Mandelbrot, 1983; Chapter 4). A fractal geo-object is one which can be fragmented into various parts, and each part has a similar geometry as the whole geo-object.

The geometry of geo-objects can be defined according to either amount of sampling data or certain criteria (Raper, 1989). If the geometry of certain geo-objects is defined by amount of sampling data, then they are called *sampling-limited* geo-objects. Examples of sampling-limited geo-objects are porphyry stocks, quartz veins, lithologic contacts, etc., because they cannot be sampled or mapped completely if they are only partially exposed. If the geometry of certain geo-objects is defined by certain criteria in order to delimit their spatial extents, then they are called *definition-limited* geo-objects. The best example of a definition-limited geo-object is an orebody, the spatial extents of which are defined by cut-off grade at prevailing economic conditions. Significant geochemical anomalies and prospective zones are both definition-limited geo-objects, although they are also both sampling-limited geo-objects.

Vector Model

In a vector model, geo-objects are represented as components of a graph. That means the geometric elements of point, linear and polygonal geo-objects are interpreted in 2-D space as in a map (Fig. 2-1). Point geo-objects are nodes defined by their graph of map (x,y) coordinates. Linear geo-objects are defined by arcs with start-nodes and end-nodes or by a series of arcs inter-connected at nodes called vertices. Polygonal geo-objects are defined by inter-connected arcs that form a closed loop.

The so-called *spaghetti model* is the simplest type of vector model (Fig. 2-2), which represents geo-objects in spatially less structured forms. That means, intersections between linear geo-objects are not recorded, whereas boundaries between polygonal geo-objects are represented separately. The latter is usually not without error and could result in so-called false or sliver polygons. The spaghetti vector model leads to inefficient data storage and is not amenable to true GIS functions (e.g., neighbourhood

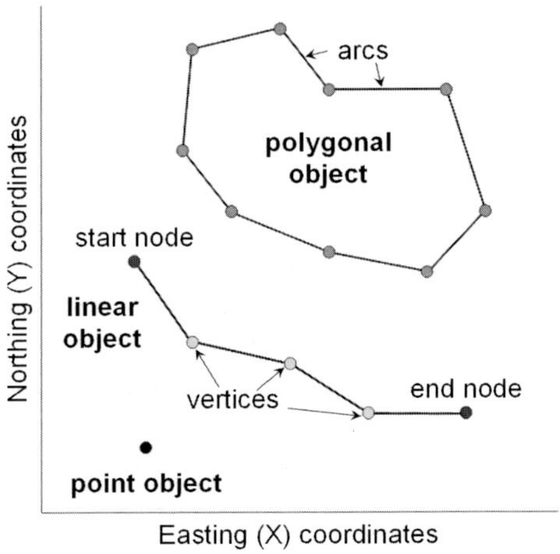

Fig. 2-1. Vector model of geo-objects.

operations; see further below). This type of vector model has, nonetheless, the advantage that geo-objects can be readily scaled, transformed to other map projections and displayed using inexpensive systems for visualisation.

The *topological model* offers vector representation of geo-objects in a spatially-structured form (Fig. 2-2) Topology is concerned with spatial relationships between geo-objects in terms of containment, connectivity, adjacency or proximity. In a topological model, linear geo-objects (including boundaries of polygonal geo-objects) are recorded in node-arc structures such that nodes represent intersections between linear geo-objects and form polyline segments or arcs and then arcs form polygons. Boundaries of polygonal geo-objects are not recorded separately. The model results in efficient storage of attributes of geo-objects and, more importantly, in explicit definition of spatial relationships between nodes, arcs and polygons. That means geo-objects on either left or right of another geo-object are explicitly defined. The topological model is amenable to true GIS functions (e.g., neighbourhood operations; see further below), not only because spatial relationships between geo-objects are defined but also because such spatial

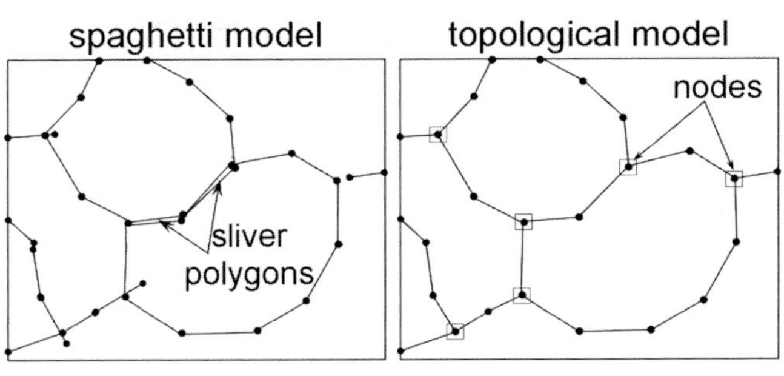

Fig. 2-2. Spaghetti model versus topological model of vector data.

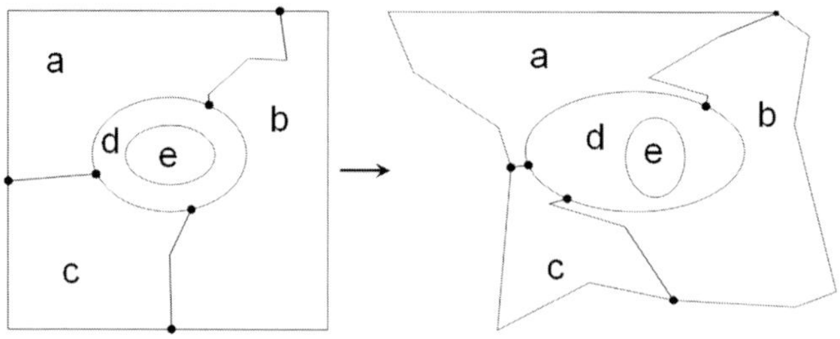

Fig. 2-3. Topological relationships between geo-objects remain unchanged under transformation.

relationships between geo-objects are independent of map scale or measurement scales and are preserved even under transformations to various map projections (Fig. 2-3). A disadvantage of the topological model is that defining spatial relationships between geo-objects during spatial data capture and map editing can be time-consuming.

Because a vector model represents geo-objects in 2-D space, it is not an appropriate model for surface variables such as topographic elevations, element concentrations of surficial materials, geophysical properties, etc. Although data for surface variables can be stored as a series of multi-valued points or a series of isoline contours in a vector model, a vector model does not adequately represent nor readily support calculation of surface characteristics (e.g., slope). Data of surface variables require 2.5-D representation such as tessellations of polygonal planar patches called triangulated irregular networks (TIN), which are usually treated as a vector model.

A TIN is constructed by connecting points of data (with x,y coordinates and z-values) to form a continuous network of triangles (Fig. 2-4). Note that a TIN can also be generated from points derived from isoline contours. There are various triangulation methods, but the most favoured is the *Delaunay* triangulation technique, which is a dual product of *Thiessen* or *Voronoi* or *Dirichlet* tessellations of polygons. The triangular facets defined represent planes with similar surface characteristics such as slope and aspect. A TIN model is adequate to represent geometry and topology of a surface, is efficient in data storage and can be locally manipulated to represent surface complexity by using breaklines (e.g., terrain discontinuities such as rivers or ridges on topographic surfaces). It is a significant alternative to surface representations based on regular grids.

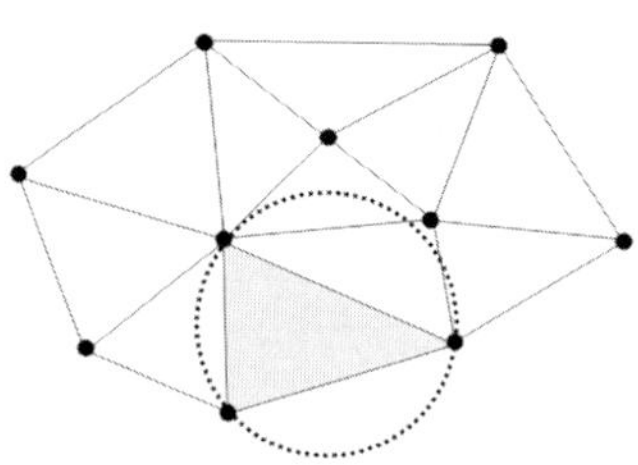

Fig. 2-4. A triangulated irregular network (TIN) by Delaunay triangulation. Triangles are defined by three points forming circumcircles not containing another point.

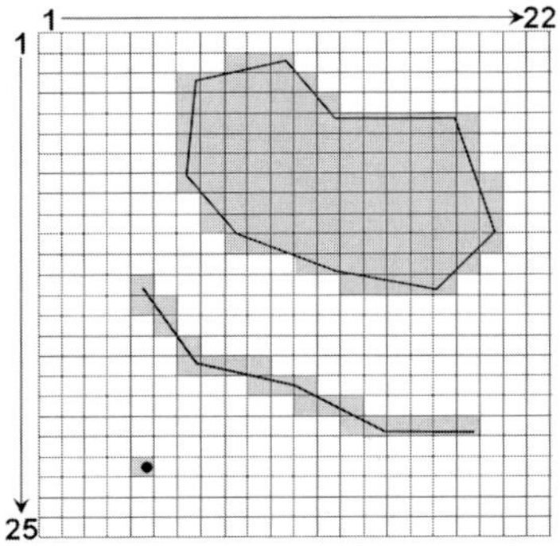

Fig. 2-5. Raster representation (grey cells) of geo-objects in Fig. 2-1.

Raster Model

In a raster model, geo-objects are represented by means of subdividing an area into a regular grid of cells or pixels organised along columns and rows (Fig. 2-5). The column/row organisation provides reference for the positions of pixels, which can be linked or *georeferenced* to a particular spatial coordinate system. The pixels are usually squares, but can also be of various equilateral shapes with a fixed size denoting the spatial resolution of a raster model. For example, the 30-m spatial resolution of Landsat Thematic Mapper imagery means that each pixel in an image measures 30 m by 30 m on the ground. The raster model is thus concerned with both location and accuracy of representing geo-objects.

In a raster model, a point geo-object (e.g., mineral deposit occurrence at a regional scale) is represented by only one pixel, a linear geo-object (e.g., a fault zone at a regional scale) as inter-connected pixels depicting length, a polygonal geo-object (e.g., a lithologic unit that is mappable at a certain scale) as adjoining pixels depicting shape. The choice of a pixel size can make a raster model an unrealistic representation of geo-objects; thus, it requires compromise between maximising spatial accuracy and minimising data storage and processing. There are a number of techniques for encoding data in raster format (e.g., run-length encoding, quadtrees, etc.), which address concerns in efficiency of data storage and processing (Holroyd and Bell, 1992).

The raster model is efficient for analysis of data within an attribute layer or between attribute layers (see further below). For a raster layer of an attribute data, operations that could be performed include neighbourhood analysis, interpolation, proximity analysis, etc. The column/row organisation of pixels in a raster model allows efficient analysis of relationships between two or more attribute data layers by overlay operations. The raster model is also satisfactory for representing surface entities. The grid structure of surface representations is simple to understand. The raster model, however, is inflexible in representing surface complexities due to constant pixel size and imparts a global directionality to a surface model as influenced by the two principal axes of a grid.

Calculation of surface characteristics such as slope is, however, more realistic with a raster model than with a TIN model.

Attribute Data Models

In a vector model, each point, line or polygon is assigned the corresponding attributes of geo-objects they represent. In a raster model, each pixel is assigned the corresponding attributes of geo-objects it represents. A map layer in a vector or raster model represents attribute data for a particular variable. For example, a map layer of stream sediment sample points could represent labels of each point, whilst another map layer of the same points could describe concentrations of a specific element at each point. Parts of a study area without data for certain attribute are, in either vector or raster model, null or undefined.

In a computer system, or more specifically in database terminology (see below), attribute data represent observed and measured properties of geo-objects. Attribute data have spatial, temporal or thematic characteristics. Spatial attributes pertain to properties that vary in space and their variations can be characterised by location, topology and geometry. Temporal attributes pertain either to age of geo-objects or to a period of data acquisition. Thematic attributes are neither spatial nor temporal properties but pertain to some forms of classifications to which geo-objects can be related, for example, lithology, mineral deposit-type or mineral occurrence (i.e., presence or absence), faults of certain orientations, etc. In many GIS studies related to mineral exploration, temporal are thematic attributes are usually considered to be non-spatial.

Attribute data can be classified as either *continuous* or *discrete* variables. Continuous variables take on any value (i.e., real values), whilst discrete variables take on only certain values (i.e., real integers). Spatial attribute data are mostly continuous variables, whilst non-spatial data are mostly discrete values. Element concentrations are examples of a continuous variable, whereas stream order is a discrete variable. Because modeling of geo-objects involves *discretisation* of continuous variables and *quantisation* (i.e., numerical representation for quantitative integration) of discrete variables, it is more didactic to classify attribute data as *quantitative* or *qualitative* variables.

Attribute data representing numerical magnitude of geological, geochemical or geophysical properties are quantitative variables. Data of quantitative variables are usually measured on either *ratio* or *interval* scales and are mostly represented by continuous values but can also take on discrete values. For example, element concentrations and temperature are continuous variables measured on ratio scales and on interval scales, respectively. In contrast, surface reflectance/absorption properties are continuous variables measured on ratio scales but can be represented discretely as real integers [0,255] in raster images. Quantitative variables are important forms of attribute data because they can be manipulated by mathematical operations or transformations, which are essential to spatial analysis.

Data of qualitative variables usually take on discrete values or labels according to either *ordinal* or *nominal* scales of measurements. A percentile classification of element concentrations is an example of ordinal measurement scale. Lithology is an example of nominal or *categorical* scales of measurements. Qualitative variables, unlike quantitative variables, usually have to be represented numerically before they can be integrated in mathematical operations. For example, lithology, which is a qualitative variable, can be integrated quantitatively in the estimation of local background by representing it numerically as an areal proportion of a drainage sample catchment basin (see Chapter 5). A simple numerical representation of qualitative variables is the assignment of discrete values in either *binary* or *ternary* scales of measurement according to a particular proposition. For example, for a proposition that "this site contains a mineral deposit", lithologic units that are unfavourable and favourable host rocks according to genetic models of the deposit-type sought can be assigned a value of [0] and [1], respectively. Other examples of types of numerical representations of qualitative variables (but also of quantitative variables) are fuzzy membership and probability, which range in the interval [0,1] reflecting degrees of non-ambiguity and certainty, respectively, with respect to a proposition (see Chapter 7).

MANAGEMENT OF SPATIAL DATA

In a GIS, management of spatial data is concerned with (1) storing data in the computer (i.e., spatial data capture) and (2) organising data in the computer (i.e., spatial database creation). Management of spatial data takes a major proportion of resources (personnel, time and money) in any GIS-based project.

Spatial Data Capture

The first fundamental step in spatial data capture is to choose a *coordinate system*, into which all geo-objects or data are geographically-registered or georeferenced. A coordinate system consists of a *spheroid* (or an *ellipsoid*) representing the Earth's surface and a *map projection* to convert spherical or geographical coordinates (latitudes, longitudes) to planar or map (metric) coordinates. The choice of an appropriate coordinate system can benefit from the authoritative discussions on spheroids and map projections given by Maling (1992) and Snyder (1993). On the one hand, the choice of an ellipsoid depends on global surface curvatures, such that for every region or country there is a commonly used 'best fit' ellipsoid (Table 2-I). On the other hand, the selection of a map projection depends on (a) geographic position of region or country, (b) size and shape of region or country where a study area is situated and (c) requirements or objectives of the study. These three factors must be considered together if the primary aim is to obtain minimum geometric distortions in terms of either shape or area.

There are different types of map projections and each map projection creates geometric distortions but guarantees a known relationship between locations on a map and their true locations on the Earth. It is essential to use a map projection because geographical coordinates are not planar coordinates and most spatial data are visualised as 2-D features using planar coordinates. Although spatial data can be stored and manipulated using geographical coordinates, storing spatial data using map projections

TABLE 2-I

Commonly used spheroids of coordinate systems: year developed, length of semi-axes and usage.

Name	Year	Lengths of semi-axes (m)		Geographical usage
		Major	Minor	
Everest	1830	6377276	6356075	India, Sri Lanka, Myanmar
Bessel	1841	6377397	6356079	Central Europe, Chile, China, Indonesia
Airy	1849	6377563	6356257	Great Britain
Clarke	1866	6378206	6356584	North America, Philippines
Clarke	1880	6378249	6356515	France, Africa (parts)
International	1924	6378388	6356912	Whole word except North America and Africa
Krasovsky	1940	6378245	6356863	Russia, Eastern Europe
IAU65	1965	6378160	6356775	Australia
GRS80	1980	6378137	6356752	North America
WGS84	1984	6378137	6356752	Whole world

avoids iterative conversion from geographic coordinates to planar coordinates each time spatial data are visualised.

A cylindrical projection is appropriate for countries lying in the tropical latitudes about the equator, whilst a conical projection is appropriate for countries lying in temperate latitudes and an azimuthal projection is appropriate for countries lying in the polar regions. A cylindrical projection is suitable for large-size rectangular regions such as Russia, whilst a conical projection is suitable for medium-size triangular regions such as India, and an azimuthal projection is suitable for small-size circular regions such as the Netherlands. The property of a map projection must be considered depending on the purpose or requirements of mapping. Equal-area projections are appropriate for mapping in which accurate area measurements are required, whilst orthomorphic (or conformal) projections are appropriate for mapping in which accurate angular measurements are required, and equal-distance projections are appropriate for mapping in which accurate distance measurements are required.

Once a suitable coordinate system is chosen, the capture of spatial data to a database can be initiated. Table 2-II lists possible methods of capture of analogue or digital spatial data. Analogue data are usually in paper form (e.g., paper maps, positive prints of aerial photographs, etc.) and thus not computer-readable in contrast to digital data. Analogue data can be either encoded directly into a GIS database or encoded digitally using different software and imported later into a GIS. Capturing data from analogue maps or images is often carried out either by (a) manual digitisation of maps or images mounted on digitising tables, (b) manual on-screen digitisation of optically (raster) scanned maps or images or (c) automatic feature extraction from scanned maps or images (Ansoult et al., 1990). The accuracy of digitising maps or images depends upon the accuracy of carefully-selected control points on the map/image being digitised and upon the skill of the operator. The selection and digitisation of control points is vital for geocoding the data – the process of geographic registration of locations of geo-objects. Specific

TABLE 2-II

Methods of capturing geoscience spatial data from different sources.

Example data	Data source	Source type	Possible encoding methods
Analytical results for geochemical samples	Tables	Analogue	Keyboard entry Text scanning
		Digital	File transfer File format conversion on import
Lithologic units Faults/fractures Drainage lines	Maps	Analogue	Manual digitising Raster scanning then on-screen digitising Automatic feature extraction
		Digital	File transfer File format conversion on import
	Air-photos	Analogue	Manual digitising Raster scanning then on-screen digitising Automatic feature extraction
		Digital	File transfer File format conversion on import
Lithologic units Altered rock units Faults/fractures Drainage lines	Satellite imagery	Digital	File transfer File format conversion on import Image processing and reformatting

procedures for manual digitising vary depending on software used. Digital data can be imported directly into a GIS database; however, many digital data often require conversion to correct format for use in a specific GIS. Most GIS software packages have routines for importing and converting digital data through a variety of interchange formats. Again, it is important that the original geographical coordinates of digital data are converted into the working coordinate system.

Spatial Database Creation

A spatial database is an organised collection of geographically-referenced data and their attributes. A database approach is favoured in GIS because it overcomes problems with traditional methods of data management and it provides or supports the following functions (Stern, 1995):

- storage and manipulation of very large data sets;
- control over data redundancy;
- data security and integrity;
- database models;
- concurrent use of data;
- back-up and recovery functions;
- spatial query languages; and
- low maintenance costs.

In order to achieve the useful functions of a spatial database, data are handled through a database management system (DBMS), which is a collection of software for creation and maintenance of a database. Creation of a spatial database takes a major proportion of resources in a GIS-based mineral exploration project. There are four major stages in database creation (Oxborrow, 1989).

- *Data investigation*. The task in this stage is to consider, according to needs or objectives of a project, the type, quantity and qualities of spatial data to be included in a database.
- *Database modeling*. The task in this stage, which can be performed independently of the software used, is to develop a conceptual or logical database model by analysing characteristics of entities and relationships between entities and attributes.
- *Database design*. The task in this stage, which depends on the software and its data model, is to create a practical design for a database. That means, the logical design of a database produced in the data modeling stage is translated into a practical design using a DBMS.
- *Database implementation*. The task in this stage is to populate the database with attribute data. This is followed always by monitoring and upkeep, including fine tuning, modification and updating of attribute data.

The database models that have been used in different GIS applications are the *hierarchical*, *network*, *relational* and *object-oriented* models. The hierarchical model was useful in geochemical applications around the mid-1980s (Mattiske, 1983). The hierarchical and network models have some importance in GIS applications to geological studies (e.g., Lopez-Blanco and Villers-Ruiz, 1995; Wielemaker et al., 2003). The object-oriented model became an emerging trend and topic of many researches in GIS applications to geological studies (e.g., Alms et al., 1998; Breunig et al., 1999). The relational model is now predominantly used in GIS and is most widely used in GIS applications for geological exploration (e.g., Schmatz et al., 1995; Laxton and Becken, 1996; Colman-Sadd et al., 1997; Pouch, 1997; Diepenbroek et al., 2002). The relational database model is given some supplementary treatment here; the other database models are not discussed further.

The relational database model was developed by Codd (1970) based on the relational algebra. In such a database model, data and relations between them are organised in a 2-D structure (i.e., table). The terminology of relational databases as compared to the terminology of traditional tables or simple spreadsheets is given in Table 2-III. Each table represents data for one entity or theme (e.g., stream sediment geochemical data). Each row or tuple is a record of different properties of a geo-object (e.g., a sample). Each column represents an attribute or property. The order of tuples or attributes is trivial as long as the contents of a table are not changed or the significance of relations is not altered. Each cell (i.e., intersection between a row and a column) can have only one value or is left blank if values are unknown or unmeasured. Where relations exist between different data sets, tables representing such data sets are linked by certain

TABLE 2-III

Relational database terminology.

Analogue version	Digital version	Relational database
Table	File	Relation
Row	Record/case	Tuple
Column	Field	Attribute
Number of rows	Number of cases	Cardinality
Number of columns	Number of fields	Degree
Unique ID	Primary key	Index
	Possible values	Domain

(Adapted from Date, 1990)

common attributes designated as keys, through which retrieval, search or query (or so-called join operations) of data from one table to another is made possible and efficient.

Attributes designated as keys must have entries that are unique; however, there is no standard to predefine keys. In GIS, location (i.e., coordinate pairs) is usually but tacitly designated as primary key. On the one hand, for data or tables pertaining to point geo-objects such as sample locations, unique identifiers of each location (e.g., sample number) can be designated as key. On the other hand, for data or tables pertaining to linear or polygonal geo-objects, unique identifiers to be used as keys must first be created and assigned to each linear or polygonal geo-object. The assignment of unique identifiers to geo-objects usually takes place during spatial data capture and is considered during the database modeling stage of spatial database creation. This allows for an efficient link between maps of different types of geo-objects and their corresponding attribute tables and therefore allows for efficient analysis of relationships between geo-objects of interest through suitable spatial data operations. Extensive explanations about operations on relational databases (or tables) of geoscience exploration data and their relationships with maps can be found in Bonham-Carter (1994).

OPERATIONS ON SPATIAL DATA

The focus of the following introductory discussions is on generic concepts of certain operations rather than on algorithms behind such operations, which could vary in vector- or raster-based GIS software packages.

Spatial query and selection

In the early stages of spatial data analysis, it is common to (a) visualise and query the spatial database and then (b) select locations of spatial data or geo-objects of interest. Fig. 2-6 shows an example how spatial data or information can be visualised and queried. Clicking the cursor on any portion of a map view would indicate data attributes at specific locations, meanwhile an interactive table view could indicate map coordinates

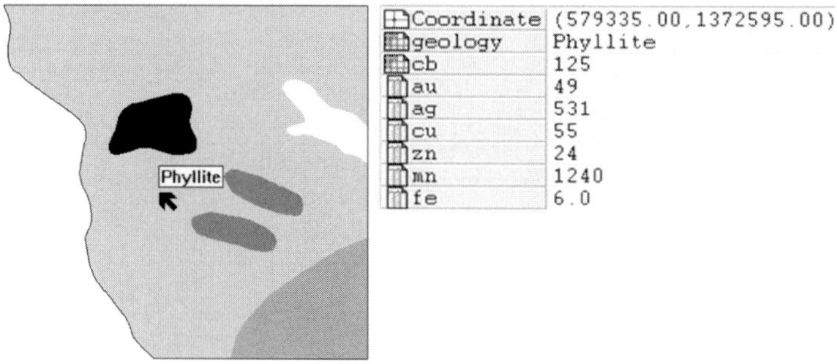

Fig. 2-6. Spatial query of lithologic information by clicking cursor on any part of a lithologic map or through an interactive table view that indicates, at cursor position, map coordinates and attributes in the displayed map ('geology'), and in other maps in the spatial database. The 'cb' is drainage catchment basin number with attributes for certain elements.

and data attributes at the cursor position. The process of selecting spatial data or geo-objects for further analysis can then be performed by using either data attribute conditions or topological relationships of geo-objects.

Criteria for selection of certain spatial data by using data attribute conditions are specified in a query language, which can be a software specific language or a structured query language (SQL) if the spatial data sets are stored in a relational database. A data attribute condition criterion to select certain spatial data is called an *atomic condition*, which makes use of relational operators (=, <, >, ≤, ≥, etc.). Fig. 2-7A shows a map of spatial data of interest, whilst Fig. 2-7B shows an output map of selected spatial data with values above a certain threshold. Two or more attribute condition criteria to select certain spatial data make up a *composite condition*, which make use of Boolean logic operators (AND, OR, NOT, XOR, etc.). The AND operator selects spatial data that meet all of the attribute condition criteria. The OR operator selects spatial data that meet either one or all of the attribute condition criteria. The NOT operator selects spatial data that do not meet all of attribute condition criteria. The XOR operator selects spatial data that

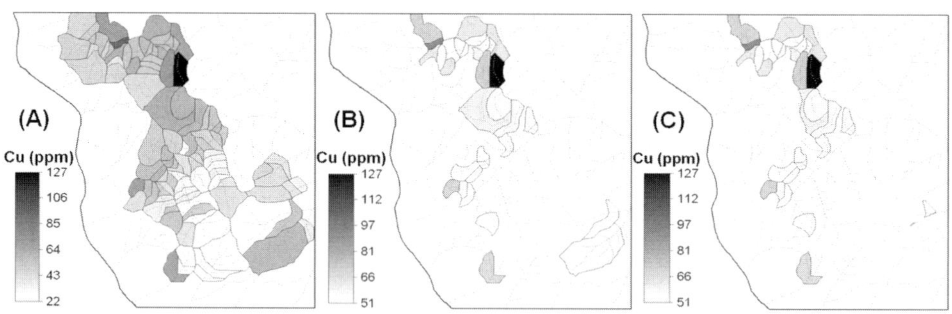

Fig. 2-7. Selection of spatial data by using attribute conditions. (**A**) Input map of Cu data. (**B**) Output map using an atomic condition of "Cu>50". (**C**) Output map using composite conditions of "Cu>50" AND "lithology=phyllite" (see Fig. 2-6).

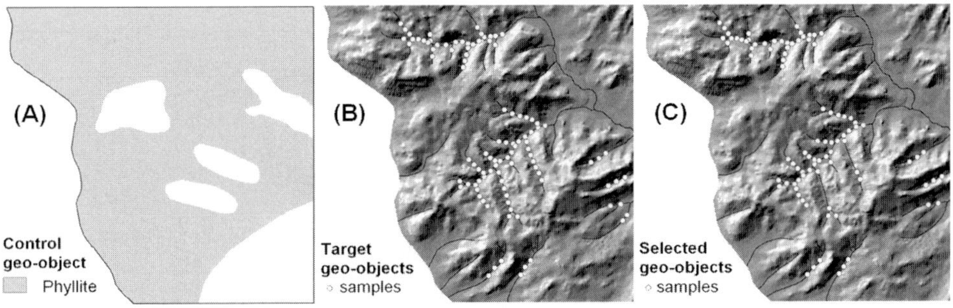

Fig. 2-8. Selection of spatial data by using topological relationships. (**A**) Map of control geo-object. (**B**) Map of target geo-objects for selection. (**C**) Map of selected geo-objects contained only in the control geo-object.

meet either one but not all of the attribute condition criteria. Fig. 2-7C shows an output map generated by selecting from an input map values above a certain threshold and pertaining to a certain thematic class. This way of selecting spatial data by using data attribute conditions can be useful, for example, in analysis of geochemical data pertaining to different lithologies.

Selection of certain geo-objects of interest based on their topological relationships (containment, connectivity, adjacency or proximity) with other types of geo-objects is carried out in two steps. The first step is to decide on a control geo-object for the selection of target geo-objects. The second step is to apply a spatial relationship of target geo-objects with the control geo-object. Fig. 2-8 shows a map of a control geo-object, a map of target geo-objects and a map of selected target geo-objects *contained* in the control geo-object. Selection of spatial data by using attribute conditions is useful, for example, in analysis of geochemical data pertaining to different lithologies. This method of selecting geo-objects of interest can be useful, for example, in analysis of subsets of geochemical data pertaining to different lithologies (see Chapter 3).

Classification and re-classification

The objective in classification or re-classification is to reduce detail in input data in order to reveal patterns or spatial distributions of interest. Classification is applied to spatial data measured on ratio scales. For example, one might be interested to identify zones of different levels of concentrations of certain metals. This requires partitioning or classification of geochemical data (point or surface) into different ranges (Fig. 2-9). Re-classification is applied to spatial data of interval, ordinal and nominal scales. For example, one might be interested in zones underlain by certain lithologies according to their mode of genetic formation. This requires re-classification of different lithologic units according to whether they are igneous, sedimentary or metamorphic rocks (Fig. 2-10). Classification or re-classification can be carried out via either map or attribute table calculations. Quantitative attribute data, particularly continuous variables, are more adequately re-classified via calculations with raster maps (e.g., classification of multi-

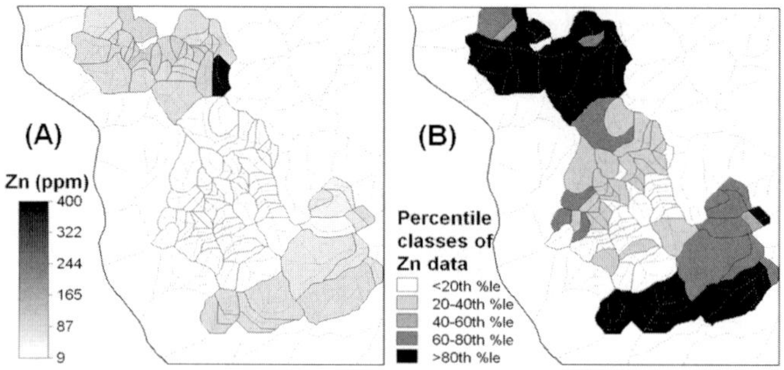

Fig. 2-9. Stream sediment geochemical data displayed as sample catchment basins. (**A**) Raw data of Zn (ppm). (**B**) Classification of Zn data into classes of 20-percentile intervals.

spectral satellite images), whereas qualitative attribute data are adequately re-classified via table calculations.

Measurements

Various types of measurement information on vector and/or raster maps can be obtained: (a) number of point, lines, polygons or pixels; (b) length of lines; and (c) perimeter and area of polygons. These measurements can be calculated by means of creating a *map histogram* of any type of map. The usefulness of measurements depends on the concern of the analyst. For example, one analyst might wish to check the actual number of sample location points in order to monitor and update a spatial database. The number and length of a particular set of lines (e.g., faults) are useful in structural analysis by way of rose diagrams. Area calculations are useful in fractal analysis and catchment basin analysis of geochemical anomalies (Chapters 4 and 5, respectively) and in the analysis of spatial associations between mineral deposits and geological features (Chapter 6). Areas of polygons (e.g., prospective zones) could also be an important criterion, in combination with other criteria, for prioritisation of follow-up exploration work. Note that in raster maps the number of pixels multiplied by the square of the pixel

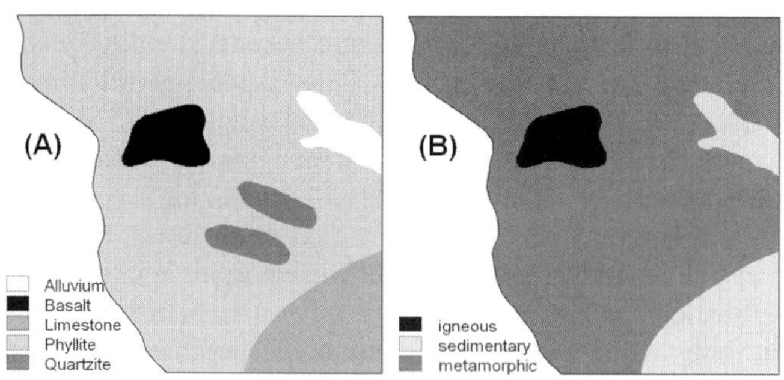

Fig. 2-10. Re-classification of lithologic units (**A**) according to their genetic mode (**B**).

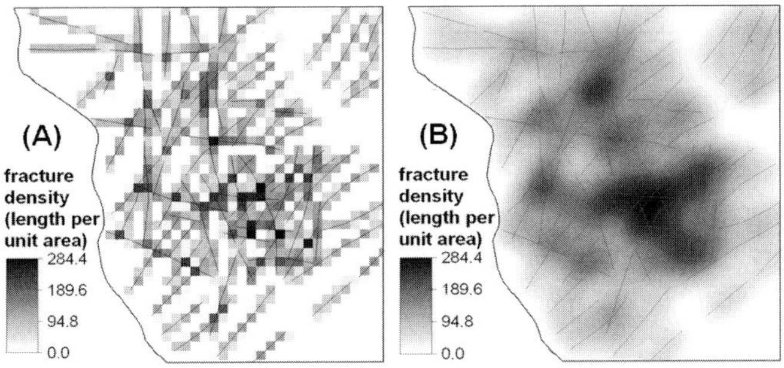

Fig. 2-11. Fractures (linear features) and fracture density estimated as total length per unit area or pixel (**A**) and then smoothed via a gridding method (**B**).

size is equivalent to area.

Distance calculation is important in many GIS analyses for mineral exploration. For example, Seoane and De Barros Silva (1999) introduced a drainage sinuosity index to rank gold-anomalous catchment basins. They calculated sinuosity index as ratio of total length of drainages in a segment basin to distance between a sampling point represented by a segment basin and the most upstream point of that segment basin. Distance calculations are also important in quantifying spatial associations between mineral deposits and certain geological features (see Chapter 6).

Another form of measurement is point or line density, which is the number of points or the total length of lines, respectively, per unit area. Fig. 2-11A shows a map of fractures and the corresponding fracture density created by a simple method of measuring total length of fractures per unit area. Note the blocky character of the simple fracture density map, from which it is evident which pixels contain a facture segment. A smoother facture density map (Fig. 2-11B) could be achieved via a gridding method (see further below). It is clear in the example that a fracture density map, for example, is a form of transformation of line or point geo-objects into area or surface geo-objects. Point density calculation is an important concern in the analysis of spatial association between mineral deposits and certain geological features (Chapter 6). In such analysis, linear geological objects are represented as or transformed into polygonal features.

Transformations

Most GIS operations on spatial data can be considered transformations. In fact, calculation of density of point or linear geological features is transformation of the 0-D or 1-D, respectively, of these geo-objects into 2-D. Perhaps the most important type of transformation in a GIS is conversion of geographical coordinates, in which most spatial data are probably originally stored, into planar coordinates of suitable map projections (Maling, 1992). Geometric corrections of satellite imagery and transformations of various raster maps into a common pixel size are also important in GIS studies. Such transformations, known as *resampling* (Mather, 1987), ensure that map layers in a GIS are properly georeferenced. Transformations to derive digitally-encoded data, such as line generalisations, are important in capturing spatial data (McMaster and Shea, 1992; Garcia and Fdez-Valdivia, 1994). There are many other types of transformations in a GIS. Bonham-Carter (1994) provides elaborate discussions on the concepts and algorithms of many different types of spatial data transformations that are applicable to geoscience modeling in general.

The discussion here concentrates on spatial data transformations that are more directly and usually involved in mapping of geochemical anomalies and mineral prospectivity. These are point-to-area, point-to-surface, line-to-area, line-to-surface, area-to-point, area-to-area and surface-to-area transformations. The last two transformations are handled via re-classification operations (see above). Some of these transformations may require conversion from a vector data model to a raster data model and vice versa. Detailed discussions on vector-to-raster and raster-to-vector conversions can be found in Clarke (1995), Mineter (1998) and Sloan (1998). Area-to-point transformation, for example, can be handled by vector-to-raster conversion, whereby polygonal geo-objects are converted to pixels and each pixel can treated as a point.

Most geoscience spatial data used in mapping of geochemical anomalies and mineral prospectivity are recorded as attributes of sampling points (Fig. 2-12A). Because the objective of most mineral exploration activities is to define anomalous zones rather than points (except in defining locations for drilling), point-to-area and/or point-to-surface transformations are required to analyse and model spatial information from point data. The types of transformations performed depend on the type of geo-objects represented by point data and on the nature of attribute data. On the one hand, point-to-area transformations of point data representing geo-objects such as intersections of curvi-linear structures or locations of mineral deposits can be modeled appropriately by, for example, point density calculations. On the other hand, point-to-area and point-to-surface transformations of point data representing qualitative or quantitative attributes can be modeled by, respectively, non-interpolative transformations or spatial interpolations. The objective of such transformations is to reconstruct the *continuous field*, respectively, which was measured at the sampling points.

Non-interpolative transformations are suitable for point data measured on a nominal scale. In some cases, such transformations are also applicable to point data measured on ordinal, interval or ratio scale. Non-interpolative transformations involve creation of zones of influence around points with assumption of homogeneity of attribute data in each zone. Bonham-Carter (1994) describes a number of methods of non-interpolative point-to-area transformations, which are briefly reviewed here. The simplest method is to associate attributes of each point to a rectangular cell in a regular grid. Cells with more than one point are assigned attributes that are aggregated according to some rule and depending on measurement scale, whilst cells without points are assigned null attributes (Fig. 2-12B). This method has been used for regional geochemical mapping (e.g., Garrett et al., 1990; Fordyce et al., 1993). A modification of representing point data as rectangular cells is to draw equal-area circular cells centred on points and to assign attributes of each point to the corresponding circle; zones outside the circles are assigned null attributes (Fig 2-12C). An advantage of this method is that the size of the circle can

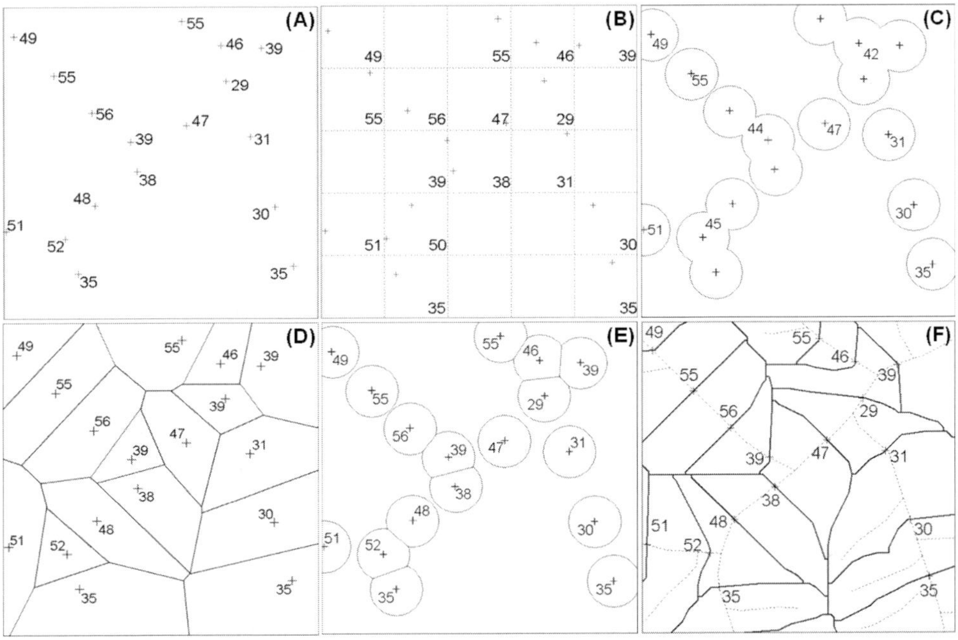

Fig. 2-12. Point-to-area transformations (adapted from Bonham-Carter, 1994, pp. 146). (**A**) Distribution of point data on a map. (**B**) Points transformed to regular cells. Cells with more than one point are assigned aggregated (e.g., mean) attributes whilst cells without points are null attributes. (**C**) Points transformed to circular cells. Zones defined by overlapping circles are assigned aggregated attributes. (**D**) Points transformed to Thiessen polygons. (**E**) Points transformed to areas defined by overlap of Thiessen polygons and circular cells. (**F**) Stream sediment sample points transformed to sample catchment basins. Dotted lines are streams. Solid lines are outlines of drainage catchment basins.

be chosen with subjectivity to represent zone of influence of a point. A disadvantage of this method is that some circles will overlap, which provides difficulty in deciding assignment of attributes to overlapping zones. This problem can be overcome by creating Thiessen or Voronoi or Dirichlet polygons around each point (Fig 2-12D) (Burrough and McDonnell, 1998). The points can then be represented by Thiessen polygons restricted to circular zones (Fig 2-12E). Bartier and Keller (1991) represented stream sediment point data as Thiessen polygons to integrate such data with bedrock geological data in a GIS analysis. They recognise, however, that representation of stream sediment data as Thiessen polygons is less appealing intuitively than representation of such data as sample catchment basin polygons, which is another method of point-to-area transformation (Fig 2-12F).

Spatial interpolation is involved in point-to-surface transformations of point data representing continuous variables. Surface models produced by any interpolation method can be symbolised and visualised by contouring, a subject that is treated thoroughly by Watson (1992). For a given set of irregularly- or regularly-spaced point data, there are

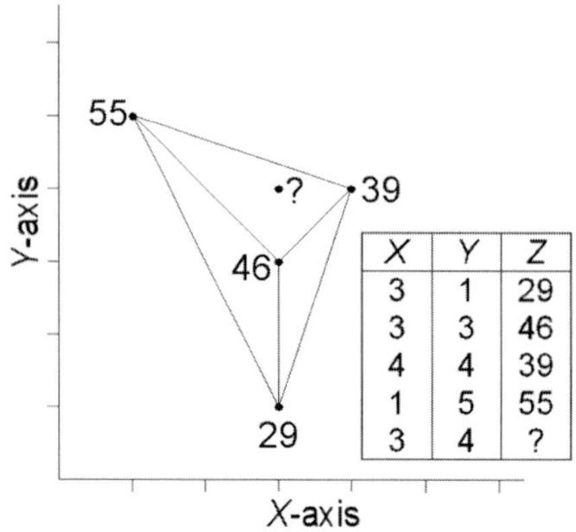

The equation for a triangular facet is:

$$Z = a + bX + cY$$

where X and Y are map coordinates of points with attribute values Z, and a, b and c are coefficiens.

The vertices of the triangle, with a point of unknown Z value, are given by the equations:

$$55 = a + b + 5c$$
$$39 = a + 4b + 4c$$
$$46 = a + 3b + 3c$$

whose solutions are a=67, b=-5.75 and c=-1.25.

The estimated Z value at the point of interest is

$$\hat{Z} = 67 - 5.75(3) - 1.25(4) = 44.75$$

Fig. 2-13. Transformation of data points (at vertices of triangles) into a surface via Delaunay triangulation. With assumption that an unknown point lies on the plane of a triangular facet, the value at that point can be estimated based on the equations of the triangle's vertices.

several spatial interpolation methods, which can generally be classified as either triangulation (i.e., TIN generation) or gridding methods. In triangulation methods, given control points form vertices of triangles and values at any point are estimated according to the equation for a triangular facet containing such points (Fig. 2-13). Triangulation methods are suitable for modeling of topographic, stratigraphic or structural surfaces. In gridding methods, values of the surface to be modeled are estimated at locations, called 'grid nodes', arranged in a regular pattern completely covering area of interest. Grid nodes are usually arranged in a square pattern and a zone enclosed by four nearest neighbouring grid nodes is called a 'grid cell'. The choice of a grid cell size, which determines accuracy and computing time of a surface model, is largely a subjective judgment but depends primarily on density and distribution of a given point data set. Generally, values of the surface at grid nodes are unknown and are estimated using control points where values of the surface are known. Various gridding methods exist and their detailed descriptions can be found in several textbooks (e.g., Burrough and McDonnell, 1998). For each gridding method, the estimation process involves three essential steps. Control points are first sorted according to their geographic coordinates. From the sorted controls points, a search is made for control points within a neighbourhood surrounding a grid node to be estimated. The value of a grid node is finally estimated by some mathematical function of values of control points within a search neighbourhood (which is usually circular or elliptical). An example of a mathematical function is *moving average*, whereby for each grid node the average of values at control points within a search neighbourhood that is 'moved' from one grid node to another is estimated (Fig. 2-14). Values at control points are projected horizontally to a grid node, where they are weighted and averaged. Weights are calculated because control points closer to a grid node to be estimated should have more influence on the estimated value than control points farther away. Of the different

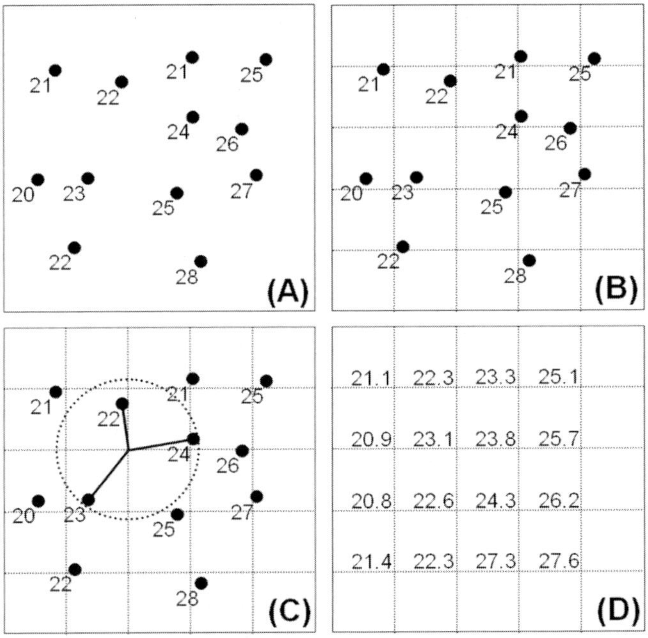

Fig. 2-14. Transformation of data points into a surface grid. (**A**) Control data points; each point is characterised by its x-coordinate (east-west or across page width), y-coordinate (north-south or down page length), and z-coordinate (value beside a point). Point data are identified by numbering them sequentially, as read by software, from 1 to i. Thus, a control point i has coordinates x_i and y_i and a value z_i. (**B**) A regular grid of nodes is superimposed on the map. These grid nodes are also numbered sequentially from 1 to k. Each grid node has coordinates x_k and y_k, and a value to be estimated z_k. (**C**) The value z_k at a grid node k is estimated from n control points found within a search neighbourhood, of specified area of influence, centred at k. (**D**) Completed grid with estimated values of z_k.

'weighted moving average' methods, inverse distance method and kriging have been usually applied to model geochemical surfaces. Point-to-surface transformations of geochemical data by spatial interpolation are applicable in fractal analysis of geochemical anomalies (see Chapter 4).

Methods of point-to-surface transformations are also applicable to line-to-surface transformations of linear geo-objects representing continuous variables (e.g., isolines of elevation). That means all points or samples of points along linear geo-objects are used in triangulation or gridding methods. In contrast, methods of point-to-area transformations are not readily applicable to line-to-area transformations, particularly if linear geo-objects represent qualitative variables. For linear geo-objects representing qualitative variables (e.g., curvi-linear structures, lithologic contacts, etc.), the idea of line-to-area transformation is to generate zones of proximity to linear features through an operation known as *buffering* or *dilation*, which depends on distance calculations. Points or polygons can also be buffered (i.e., point-to-area or area-to-area transformations) if they represent qualitative variables. Buffering is performed when one intends to determine spatial associations between locations of mineral deposits and various geological features (e.g., structures, anomalies). Buffering is among the most common neighbourhood operations, which are treated briefly below and treated in more detail in Chapter 6.

Neighbourhood operations

The general objective of a neighbourhood operation is to analyze the characteristics and/or spatial relationships of locations surrounding some specific (control) locations. Note that control locations are actually part of the neighbourhood to be analyzed. Thus, in fact, spatial interpolation techniques are a type of neighbourhood operation, because they aim to estimate values at unsampled locations based on values at sampled locations. Most types of neighbourhood operations applied in mapping of geochemical anomalies and mineral prospectivity are performed using a raster data model, because this ensures spatial adjacency of control pixels to neighbouring pixels. Buffering, however, may be performed using either vector or raster data.

Neighbourhood operations applied to raster maps are basically *filtering* operations. Filtering can be performed in the time domain, frequency domain or spatial domain. Filtering in the spatial domain is a basic function in GIS, which is further discussed here. Filtering operations in the time domain and frequency domain are beyond the scope of this volume; Davis (2002) provides a clear discussion of filtering operations in the time and frequency domains as applied to geological data analysis.

Filtering of a raster map involves an equal-sided filter window, also called a "kernel" or "template", which moves across a raster map one pixel at a time. A filter has an odd number of pixels on each of its sides so that it defines a symmetrical neighbourhood about the central pixel (Fig. 2-15). The simplest filter is a square of 3x3 pixels. Each pixel visited by a filter becomes the control location in a neighbourhood and a new value is calculated for that pixel according to certain mathematical 'search' functions that are desired to characterise that neighbourhood.

There are three basic elements in a neighbourhood operation – the control pixel, the neighbouring pixels and the search function to be applied to the neighbourhood. Because there are four general types of spatial data – ratio, interval, ordinal, nominal – the choice of mathematical search functions used in filtering operations depends on the type of data being studied. Note also that if a function can be applied to any type of data in the order as listed above, then that function can also be applied to the preceding type of data. The following discussion details examples of eight mathematical search functions, which are described along with their results for data in Fig. 2-15A.

The data in Fig. 2-15A can be ratio, interval or ordinal. A MINIMUM function returns to the central pixel the value of the pixel in the neighbourhood with the lowest value (Fig. 2-14B). The MINIMUM function is often used in a BOOLEAN search query (true or false) to find 'false pits' (single pixel depressions) in a digital elevation model (DEM) before performing runoff simulations. A MAXIMUM function returns to the central pixel the value of the pixel in the neighbourhood with the highest value (Fig. 2-

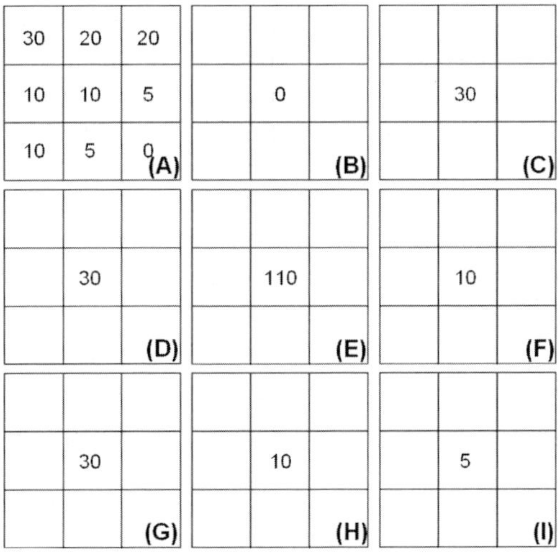

Fig. 2-15. Neighbourhood operations with mathematical search functions. (**A**) Example raster map data. (**B**) Result of a MINIMUM function. (**C**) Result of a MAXIMUM function. (**D**) Result of a RANGE function. (**E**) Result of a SUM operation. (**F**) Result of a MEDIAN function. (**G**) Result of a MINORITY function. (**H**) Result of a MAJORITY function. (**I**) Result of a DIVERSITY function.

15C). The MAXIMUM function can be used to find, for example, locations with highest probability of deposit-type occurrence in a mineral prospectivity map. A RANGE function returns to the central pixel the difference between highest and lowest values in the neighbourhood (Fig. 2-15D). The RANGE function can be used, for example, to find the range of metal concentrations surrounding every location in a soil geochemical map, which may indicate locations with high geochemical contrast. A SUM function returns to the central pixel the sum of the pixel values in the neighbourhood (Fig. 2-15E). The SUM function is useful in measuring density of geo-objects in a map. A MEDIAN function returns to the central pixel the median of pixels values in the neighbourhood (Fig. 2-15F). The MEDIAN function can be used, for example, to smooth values in a map; it serves a similar purpose to using an AVERAGE function. The AVERAGE function usually returns an integer value rather than a real value, and thus it is not suitable for interval or ordinal data. A MINORITY function returns to the central pixel the pixel value that occurs least frequently in the neighbourhood (Fig. 2-15G). The MINORITY function is rarely used. A MAJORITY function returns to the central pixel the pixel value that occurs most frequently in the neighbourhood (Fig. 2-15H). The MAJORITY function can be used to replace missing values in a map; for example, to assign the most common lithologic unit in a neighbourhood. Note that unique interval or ordinal values can be assigned to lithologic units instead of their names, because the latter type of data are not amenable to mathematical operations. A DIVERSITY (or VARIETY) function returns to the central pixel the number of different pixel values in the neighbourhood (Fig. 2-15I). The DIVERSITY function can be used, for example, to find edges of polygons of different lithologic units whereby it returns a value of [1] for interior of a lithologic unit, [2] along the contacts of two lithologic units, and [3] or higher values where three or more lithologic units join (cf. Mihalasky and Bonham-Carter, 2001).

The preceding examples belong to the aggregation type of neighbourhood operations. Clear introductory discussions of other different types of spatial filters, particularly those used in raster image analysis, and the functions associated with such filters can be found in Bonham-Carter (1994, p. 204-212). Other types of neighbourhood operations involve 'spread' or 'seek' computations. Spread computations are applicable, for example, to flood inundation studies (e.g., Peter and Stuart, 1999) or environmental pollution studies (e.g., Haklay, 2007). Seek computations are applicable, for example, to hydrological studies (e.g., Vieux, 2004).

Map overlay operations

The previously discussed operations on spatial data – spatial query and selection, classification and re-classification, measurements, transformations, neighbourhood – are usually applied to analyze spatial patterns of interest in single maps of geoscience spatial data sets. However, the previously discussed operations could also actually involve at least a pair of maps. For example, selection of stream sediment samples in zones underlain by certain lithologic units (Fig. 2-8) involves a map of stream sediment sample locations and a lithologic map. In addition, mapping of stream sediment sample catchment basins via a neighbourhood operation involves a map of stream sediment sample locations, a map of drainage lines and a DEM. These examples show that map overlay operations are implicitly involved in some of the previously discussed operations on spatial data. Map overlay operations are perhaps the most important of all GIS functionalities. There are two important conditions that must be fulfilled in order to perform overlay operations: (1) maps are georeferenced to the same coordinate system; (2) maps must overlap and thus pertain to the same study area. The principle in overlay operations is to integrate maps of certain attributes of every location in order to produce a map of new attributes for every location.

The three most common overlay operations are *clip*, *overwrite* and *intersect* (Fig. 2-16). The clip operation, which is also called an *impose* operation, restricts the spatial extent of the first map to the spatial extent of the second map (the clip map) (Fig. 2-16A). The clip operation is useful, for example, to retrieve from a source thematic map spatial data pertaining to a study area. The clip operation does not result in a new attribute table; the output map adopts the attribute table of the first map. The overwrite operation, which is also called a *stamp* operation, adopts the data from the first map except where there are data in the second map; data in the second map take priority in the output. The overwrite operation is useful, for example, to update an existing lithologic map with recent results of lithologic mapping (Fig. 2-16B). The overwrite operation results in an attribute table for the output map only if the second map has new data attributes. Creation of a new attribute, however, is not necessary if the attribute table of the first map is updated initially so that it can be associated with the second map.

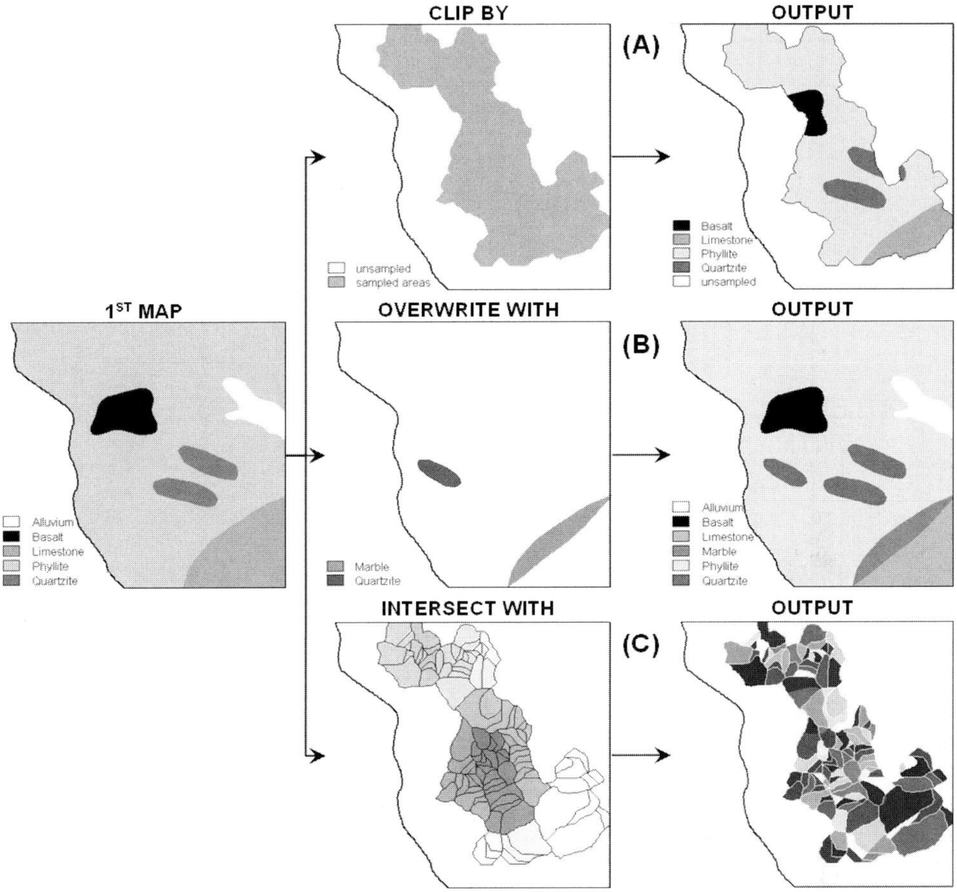

Fig. 2-16. Two-map overlay involving a geological map as common first map and different second maps. (**A**) Geological map is *clipped by* map of an area sampled with stream sediments (see Figs. 2-7A and 2-9A). (**B**) Geological map is *overwritten with* a map of recently delineated lithologic units. (**C**) Geological map is *intersected with* (or *crossed with*) a map of stream sediment sample catchment basins. Further illustration of the intersect operation is shown in Fig. 2-17.

The intersect operation, which is also called the *cross* or *spatial join* operation, is perhaps the most standard of all overlay operations. The intersect operation results in a collection of all possible intersections between geo-objects in the two input maps (Figs. 1-4 and 2-16C). It is useful, for example, in the process of integrating lithologic information in catchment basin analysis of stream sediment geochemical anomalies (Chapter 5). The intersect operation is applicable not only to polygonal geo-objects but also to linear and point geo-objects. If a map of polygonal geo-objects is intersected with a map of linear geo-objects, then the output map contains only linear geo-objects. The output geo-objects in an intersect operation adopt the geometry of geo-objects with the

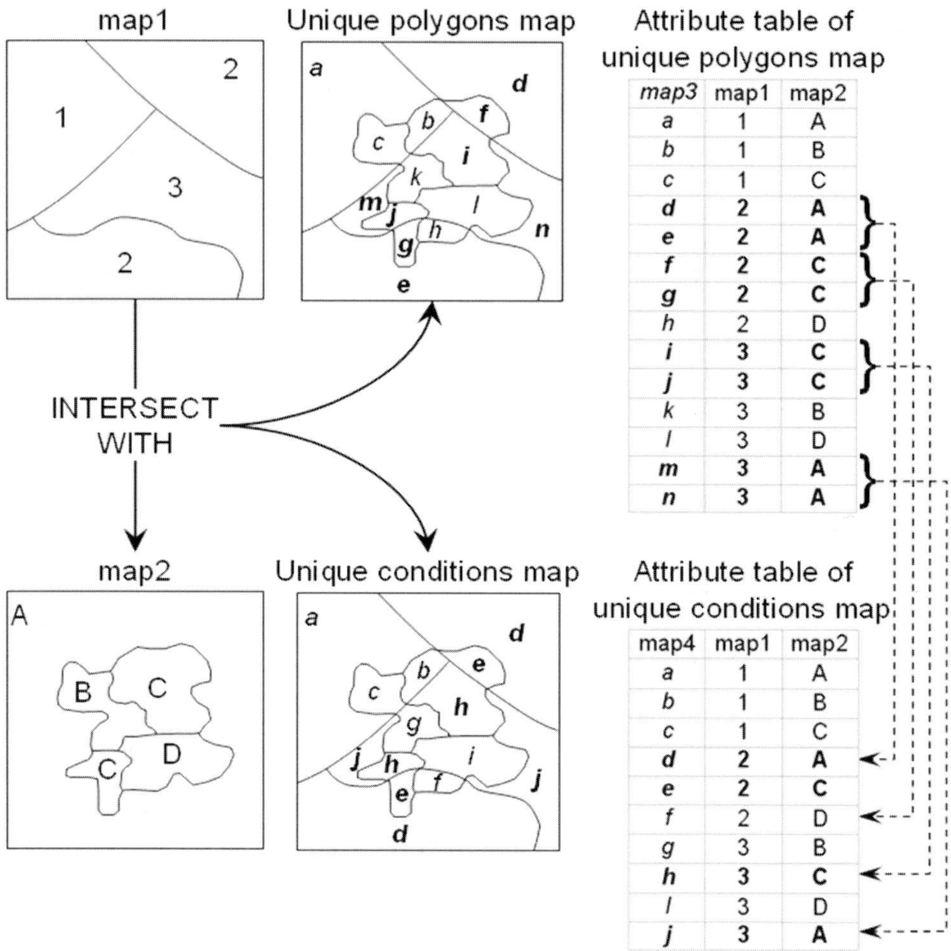

Fig. 2-17. Illustrations of result of intersect operation into either a unique polygons map or a unique conditions map. Individual geo-objects in a unique polygons map having the same combinations of attributes of the input maps are considered geo-objects of the same class in a unique conditions map. The latter map has fewer classes than the former map.

lowest spatial dimension in one of the input maps. The intersect operation always results in an attribute table that is associated with the output map.

There are two types of intersect operations, one type results in a *unique polygons* map and the other type results in a *unique conditions* map (Fig. 2-17). An intersect operation that results in a unique polygons map assigns a unique attribute (e.g., geo-object number or ID) to each polygon, even if there are polygons having the same or unique combinations of attributes of the input maps. This type of intersect operation is common in vector-based GIS software packages. An intersect operation that results in a unique conditions map assigns a unique attribute (e.g., geo-object class or ID) to polygons with unique combinations of attributes of the input maps. This type of intersect operation is common in raster-based GIS software packages. The number of unique conditions is usually less than but could be equal to the number of unique polygons. For example, in Fig. 2-17 there are eight pairs of unique polygons and each pair has unique combinations of attributes of the input maps. Thus, the attribute table associated with the unique conditions map has four records less than the attribute table associated with the unique polygons map. The reduced number of classes in a unique conditions map compared to a unique polygons map, or the reduced number of records in the attribute table associated with the former map compared to the latter map, is favourable in speeding up modeling computations. This is a reason why the intersect operation that results in a unique conditions map is usually preferred over the intersect operation that results in a unique polygons map.

Of the different overlay operations, the intersect operation is widely used in the analysis of at least two maps in order to obtain spatial information of interest. The objective of overlay operations in the analysis of either a pair of maps or multiple maps could be either (a) to combine maps according to certain conditions or rules or (b) to examine spatial relationships between two different geological features. The conditions or rules applied in combining a pair of maps represent a conceptual model of theoretical relationships between factors of or controls on how and where certain geochemical anomalies and mineral deposit-types occur (see Chapter 1).

SUMMARY

A GIS facilitates efficient capture, storage, organisation, management, query, retrieval, transformation, analysis and integration of geoscience spatial data sets used in mineral exploration. Such functionalities of a GIS, in turn, facilitate efficient modeling of spatial geo-information such as geochemical anomalies and prospective areas. The registration of spatial data to a common coordinate system, the representation of geo-objects and their data attributes as either vector or raster data models, and the organisation of spatial data attributes in relational databases all contribute to facility in analysis and integration of various spatial data or geo-information as a series of data layers. The way of analysing and integrating spatial data or geo-information in a GIS is actually an extension of the traditional light table method employed by mineral explorationists in the past. Spatial data operations on single or multiple data layers provide efficient tools for analysis of inter-relationships between data, which are important in modeling of geochemical anomalies and prospective areas. The various functionalities of a GIS that are relevant in predictive modeling of geochemical anomalies and prospective areas are demonstrated further in the succeeding chapters.

PART II. GEOCHEMICAL ANOMALY MAPPING

Geochemical Anomaly and Mineral Prospectivity Mapping in GIS
by E.J.M. Carranza
Handbook of Exploration and Environmental Geochemistry, Vol. 11 (M. Hale, Editor)

Chapter 3

EXPLORATORY ANALYSIS OF GEOCHEMICAL ANOMALIES

INTRODUCTION

Among the traditional methods for modeling of uni-element geochemical anomalies (see Chapter 1), the estimation of threshold as the mean plus (or minus) twice the standard deviation (hereafter denoted as mean±2*SDEV*) of a data set is based on classical statistics and hypothesis testing. The application of classical statistics fundamentally assumes that data consist of independent samples and have a normal distribution. These assumptions also apply to probabilistic data analysis (e.g., testing significance or probability levels of independence or normality). The assumptions in probabilistic and classical statistical data analyses are rigorous and require that data have been collected under rather carefully controlled conditions as in physical experiments. Whilst mineral explorationists strive to collect precise and accurate geochemical data, there are several uncontrollable factors that influence the values and variations of element contents in Earth materials that they sample. Such factors include not only geogenic (e.g., metal-scavenging by Fe-Mn oxides, lithology, etc.) and anthropogenic (i.e., man-induced) processes but also sampling and analytical procedures. Thus, uni-element geochemical data sets invariably contain more than one population, each of which represents a unique process. In addition, because geogenic processes are spatially dependent on one another and invariably explain the highest proportion of variations in uni-element contents in geochemical samples, it follows that geochemical data are invariably not spatially independent. Thus, many uni-element geochemical data sets invariably do not follow a normal distribution model (e.g., Vistelius, 1960; Reimann and Filzmoser, 1999). Certain transformations are usually applied to 'normalise' the values in a uni-element geochemical data set (Miesch, 1977; Joseph and Bhaumik, 1997), but even then most, if not all, transformed uni-element geochemical data sets only approximate a normal distribution (e.g., McGrath and Loveland, 1992). If a geochemical data set contains more than one population and does not follow a normal distribution model, then estimation of threshold as the mean±2*SDEV* can lead to spurious models of geochemical anomalies.

As an example here, Fig. 3-1 shows that the distribution of Fe concentrations in soil displayed in Fig. 1-1 clearly deviates from normality and consists of at least two populations (Fig. 3-1). Based on the given statistics in Fig. 3-1, the threshold estimated as mean+2*SDEV* of the data is greater than the maximum value in the data. The $\log_e$-transformed soil Fe values also do not approximate a log-normal distribution model and highlight the presence of at least two populations (Fig. 3-2). Based on the given statistics

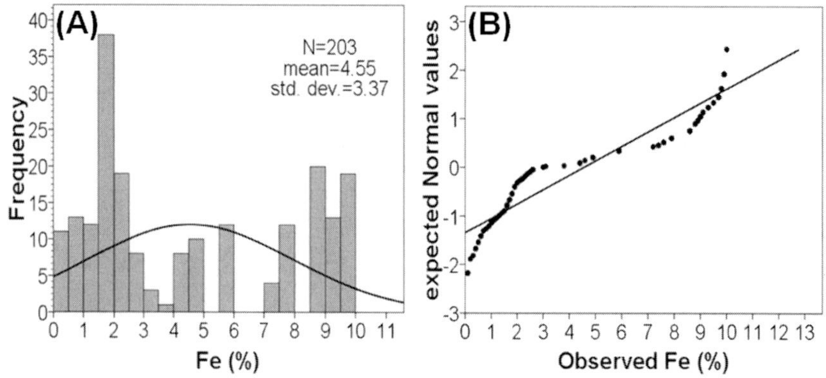

Fig. 3-1. (**A**) Histogram depicting the empirical density distribution of soil Fe (%) values (see Fig. 1-1) and the bell-shaped estimated density distribution curve based on a normal distribution model. (**B**) Normal Q-Q plot of soil Fe (%) values versus expected values based on a normal distribution model. If the soil Fe values have a normal distribution, the points should fall exactly on the line.

in Fig. 3-2, the estimated threshold as (anti-log of the) mean+2*SDEV* of the $\log_e$-transformed data is less than the maximum data value, but the value of the (anti-log of the) mean–2*SDEV* of the $\log_e$-transformed data is negative. Clearly, the application of classical statistics should be avoided in characterising empirical density distributions and mapping spatial distributions of uni-element geochemical data sets that do not follow a normal distribution model.

In the late 1970s, Tukey (1977) introduced the paradigm of exploratory data analysis

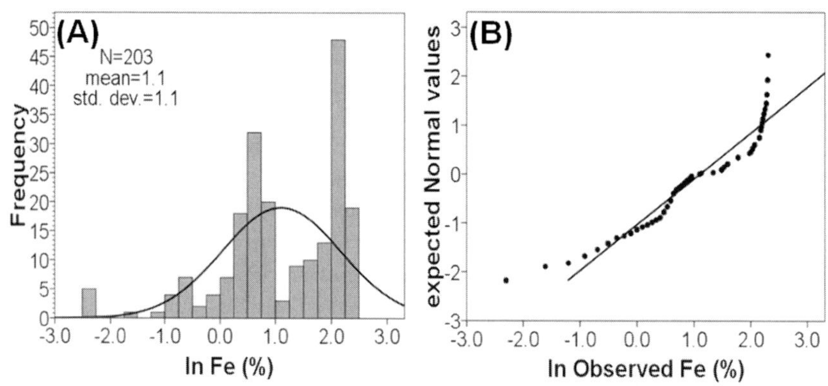

Fig. 3-2. (**A**) Histogram depicting the empirical density distribution of $\log_e$-transformed soil Fe (%) values (see Fig. 1-1) and the bell-shaped estimated density distribution curve based on a normal distribution model. (**B**) Normal Q-Q plot of $\log_e$-transformed (ln) soil Fe (%) values versus expected values based on a log-normal distribution model. If the $\log_e$-transformed soil Fe (%) values have a normal distribution, the points should fall exactly on the line.

(EDA), which was then considered as an unconventional and informal approach to analyse and interpret univariate data that do not follow a normal distribution model. Since the early 1980s, the EDA approach has gained attention in analysis and modeling of uni-element geochemical anomalies (e.g., Campbell, 1982; Smith et al., 1982; Howarth, 1983a, 1984; Garrett, 1988; Kürzl, 1988; Rock, 1988b; Chork and Mazzucchelli, 1989; Cook and Fletcher, 1993; Yusta et al., 1998; Bounessah and Atkin, 2003; Reimann et al., 2005; Reimann and Garrett, 2005; Grunsky, 2006). This chapter (a) reviews the concept and methods of EDA that are relevant in modeling of uni-element geochemical anomalies and (b) demonstrates a GIS-based case study application of EDA in modeling of significant geochemical anomalies.

EXPLORATORY DATA ANALYSIS

EDA is not a method but a philosophy of or an approach to robust data analysis (Tukey, 1977). It consists of a collection of descriptive statistical and, mostly, graphical tools intended to (a) gain maximum insight into a data set, (b) discover data structure, (c) define significant variables in the data, (d) determine outliers and anomalies, (e) suggest and test hypotheses, (f) develop prudent models, and (g) identify best possible treatment and interpretation of data. Whereas the sequence of classical statistical data analysis is problem→data→model→analysis→conclusions and the sequence of probabilistic data analysis is problem→data→model→prior data distribution analysis→conclusions, the sequence of EDA is problem→data→analysis→model→conclusions. Thus, classical statistical data analysis and probabilistic data analysis are *confirmatory* approaches to data analysis (being based on prior assumptions of data distribution models), whilst EDA, as its name indicates, is an *exploratory* approach to data analysis.

The goal of EDA is to recognise 'potentially explicable' data patterns (Good, 1983) through application of resistant and robust descriptive statistical and graphical tools that are qualitatively distinct from the classical statistical tools. From a statistical point of view, a statistic is resistant and robust (Huber, 1981; Hampel et al., 1986) (a) if it is only slightly affected either by a small number of gross errors or by a high number of small errors (resistance) and (b) if it is only slightly affected by data outliers (robustness). The descriptive statistical and graphical tools employed in EDA are based on the data itself but not on a data distribution model (e.g., normal distribution), yet they provide resistant definitions of univariate data statistics and outliers.

Graphical tools in EDA

The emphasis in EDA is interaction between human cognition and computation in the form of statistical graphics that allow a user to perceive the behaviour and structure of the data. Among the several types of EDA graphical tools (Tukey, 1977; Velleman and Hoaglin, 1981; Chambers et al., 1983), the density trace, jittered one-dimensional scatterplot and boxplot are most commonly used in uni-element geochemical data analysis (Howarth and Turner, 1987; Kürzl, 1988; Reimann et al., 2005; Grunsky, 2006). These three EDA graphics, which can be readily stacked on one another (Fig. 3-3), are

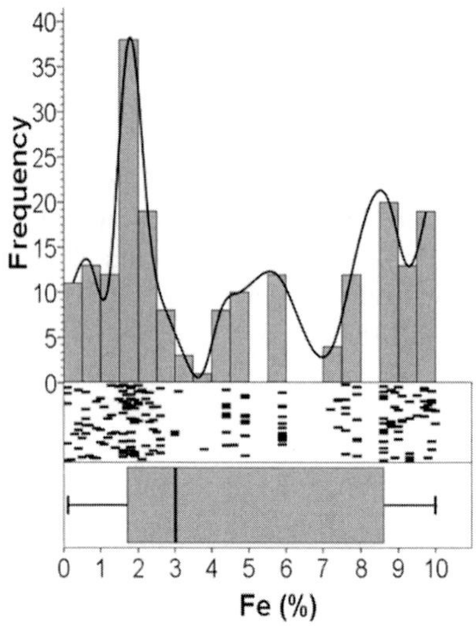

Fig. 3-3. Histogram and EDA graphics (density trace, jittered one-dimensional scatterplot, boxplot) depicting the empirical density distribution of the soil Fe data in Fig. 3-1.

often used together with a histogram. That is because the visual impression one obtains about univariate data behaviour and structure from a histogram alone is influenced by the choice of number of classes for constructing a histogram. The combination of the three EDA graphics with a histogram provides better insight to univariate data behaviour and structure than from a histogram alone. Unlike a histogram, the three EDA graphics can readily indicate any 'abnormalities' in a univariate data set.

A density trace is similar to a histogram, but it describes the empirical density distribution of univariate data in a much more realistic manner and its shape does not change significantly with change in the number of classes. The appropriate number of classes for constructing a histogram can be controlled by constructing a jittered one-dimensional scattergram, in which the univariate data are plotted at random positions across a narrow band (usually with a range of 0-1) orthogonal to the univariate data axis. Because a jittered one-dimensional scattergram is not based on classes of univariate data, it provides additional pieces information about data (i.e., structure, behaviour, local densities, gaps, outliers) that should be portrayed by a histogram and density trace. A boxplot displays information about the characteristics (e.g., central tendency, spread, etc.) of the empirical density distribution of a univariate data set, such that it is probably the most useful of the EDA graphical tools applied in geochemical data analysis and so it is given special attention here.

Boxplot and 5-number statistics of univariate data

A boxplot is created by first ordering data values from the minimum to the maximum, or vice versa (Fig. 3-4). The median value is then determined by counting

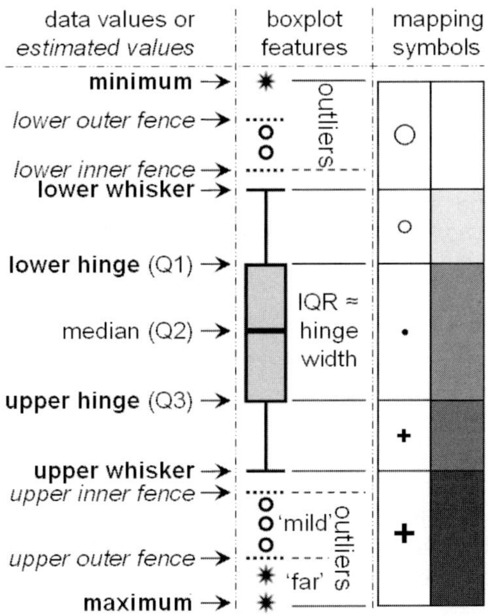

Fig. 3-4. Boxplot features representing characteristics of a univariate data set. Text in italics represents estimated values based on the inter-quartile range (IQR) or hinge width. Text in bold represents data values at which a univariate data set may be divided into five robust classes. Each class is represented by EDA-mapping symbols or grey-scale colours as shown here.

half way through the data values from the minimum to the maximum or vice versa, thereby dividing the univariate data set into two equal parts. By counting half way from the minimum to the median and from the maximum to the median, the lower hinge (*LH*) value and the upper hinge (*UH*) value are determined, respectively. The lower hinge, median and upper hinge, thus, divide a data set into four approximately equal parts known as quartiles. Values from the minimum to the lower hinge represent the 1st quartile (Q1) of a data set, values from the lower hinge to the median value represent the 2nd quartile (Q2), values from the median to the upper hinge represent the 3rd quartile (Q3) and values from the upper hinge to the maximum represent the 4th quartile (Q4). A box is then drawn between the lower and upper hinges. The box is then divided usually by a line at the median value. The absolute difference between the values at the lower and upper hinges represents the inter-quartile range (*IQR*) or hinge width:

$$hinge\ width = IQR = |lower\ hinge - upper\ hinge|\ . \tag{3.1}$$

A lower inner fence (*LIF*) and a lower outer fence (*LOF*) are defined at $1\frac{1}{2}\times IQR$ and $3\times IQR$, respectively, away from the lower hinge towards the minimum value. Algebraically, values (*X*) at the *LIF* and the *LOF* can be defined, respectively, as:

$$X_{LIF} = X_{LH} - (1.5 \times IQR) \tag{3.2}$$

$$X_{LOF} = X_{LH} - (3 \times IQR) \tag{3.3}$$

An upper inner fence (*UIF*) and an upper outer fence (*UOF*) are also defined at 1½×*IQR* and 3×*IQR*, respectively, away from the upper hinge toward the maximum. Algebraically, values (*X*) at the *UIF* and the *UOF* can be estimated, respectively, as:

$$X_{UIF} = X_{LH} + (1.5 \times IQR) \tag{3.4}$$

$$X_{UOF} = X_{LH} + (3 \times IQR) \tag{3.5}$$

For log-transformed data, the fences are defined using log-transformed values in either equation (3.4) or (3.5).

The lower whisker (*LW*) and the upper whisker (*UW*) are drawn from each of the hinges toward the most extreme data values within the inner fences. Algebraically, values (*X*) of the *LW* and the *UW* can be determined, respectively, as:

$$X_{LW} = \min(X[X > X_{LIF}]), \text{ and} \tag{3.6}$$

$$X_{UW} = \max(X[X < X_{UIF}]). \tag{3.7}$$

where the values in brackets are those within the inner fences and the hinges. For log-transformed data, the log values of the inner fences must be anti-logged for use in either equation (3.6) or (3.7). Data values beyond the inner fences are considered outliers. Data values between the inner and outer fences are considered 'mild' outliers, whilst data values beyond the outer fences are considered 'far' or extreme outliers, i.e., very unusual values (Kotz and Johnson, 1985, pp. 136-137). Mild and extreme outliers beyond the inner fences are marked by different symbols (e.g., open circles and asterisks, respectively (Fig. 3-4)).

A boxplot thus defines the 5-number summary statistics (minimum, *LH*, median, *UH* and maximum) and describes the most important characteristics of a univariate data set, namely (Tukey, 1997; Hoaglin et al., 2000): (a) location or central tendency; (b) spread; (c) skewness; (d) lengths of tails; and (e) outliers. As the box represents approximately 50% or two quartiles of a univariate data set, it means that at most 25% of data can be outliers but these values do not significantly affect the median and the hinges. In addition, because the inner fences are defined by the *IQR* or the hinge width, they are not seriously affected by outliers. These imply that the boxplot (or box-and-whiskers) plot is resistant and robust against extreme outliers in a univariate data set.

Classification of uni-element geochemical data

Based on a boxplot, an exploration uni-element geochemical data set can usually be divided into five robust classes (Fig. 3-4): (1) minimum–*LW*; (2) *LW*–*LH*; (3) *LH*–*UH*; (4) *UH*–*UW*; and (5) *UW*–maximum. The *UIF* is usually considered the threshold separating background values and anomalies (e.g., Bounessah and Atkin, 2003; Reimann et al., 2005), although the *UOF* can also be used as the threshold (e.g., Yusta et al., 1998). However, an estimated value representing the *UIF* (equation (3.4)) may not be actually part of an exploration uni-element geochemical data set, so that outliers beyond the *UW* represent anomalies. Thus, data values in the *UH*–*UW* class (at most 25% of a data set) can be considered high background, data values in the *LH*–*UH* class (at most 50% of a data set) are background, data values in the *LW*–*LH* class (at most 25% of a data set) are low background and data values in the minimum–*LW* class are extremely low background.

Aside from the boxplot-defined threshold (e.g., *UIF* or *UW*), a threshold can be defined from the EDA statistics as median+2*MAD*. The *MAD* is the median absolute deviation, which is estimated as the median of absolute deviations of all data values from the data median (Tukey, 1977):

$$MAD = median[|X_i - median(X_i)|] \tag{3.8}$$

where the values in brackets are absolute differences between values X_i and median of such values. The *MAD* is analogous to the *SDEV* in classical statistics, so the EDA median+2*MAD* threshold is also analogous to the classical mean+2*SDEV* threshold.

Standardisation of classified uni-element geochemical data

When dealing with individual uni-element geochemical data sets showing presence of multiple populations (e.g., as shown in Fig. 3-3), analysis of only the whole of a uni-element geochemical data set is inadequate for recognition of anomalies that may be associated with individual populations in the data. It is imperative to subdivide a uni-element geochemical data set into subsets representing the various populations present. The empirical data distribution of a geochemical data set as depicted in a boxplot or in a cumulative probability plot (Tennant and White, 1959; Sinclair, 1974) and in equivalent Q-Q (quantile-quantile) or Normal Q-Q plots (Figs. 3-1 and 3-2) can be useful in graphical examination of multiple populations and in defining breaks or inflection points, at which to subdivide a uni-element geochemical data set into subsets representing those populations. Alternatively, if populations present in a uni-element geochemical data set are considered to be strongly related to certain geogenic variables (e.g., lithology) that have also been recorded during the geochemical sampling, then individual uni-element geochemical data sets may be subdivided into subsets according to such variables.

For example, Fig. 3-5A shows boxplots of subsets of the soil Fe data according to rock type at the sample sites. Comparing and contrasting these boxplots of subsets of the

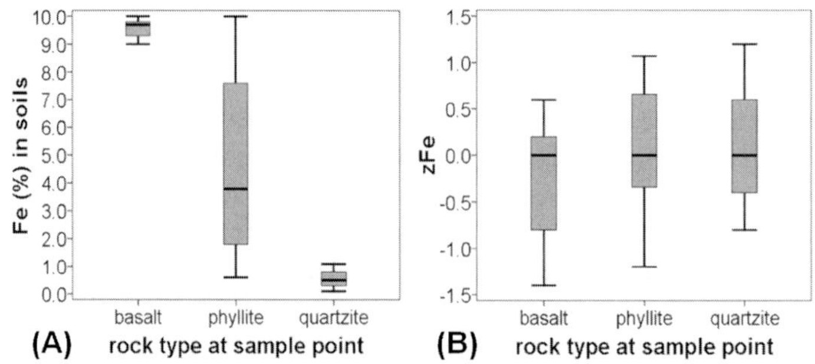

Fig. 3-5. Boxplots of subsets of the soil Fe data (shown in Fig. 1-1) based on rock type at sampling sites. (**A**) Original data values. (**B**) Data values standardised according to equation (3.10).

data with the boxplot of the whole data (Fig. 3-3) would lead one to make the following conclusions. If presence of multiple populations in the data set is not recognised (or ignored) and if such populations are not analysed individually, then the Fe data values associated with basalt could all be misclassified and mapped as high background whilst the Fe data values associated with quartzite could all be misclassified and mapped as low background. Such misclassifications based on analysis of a whole uni-element geochemical data set could lead to mapping of false positive and false negative geochemical anomalies (or Type I and Type II errors, respectively).

For proper and uniform classification of different populations that may be present in a uni-element geochemical data set, a suitable standardisation algorithm is required. A standardisation algorithm based on classical statistics is defined as:

$$Z_{ij} = \frac{X_{ij} - \overline{X}_j}{SDEV_j} \tag{3.9}$$

where Z_{ij} represents the standardised data values for population j, X_{ij} the original values i in population j, $\overline{X}_j$ the arithmetic mean of X_{ij} values and $SDEV_j$ the standard deviation of X_{ij} values. Because the mean and standard deviation in classical statistics are not resistant to outliers, the standardisation algorithm in equation (3.9) should be avoided accordingly. Yusta et al. (1998) proposed the following standardisation algorithm based on EDA statistics:

$$Z_{ij} = \frac{X_{ij} - median_j}{IQR_j}\,. \tag{3.10}$$

The standardisation algorithm in equation (3.10) makes each population j in a geochemical data set comparable to one another and consequently makes classes of the different populations in a uni-element geochemical data set comparable (Fig. 3-5B). The *MAD* can be used in lieu of *IQR* in equation (3.10), thus:

$$Z_{ij} = \frac{X_{ij} - median_j}{MAD_j} . \tag{3.11}$$

The EDA standardised values according to equation (3.11) are thus analogous to the classical standardised values according to equation (3.9).

In order to compare anomalies associated with different populations in a uni-element geochemical data set, the boxplot-defined threshold and the *IQR* defined for each population *j* can be used for standardisation (cf. Yusta et al., 1998):

$$Z_{ij} = \frac{X_{ij} - threshold_j}{IQR_j} . \tag{3.12}$$

The standardisation via equation (3.12) should make use of the same type of boxplot-defined threshold values (e.g., the boxplot *UW*). Alternatively, the median+2*MAD* and the median can be used for standardisation:

$$Z_{ij} = \frac{X_{ij} - (median + 2MAD)_j}{median_j} . \tag{3.13}$$

The standardisation algorithms in either equation (3.10) or equation (3.11) would allow representation of uni-element geochemical data from different sampling media in the same maps in order to, for example, compare spatial distributions of the same elements in rocks and soils. Equation (3.12) or equation (3.13) could be used, for example, to compare anomalies of the same (pathfinder) elements in different sampling media. Standardisation of various uni-element geochemical data sets via either equation (3.10) or equation (3.11) can be an important step prior to modeling of multi-element signatures through application of multivariate analytical techniques, which require proper estimation of the multivariate covariance (or correlation) matrix.

Mapping of classified uni-element geochemical data

EDA-mapping symbols (Tukey and Tukey, 1981; Kürzl, 1988; Reimann, 2005), such as those shown in Fig. 3-4, have been proposed to represent data in robust-class intervals defined by a boxplot. A boxplot can be conveniently used as a map legend explaining the symbols of classes of data values. For point-symbol representation of uni-element geochemical data (say, for stream sediment samples), circles and crosses are used because they can be readily perceived to signify low and high values, respectively. Thus, extremely low background values, which are usually more infrequent than anomalies in an exploration uni-element geochemical data set, are represented by large open circles;

whereas anomalies are represented by large crosses. Because extremely low background values and anomalies are usually fewer, if not absent, compared to other data values in an exploration uni-element geochemical data set, the large symbols for the former data values will not dominate a map. Low background values and high background values are represented by small circles and small crosses, respectively. Background values are each marked with the smallest symbol – a dot – because they are expected to dominate the data and its map.

Fig. 3-6A shows a map of spatial distribution of Fe contents in soils based on boxplot classes defined from the whole data set (Fig. 3-3). The spatial distributions of the Fe data, based on the boxplot classes of the whole data set, can be explained readily by variations in lithology. Fig. 1-1A shows very similar distribution of the Fe data, although the classes were defined based on a-priori knowledge that variations in the Fe data are influenced strongly by one of the lithologic units. Thus, geochemical data classification based on a boxplot and the EDA-mapping symbols has strong ability to portray physically meaningful spatial distributions of uni-element data without assumption of the normal distribution model or a-priori information about certain factors that influence variability in a geochemical data set.

Further exploratory analysis of subsets of a uni-element geochemical data set according to certain criteria could provide further insight into processes that plausibly influence variations in the data set. For example, based on subsets of the Fe data according to rock type at sample sites (Fig. 3-5B) and after standardisation according to equation (3.10), soils on phyllite immediately around the basalt unit are shown to be high background in Fe compared to soils on phyllite farther away from the basalt (Fig. 3-6B). The plausible explanation could be contamination of soils on phyllite by soils derived from the basalt. In contrast, soils at the outer portions of the basalt unit are shown to be low background in Fe compared to soils at the inner portions of the basalt unit. The plausible explanation could be contamination of soil on basalt by soils derived from the phyllite. Thus, uni-element geochemical maps based on resistant classes defined by a boxplot of a whole data set or data subsets are potentially useful in interpretation of processes that control variations in the geochemical landscape.

Instead of using different EDA point-symbols, boxplot classes can be represented with the same and equal-sized point-symbols but with different shades of grey (Fig. 3-4) or different colours (e.g., Reimann, 2005). Grey-scale or colour-scale representations are appropriate for interpolated uni-element geochemical data, although the classes must be defined from a boxplot of original data at sampling points. The symbols or colours used to represent classes of an exploration uni-element geochemical data set, as defined through a boxplot, serve to objectively portray in a map the structure and spatial distribution of that data set with a balanced aesthetic (visual) impression.

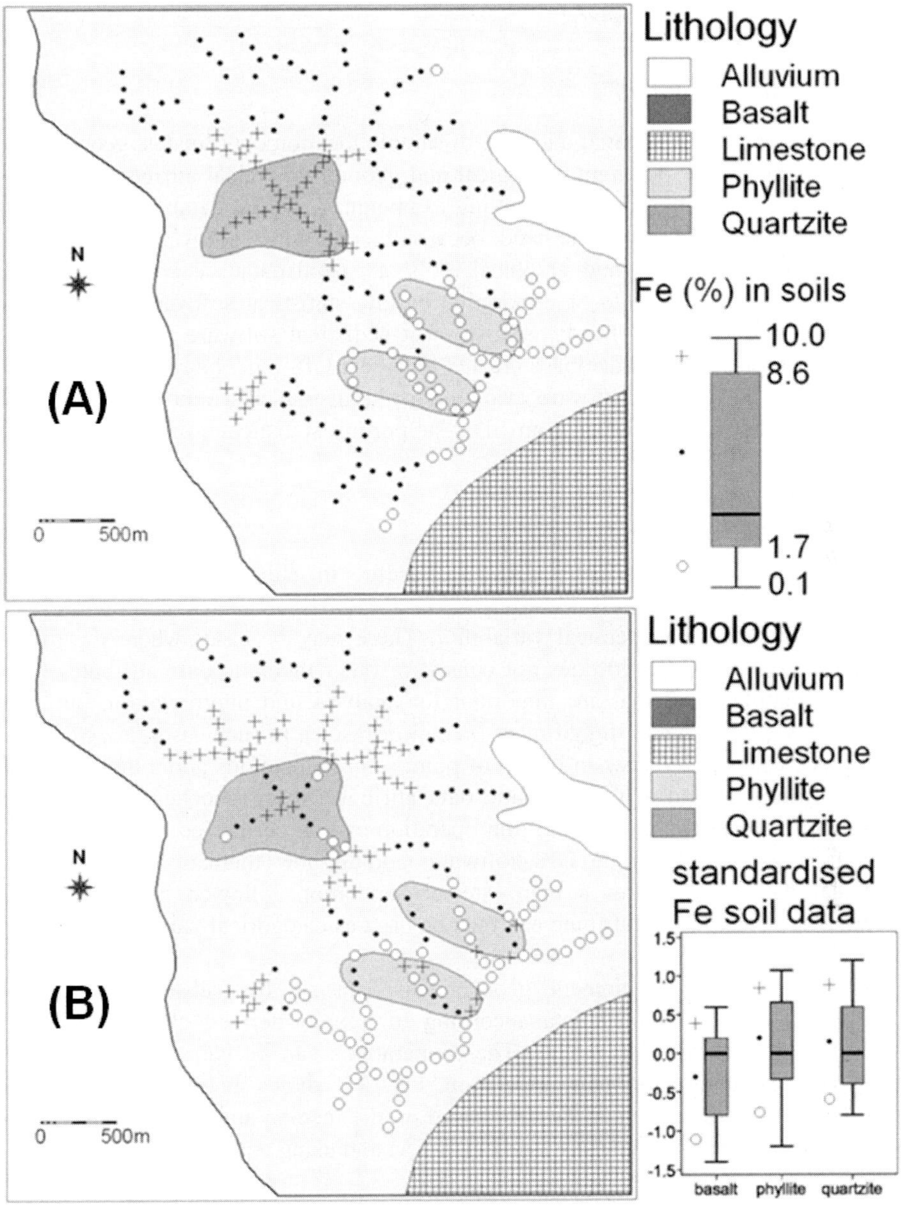

Fig. 3-6. Spatial distributions of the Fe data based on boxplot classes and EDA mapping symbols (cf. Fig. 1-1A). (**A**) Classified Fe values in the whole data set. (**B**) Classified standardised Fe values (using equation (3.10)) in subsets of the data according to rock type at sample sites.

APPLICATIONS OF GIS IN EDA

Management of spatial data

EDA clearly requires a spatial database of sample location coordinates, geochemical data attributes and other pertinent geological and geomorphological attributes that may have been recorded during the field sampling campaign. A spatial database may or may not be created in a GIS. On the one hand, because there are GIS software packages that do not have EDA statistical and graphical tools, a spatial database created in a GIS should be readily exportable to or importable by any statistical software package that supports EDA. On the other hand, because most statistical software packages do not have mapping tools, a spatial database created outside a GIS must be readily exportable to or importable by the GIS software available to the user. Thus, inter-operability of a spatial database between a GIS and an EDA-supporting statistical software package is highly desirable.

Operations on spatial data

Spatial query operations in a GIS are useful in subdividing a uni-element geochemical data set into subsets according to certain spatial attributes of variables (e.g., lithology) that control geochemical variability. There may be cases where the digital geochemical database acquired does not consist of other thematic data attributes (e.g., lithology at sample sites) that are important for analysis and interpretation, but such thematic data are available as digital maps (vector or raster). In such cases, a *spatial join* operation (see Chapter 2) between a map of point sample locations and a thematic map can be performed to add a new thematic data attribute in the geochemical database. Depending on GIS software, a spatial join operation can be performed simply via table calculation (Fig. 3-7). In some GIS software, adding new thematic attributes in a geochemical database requires a map intersect operation followed by a table join operation. The newly added attribute can then be used as categorical variable in creating boxplots (Fig. 3-5).

Classification of a uni-element geochemical data set (or subsets) as well as standardisation of geochemical data according to boxplot-defined classes and EDA statistics can be performed in a GIS. These operations can be carried out via either attribute table calculation or map calculation. Fig. 3-8 shows an example of a table calculation to standardise the soil Fe data based on the median and *IQR* of data subsets according to rock type at sample site (see Fig. 3-5A) and using equation (3.10).

Visualisation of spatial data or geo-information

A GIS can clearly support mapping and visualisation of EDA results in order to describe or explain plausible underlying processes that govern the spatial distributions of uni-element geochemical data. Maps of geochemical attributes or derivative attributes using, say boxplot-defined classes, can be readily created in a GIS (e.g., Fig. 3-6). Most

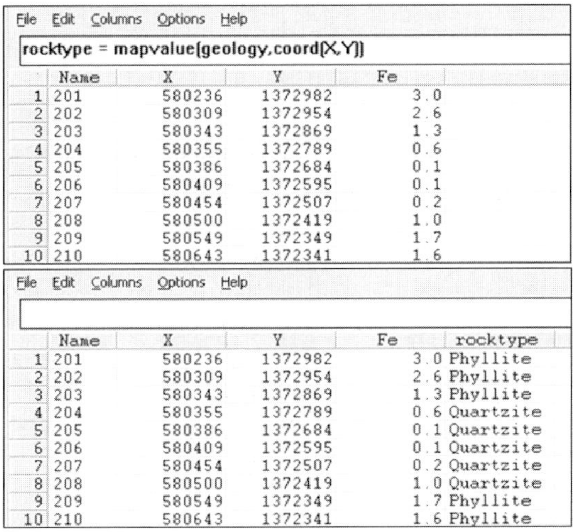

File Edit Columns Options Help

rocktype = mapvalue(geology,coord(X,Y))

	Name	X	Y	Fe
1	201	580236	1372982	3.0
2	202	580309	1372954	2.6
3	203	580343	1372869	1.3
4	204	580355	1372789	0.6
5	205	580386	1372684	0.1
6	206	580409	1372595	0.1
7	207	580454	1372507	0.2
8	208	580500	1372419	1.0
9	209	580549	1372349	1.7
10	210	580643	1372341	1.6

File Edit Columns Options Help

	Name	X	Y	Fe	rocktype
1	201	580236	1372982	3.0	Phyllite
2	202	580309	1372954	2.6	Phyllite
3	203	580343	1372869	1.3	Phyllite
4	204	580355	1372789	0.6	Quartzite
5	205	580386	1372684	0.1	Quartzite
6	206	580409	1372595	0.1	Quartzite
7	207	580454	1372507	0.2	Quartzite
8	208	580500	1372419	1.0	Quartzite
9	209	580549	1372349	1.7	Phyllite
10	210	580643	1372341	1.6	Phyllite

Fig. 3-7. Adding a thematic attribute to a geochemical database in a GIS. A table calculation formula (using software syntax) is executed in the command line (blank space at top) of the table. The particular calculation formula shown here, rocktype = mapvalue(geology,coord(X,Y)), uses an operator (mapvalue), which extracts attributes (rocktype) from a map (geology) at locations with the *x,y* coordinates [coord(X,Y)] of the geochemical samples.

GIS software packages support mapping symbols used in EDA. The map overlay functionality of GIS, which allows maps to be stacked on top of each other, clearly enhances visualisation of spatial geo-information derived from EDA (Fig. 3-6).

CASE STUDY

This case study demonstrates (a) the performance of two EDA-derived threshold values – boxplot *UW* and median+2*MAD* – against the performance of the traditional mean+2*SDEV* threshold and (b) usefulness of standardised data values based on EDA statistics in recognition of multi-element geochemical anomalies.

Study area and spatial data

The Aroroy district is located in the northwestern portion of Masbate Island in the Philippines (Fig. 3-9). The topography of the area is characterised by moderate relief and gently rolling to moderate slopes with elevations varying from zero to about 630 m and averaging about 110 m. Vegetation in the area consists mostly of grasses and shrubs. The average annual rainfall in Masbate is about 1,500 mm.

The Eocene-Oligocene Mandaon Formation, consisting mainly of andesitic-dacitic agglomerates, comprises the oldest rocks in the area (Fig. 3-9). The Miocene Aroroy Diorite, which varies in composition from quartz diorite to hornblende diorite, intruded the Mandaon Formation. Unconformably overlying the Mandaon Formation and the

File Edit Columns Options Help

zFe = iff(rocktype="basalt",(Fe-9.7)/0.5,iff(rocktype="phyllite",(Fe-3.8)/5.8,(Fe-0.5)/0.5))

	Name	X	Y	Fe	rocktype
1	201	580236	1372982	3.0	Phyllite
2	202	580309	1372954	2.6	Phyllite
3	203	580343	1372869	1.3	Phyllite
4	204	580355	1372789	0.6	Quartzite
5	205	580386	1372684	0.1	Quartzite
6	206	580409	1372595	0.1	Quartzite
7	207	580454	1372507	0.2	Quartzite
8	208	580500	1372419	1.0	Quartzite
9	209	580549	1372349	1.7	Phyllite
10	210	580643	1372341	1.6	Phyllite

File Edit Columns Options Help

	Name	X	Y	Fe	rocktype	zFe
1	201	580236	1372982	3.0	Phyllite	-0.14
2	202	580309	1372954	2.6	Phyllite	-0.21
3	203	580343	1372869	1.3	Phyllite	-0.43
4	204	580355	1372789	0.6	Quartzite	0.20
5	205	580386	1372684	0.1	Quartzite	-0.80
6	206	580409	1372595	0.1	Quartzite	-0.80
7	207	580454	1372507	0.2	Quartzite	-0.60
8	208	580500	1372419	1.0	Quartzite	1.00
9	209	580549	1372349	1.7	Phyllite	-0.36
10	210	580643	1372341	1.6	Phyllite	-0.38

Fig. 3-8. GIS-based standardisation of uni-element attributes using EDA statistics. A table calculation formula (using software syntax) is executed in the command line (blank space at top) of the table. The particular calculation formula shown uses a composite if-then expression to standardise Fe values according to values of median and IQR in subsets of the data according to rock type at sample site (see Fig. 3-5A) and using equation (3.10).

Aroroy Diorite are feldspathic wackes belonging to the Early Miocene Sambulawan Formation. Andesitic lithic tuffs (Late Miocene to Early Pliocene Lanang Formation) disconformably overlie the Mandaon Formation and the Sambulawan Formation. The Pliocene Nabongsoran Andesite consists of porphyritic stocks, plugs and dikes that intrude into the series of dacitic-andesitic volcano-sedimentary rocks (i.e., the Mandaon, Sambulawan and Lanang Formations) and the Aroroy Diorite. The Nabongsoran Andesite porphyry intrusions, many of which are not mappable at the map scale of Fig. 3-9, are probably responsible for either high-sulphidation (Mitchell and Balce, 1990) or low-sulphidation (Mitchell and Leach, 1991) hydrothermal alteration and epithermal Au mineralisation in the intruded rocks. Gold, in at least 13 mineral deposit occurrences in the area (Fig. 3-9), is associated with sulphide (mainly pyritic) minerals in wide-sheeted and manganese-bearing quartz or silicified veins in generally northwest-trending faults/fractures that cut the volcano-sedimentary rocks.

A subset of stream sediment geochemical data (135 out of more than 2,200 samples analysed for Cu, Zn, Ni, Co, Mn, As; JICA-MMAJ, 1986) pertaining to the study area is used. This subset of geochemical data represents a total drainage basin area of about 101 km^2 (i.e., a sampling density of one sample per 1-2 km^2). In the geochemical data for As, 40 stream sediment samples have censored values (i.e., half the detection limit of 0.5 ppm As). The geochemical data, lithological map and locations of the epithermal Au deposits were compiled in a GIS. These spatial data sets were used by Carranza (2004a) in a case study of GIS-based modeling of stream sediment geochemical anomalies.

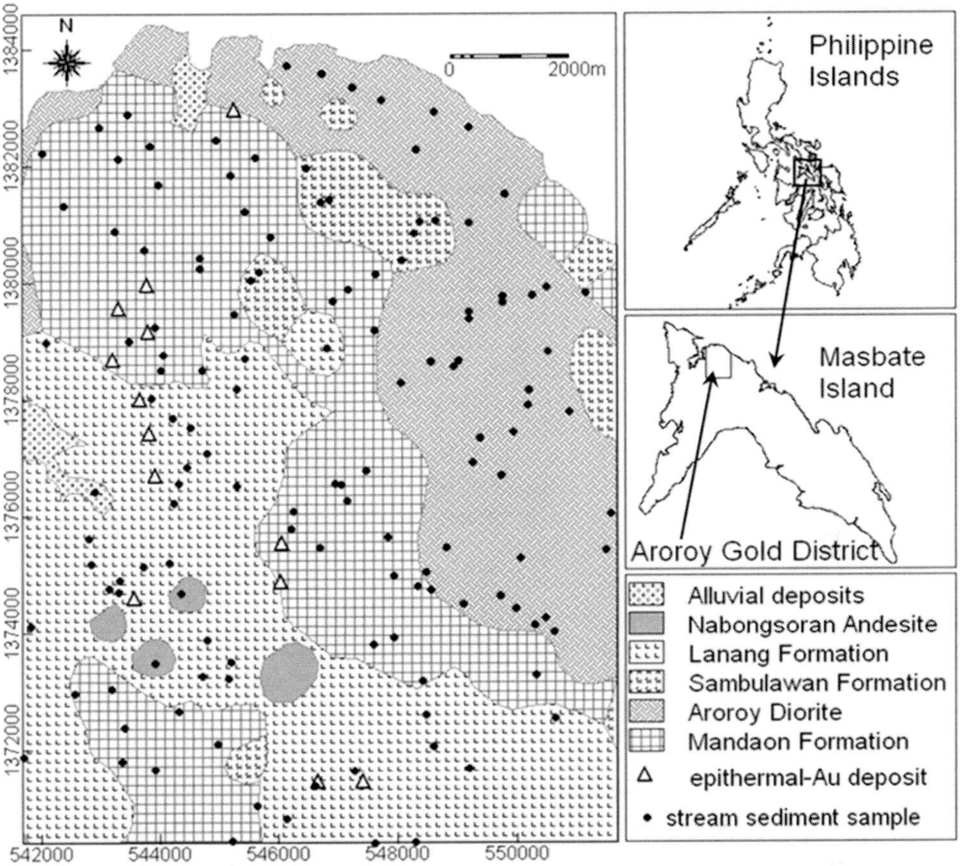

Fig. 3-9. Simplified lithological map (modified from Baybayan and Matos (1986) and JICA-MMAJ (1986)) and locations of stream sediment geochemical samples (JICA-MMAJ, 1986) in the Aroroy district, Masbate, Philippines. Map coordinates are in metres (UTM projection, zone 51).

Analysis of uni-element data distributions

In the initial stages of modelling significant anomalies, exploration geochemists "eyeball" each uni-element data set in order to obtain impressions about the data structure, behaviour, local densities, gaps and, of course, outliers. Descriptive statistics and statistical graphics are the useful tools in "eyeballing" empirical density distributions of uni-element data sets. Careful study of empirical density distributions of uni-element data sets is very important in the analysis and modeling of geochemical anomalies.

In the study area, the empirical density distributions of each of the raw uni-element data sets are all asymmetric, i.e., they are positively-skewed (or right-skewed) (Figs. 3-10). In the individual raw uni-element data sets, the means are greater than the medians

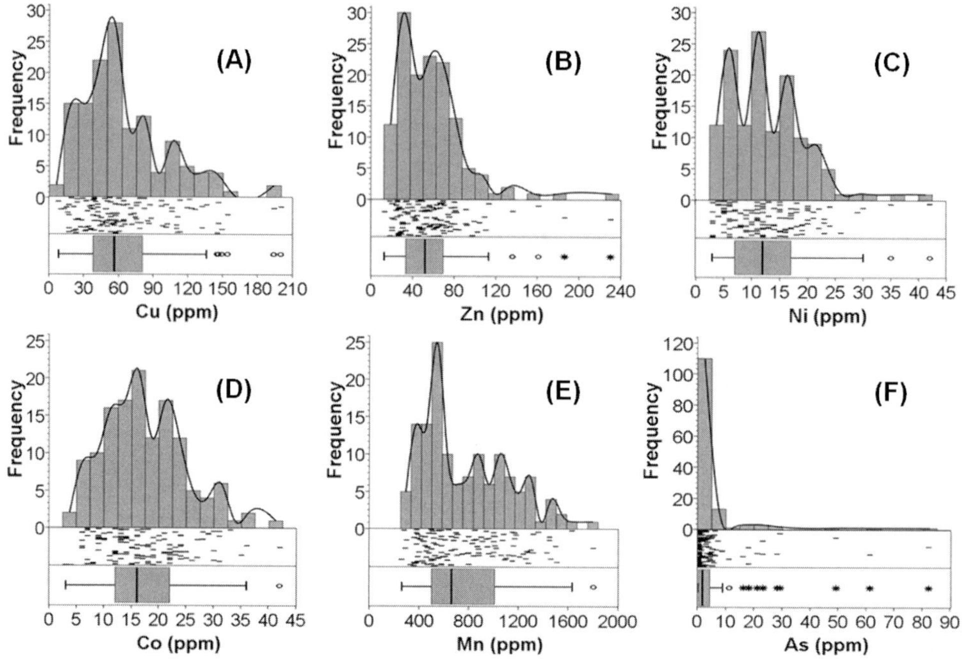

Fig. 3-10. Histograms and EDA graphics of empirical density distributions of the raw uni-element data sets, Aroroy district (Philippines). (**A**) Cu. (**B**) Zn. (**C**) Ni. (**D**) Co. (**E**) Mn. (**F**) As.

and the *SDEV*s are greater than the *MAD*s (Table 3-I). Because the individual uni-element data sets are either moderately or strongly right-skewed, the values of their mean–2*SDEV* are mostly negative whilst the values of their median–2*MAD* are mostly positive. The negative values of the mean-2*SDEV* indicate that estimates of the mean in the individual raw uni-element data sets are statistically non-significant. It follows that any estimate of threshold values according to the mean+2*SDEV* is non-meaningful. The asymmetric distributions of the individual uni-element data sets thus call for application of suitable transformations because "data should approach a symmetrical distribution before any threshold estimation methods are applied" (Reimann et al., 2005).

Several types of numerical transformation functions can be applied in order to reduce asymmetry of empirical density distribution of uni-element data (Miesch, 1977; Garrett et al., 1980; Joseph and Bhaumik, 1997). The purpose of transforming geochemical data should not be to obtain a (near) normal density distribution, as this is virtually impossible considering that most, if not all, exploration geochemical data sets are multi-modal (Fig. 3-10). For purposes of illustration, $\log_e$-transformation is applied here.

The empirical density distributions of each of the $\log_e$-transformed uni-element data sets have better symmetry compared to the respective raw data sets (Figs. 3-11) such that the anti-logs of both the *SDEV*s and *MADs* of the $\log_e$-transformed data sets are much smaller than those of the raw data sets (Table 3-I). The empirical density distributions of

TABLE 3-I

Classical and EDA statistics of uni-element concentrations (in ppm) in the raw and $\log_e$-transformed data for n=135 samples (except As*, for which n=95), Aroroy district (Philippines).

	Min.	Max.	Mean		*SDEV*		Median		*MAD*	
			Raw	$\log_e$	Raw	$\log_e$	Raw	$\log_e$	Taw	$\log_e$
Cu	8	200	64.04	3.98	37.84	0.62	56	4.03	20	0.38
Zn	13	230	56.90	3.90	32.43	0.52	52	3.95	18	0.37
Ni	3	42	12.59	2.37	6.86	0.59	12	2.48	5	0.40
Co	3	42	17.32	2.75	7.37	0.47	16	2.77	5	0.29
Mn	260	1800	769.93	6.55	345.69	0.45	660	6.49	230	0.37
As	0.25	82	4.62	0.34	10.62	1.49	1.5	0.41	1.25	1.14
As*	0.5	82	6.45	1.06	12.22	1.17	3.0	1.10	2.00	0.69

*without censored values.

the $\log_e$-transformed data sets do not, however, strictly follow a normal distribution model (Fig. 3-12). In the individual $\log_e$-transformed uni-element data sets, the medians

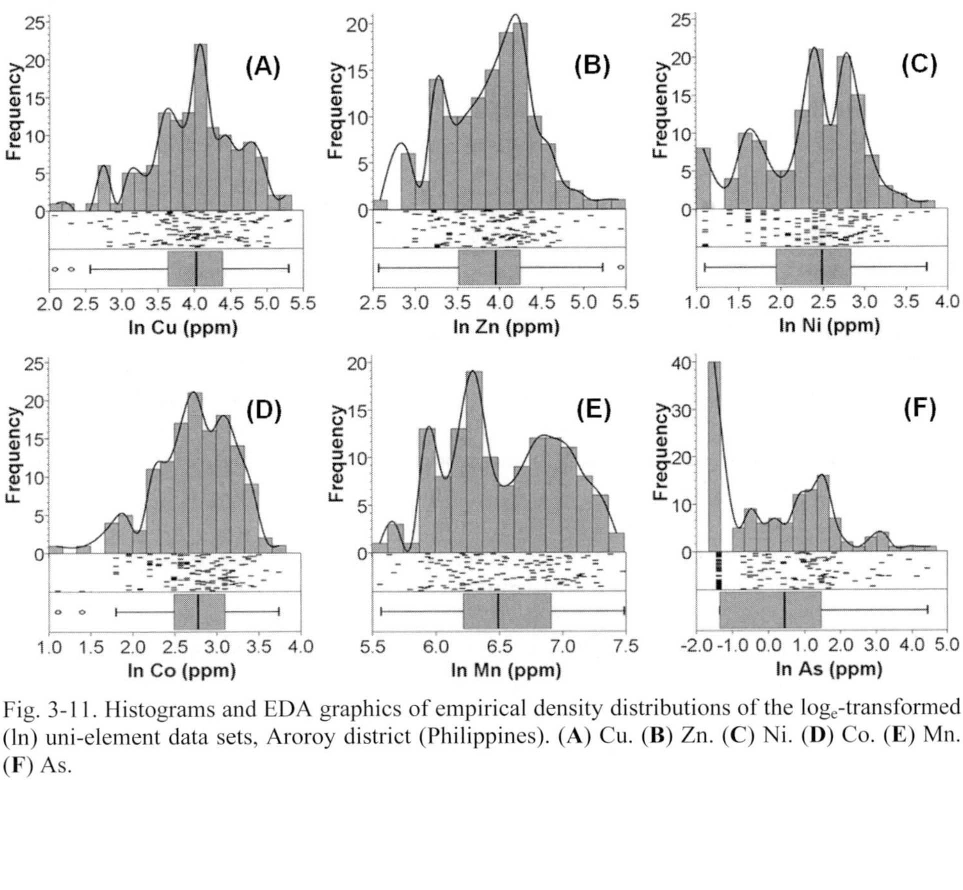

Fig. 3-11. Histograms and EDA graphics of empirical density distributions of the $\log_e$-transformed (ln) uni-element data sets, Aroroy district (Philippines). (**A**) Cu. (**B**) Zn. (**C**) Ni. (**D**) Co. (**E**) Mn. (**F**) As.

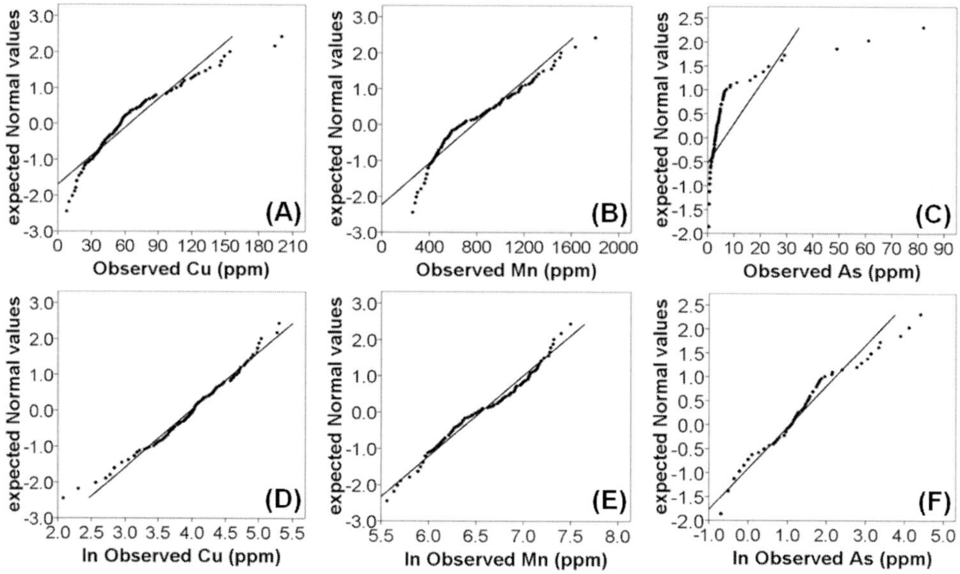

Fig. 3-12. Normal Q-Q plots of selected uni-element data (Aroroy district, Philippines) showing their deviations from the normal distribution (straight line) model. Raw data of (**A**) Cu, (**B**) Mn and (**C**) As. $\log_e$-transformed (ln) data of (**D**) Cu, (**E**) Mn and (**F**) As.

are slightly greater than the means and the *SDEV*s are all greater than the *MAD*s (Table 3-I). In addition, the values of either the mean–2*SDEV* or the median–2*MAD* in the $\log_e$-transformed data sets are mostly positive, except for the As data set. The results indicate that estimates of the classical descriptive statistics, unlike the estimates of the EDA descriptive statistics, are much more sensitive to values at/near one or both tails of any data set. The results also show that the $\log_e$-transformation has reduced the influence of very low or very high values at/near one or both tails of any of the data sets and thus improved the symmetry of their empirical density distributions. However, for the As data set, the $\log_e$-transformation is still insufficient to proceed to threshold estimation.

The individual raw uni-element data sets are all multi-modal, indicating presence of at least two populations (Figs. 3-10 and 3-11), which means that each data set must be subdivided into subsets representing each population. Graphical examination of a probability (or Q-Q) plot of a uni-element data can be useful in defining population break points (Sinclair, 1974). Identifying population break points in a probability (or Q-Q) plot is, however, highly subjective, requires experience and, thus, is not a trivial task. For example, inflection points are relatively easier to identify in the Normal Q-Q plots of the $\log_e$-transformed data sets for Mn and As than in the Normal Q-Q plot of the $\log_e$-transformed data for Cu (Figs. 3-12D to 3-12F). Nonetheless, the presence of at least two populations in each of the individual uni-element data sets is plausibly mainly due to lithology. Each of uni-element data sets was then subdivided into two subsets according to rock type at every sample location. The samples in areas underlain by diorite have

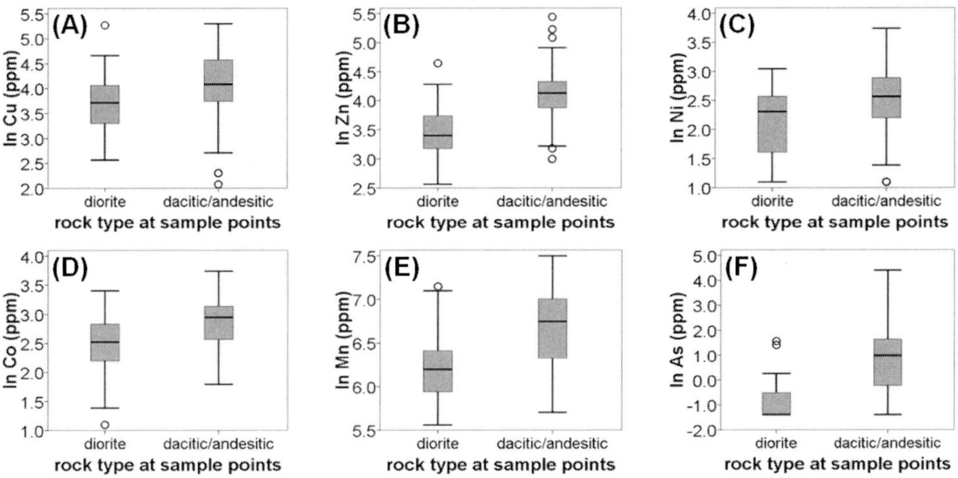

Fig. 3-13. Boxplots of subsets of the $\log_e$-transformed (ln) uni-element data according to rock type at sample points, Aroroy district (Philippines). (**A**) Cu. (**B**) Zn. (**C**) Ni. (**D**) Co. (**E**) Mn. (**F**) As.

rather lower concentrations of each of the elements under study than the samples in areas underlain by dacitic/andesitic rocks (Table 3-II, Fig. 3-13). The outliers of Cu, Zn and As in samples in areas underlain by diorite (Figs. 3-13A, 3-13B and 3-13F) would not be recognised if thresholds based on the *UW*s of boxplots of the whole $\log_e$-transformed data sets for these elements (Figs. 3-11A, 3-11B and 3-11F) were used in mapping.

Censored values must be removed especially if they form one population (Fig. 3-11F) because they result in reduced estimates of the descriptive statistics of the whole data set (Table 3-I) and potentially affect recognition of outliers. The boxplot of the $\log_e$-transformed As data inclusive of the censored values (Fig. 3-11F) does not show any outliers, whereas the boxplot of the $\log_e$-transformed As data exclusive of the censored values indicates the presence of outliers (Fig. 3-14A). The histogram and Normal Q-Q plot of the $\log_e$-transformed As data exclusive of the censored values indicates the presence of at least two populations (Fig. 3-14). Boxplots of the data subsets according to rock type at sample points show As outliers in samples in areas underlain by diorite and by dacitic/andesitic rocks. However, exclusive of the censored As data, the subset of samples in areas underlain by diorite is now very small (n=13; Table II), so probably only the descriptive statistics of the As data subset for samples in areas underlain by dacitic/andesitic rocks are meaningful.

Analysis of empirical density distributions of uni-element data sets or subsets should also be coupled with visualisation of their spatial distributions to determine whether any data treatment results in explicable or meaningful spatial patterns. Maps depicting spatial distributions of uni-element data, say based on boxplot-defined classes and EDA mapping symbols, are useful tools in eyeballing the data. For the study area, the maps in Fig. 3-15 indicate that the lithology has strong controls on the spatial distributions of most of the individual uni-element data, whereas the epithermal Au deposit occurrences

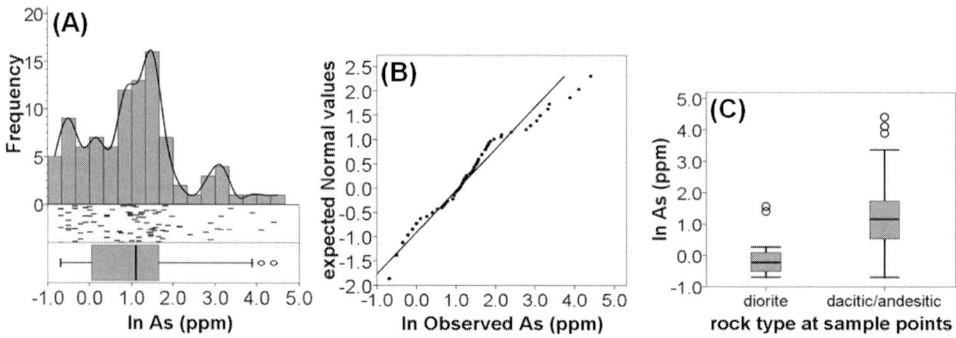

Fig. 3-14. Log_e-transformed (ln) As data exclusive of censored values, Aroroy district (Philippines). (**A**) Histogram and EDA graphics. (**B**) Normal Q-Q plot showing deviation of the data from the normal distribution (straight line) model. (**C**) Boxplots of subsets of the data according to rock type at sample points.

provide additional controls on the spatial distributions of Mn and As. High background Mn values and high background to outlying As values occur along a northwest-trending zone in the western part of the area following the same trend of the epithermal Au deposit occurrences. Note that the gold deposit occurrences are associated with Mn-stained silicified veins deposited along northwest-trending faults/fractures that cut dacitic/andesitic volcano-sedimentary rocks. The As outliers in the northwestern part of

TABLE 3-II

Classical and EDA statistics log_e-transformed data subsets according to rock type at sample points, Aroroy district (Philippines).

Rock type at sample points	Element	Mean	*SDEV*	Median	*MAD*
Aroroy Diorite (at n=38 samples, except the As data* for n=13 samples)	Cu	3.73	0.55	3.72	0.38
	Zn	3.45	0.41	3.44	0.30
	Ni	2.10	0.61	2.11	0.50
	Co	2.50	0.49	2.52	0.32
	Mn	6.22	0.37	6.19	0.25
	As	-0.92	0.78	-1.39	0.00
	As*	-0.02	0.74	-0.22	0.58
Dacitic/andesitic volcano-sedimentary rock (at n=97 samples, except the As data* for n=82 samples)	Cu	4.08	0.62	4.08	0.37
	Zn	4.08	0.45	4.13	0.24
	Ni	2.48	0.55	2.56	0.36
	Co	2.85	0.42	2.94	0.28
	Mn	6.68	0.41	6.75	0.32
	As	0.83	1.41	0.99	0.82
	As*	1.23	1.13	1.16	0.32

*Excluding samples with censored values.

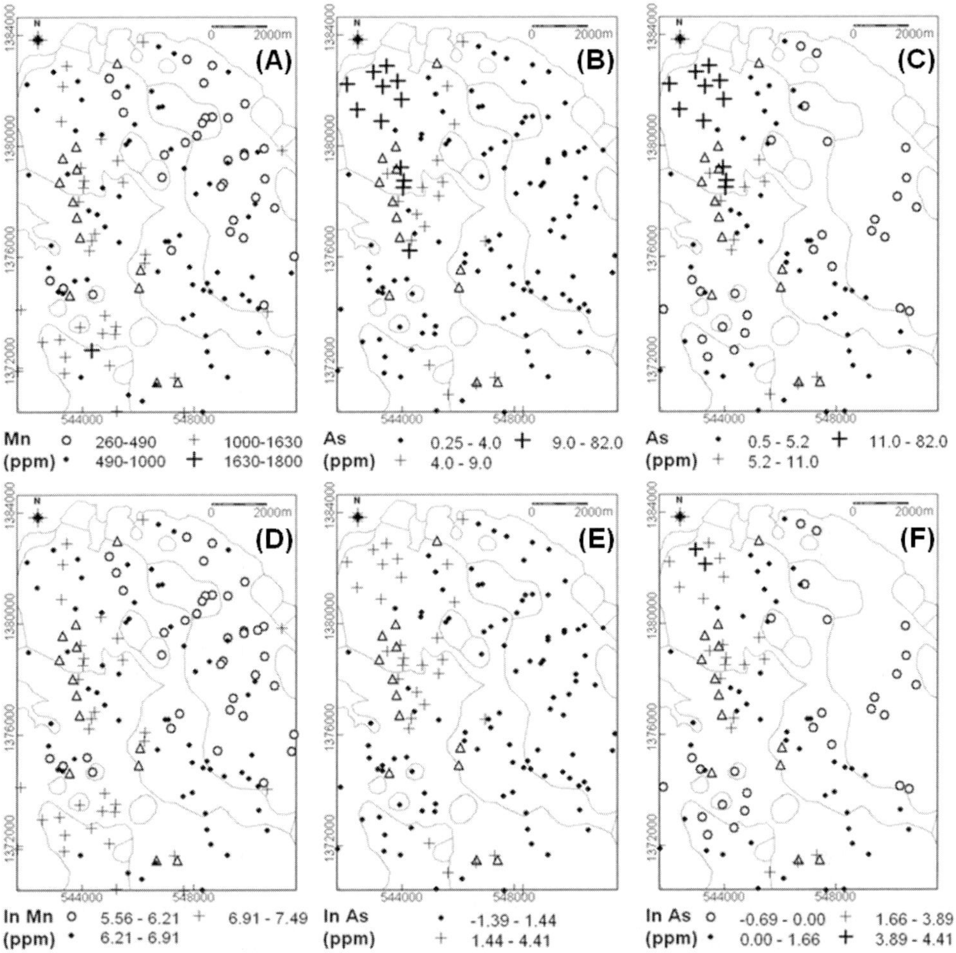

Fig. 3-15. Spatial distributions of selected uni-element data (Aroroy district, Philippines) based on boxplot-defined classes and EDA mapping symbols. (**A**) Raw Mn data (boxplot in Fig. 3-10E). (**B**) Raw As data with censored values (boxplot in Fig. 3-10F). (**C**) Raw As data without censored values (boxplot not shown). (**D**) Log_e-transformed Mn data (boxplot in Fig. 3-11E). (**E**) Log_e-transformed As data with censored values (boxplot in Fig. 3-11F). (**F**) Log_e-transformed As data without censored values (boxplot in Fig. 3-14A). Triangles represent locations of epithermal Au deposit occurrences. Light-grey lines represent lithologic contacts (see Fig. 3-9).

the study area are associated with hydrothermally altered volcano-sedimentary rocks. The maps in Fig. 3-15 also show that low background to censored values of As pertain mainly to the Aroroy Diorite, so that subdividing the uni-element data sets according to rock type at sample locations is non-trivial. Moreover, the maps of the As data in Fig. 3-15 indicate that different forms of the same data set, raw or transformed, influence the mapping of geochemical anomalies. The following section compares the performance of

TABLE 3-III

Threshold values defined as mean+2*SDEV*, median+2*MAD* and boxplot *UW* of raw and $\log_e$-transformed uni-element data set for n=135 samples (except As*, for which n=95), Aroroy district (Philippines).

	Mean+2*SDEV*		Median+2*MAD*		Boxplot *UW*	
	Raw	Antilog$_e$	Raw	Antilog$_e$	Raw	Antilog$_e$
Cu	139.72	184.93	96	120.30	136	200
Zn	121.76	139.77	88	108.85	113	187
Ni	26.31	34.81	22	26.58	30	42
Co	32.06	43.73	26	28.50	36	42
Mn	1461.93	1719.86	1120	1380.22	1630	1800
As	25.86	27.66	4	14.73	9.0	82.0
As*	30.89	29.96	7	11.94	11.0	48.9

*Excluding samples with censored values.

different threshold values (boxplot *UW*, median+2*MAD*, mean+2*SDEV*) in the $\log_e$-transformed uni-element data sets, which have more symmetrical distributions than the respective raw data sets.

Analysis of uni-element threshold values and anomalies

Table 3-III shows the threshold values defined as the mean+2*SDEV*, median+2*MAD* and boxplot *UW* in each of the raw and $\log_e$-transformed uni-element data sets. For the raw uni-element data sets, the threshold values defined by the median+2*MAD* are always the lowest, followed by those defined by either the boxplot *UW* or the mean+2*SDEV*, depending on the uni-element data set. For the $\log_e$-transformed data sets, the threshold values defined by the median+2*MAD* are also always the lowest, followed mostly by those defined by the mean+2*SDEV* and the threshold values defined by the boxplot *UW* are mostly highest, depending on the uni-element data set. These findings about the ranking of threshold values defined by each of the three methods and per type of data (raw or $\log_e$-transformed) are consistent with the findings of Reimann et al. (2005).

Because the $\log_e$-transformed data approach symmetrical distributions (Fig. 3-12), the threshold values determined from such data should be used in mapping of anomalies. The information in Table 3-III already indicates that the median+2*MAD* threshold values result in the highest number of anomalies, followed by the mean+2*SDEV* threshold values and then by the boxplot *UW* threshold values. So, for our pathfinder element for epithermal Au deposits – As – there are no anomalies based on the boxplot *UW* (Table 3-III, Fig. 3-11F), whereas anomalies based on the mean+2*SDEV* and the median+2*MAD* have, respectively, poor and good spatial associations with the known epithermal Au deposit occurrences in the case study area (Fig. 3-16). So, with respect to As anomalies, which one expects to be present and strong because of the epithermal Au deposits, the median+2*MAD* performs best, followed by the mean+2*SDEV* and then by the boxplot *UW*. The same is true even if the censored values in the As data are discarded (Table 3-

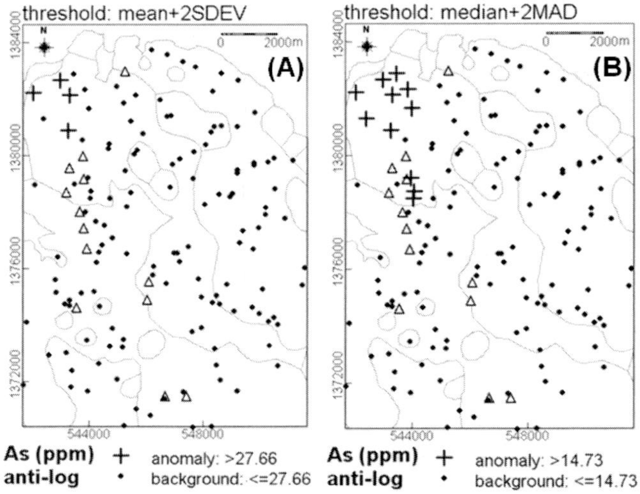

Fig. 3-16. Anomalies in the $\log_e$-transformed As data set, Aroroy district (Philippines) based on threshold defined as (**A**) mean+2*SDEV* and (**B**) median+2*MAD*. There are no As anomalies according to the boxplot *UW* (see Fig. 3-11F). Triangles represent locations of epithermal Au deposit occurrences. Light-grey lines represent lithologic contacts (see Fig. 3-9).

III). For the other elements under study, the threshold values based on the boxplot *UW* mostly indicate absence of anomalies (Fig. 3-11), whereas threshold values based on either the mean+2*SDEV* or the median+2*MAD* mostly indicate presence of anomalies. However, the threshold defined by the mean+2*SDEV* of $\log_e$-transformed Co values is greater than the maximum value in that data set (Table 3-III), suggesting that threshold values based on the mean+2*SDEV* can be misleading. In the study area, there are likely no anomalies of Ni and Co but there are likely weak anomalies of Cu, Zn and Mn associated with the epithermal Au deposit occurrences. So, with respect to Ni and Co anomalies, which one expects to be absent, the boxplot *UW* performs best, followed by the mean+2*SDEV* and then by the median+2*MAD*. Finally, with respect to Cu, Zn and Mn anomalies, which one expects to be present but perhaps weak, the mean+2*SDEV* apparently performs best, while the median+2*MAD* and the boxplot *UW*, respectively, over-estimate and under-estimate the anomalies.

The results from each of the whole $\log_e$-transformed uni-element data sets suggest that each of the three methods performs differently depending on the actual anomalies that are likely to be present (or absent) in an area. Reimann et al. (2005) pointed out that the boxplot *UW* threshold performs adequately in cases where there are 'actually' less than 10% outliers, whereas the median+2*MAD* performs adequately in cases where there are 'actually' at least 15% outliers. Although the median+2*MAD* of the whole $\log_e$-transformed As data set performed best among the three methods, in Fig. 3-16B there are only 11 (or 8.1%) anomalous samples out of the total 135 suggesting that such an

TABLE 3-IV

Threshold values defined as mean+2*SDEV*, median+2*MAD* and boxplot *UW* of the $\log_e$-transformed uni-element data subsets according to rock type at sample point, Aroroy district (Philippines).

Rock type at sample points	Element	Antilog$_e$ of Mean+2*SDEV*	Antilog$_e$ of Median+2*MAD*	Antilog$_e$ of boxplot *UW*
Aroroy Diorite (at n=38 samples, except the As data* for n=13 samples)	Cu	125.21	88.23	106
	Zn	71.52	56.83	71
	Ni	27.66	22.42	21
	Co	32.46	23.57	30
	Mn	1053.63	804.32	1210
	As	1.89	0.25	1.29
	As*	4.30	2.56	1.29
Dacitic/andesitic volcano-sedimentary rock (at n=97 samples, except the As data* for n=82 samples)	Cu	204.38	123.97	200
	Zn	145.47	100.48	137
	Ni	35.87	26.58	42
	Co	40.04	33.12	42
	Mn	1808.04	1619.71	1800
	As	38.47	13.87	82
	As*	32.76	6.05	29

*Excluding samples with censored values.

anomaly map is probably not optimal. Therefore, it is instructive to study further anomalies based on data subsets according to, say, rock type at sample locations.

Table 3-IV shows the threshold values defined as the mean+2*SDEV*, median+2*MAD* and boxplot *UW* in the individual $\log_e$-transformed uni-element data subsets according to rock type at sample locations. As in Table 3-III, the threshold values defined by the median+2*MAD* are always the lowest, followed by those defined by either the boxplot *UW* or the mean+2*SDEV*, depending on the element and data subset. The uni-element threshold values for samples in areas underlain by diorite are probably not physically meaningful because they are based on only 38 samples. For example, the median+2*MAD* threshold for As is equivalent to the minimum (i.e., censored) data value (0.25) in the data subset for samples in areas underlain by diorite, suggesting that all As values in the data subset are anomalous. For the samples in areas underlain by dacitic/andesitic rocks, the threshold values for Cu, Co and Mn based on the mean+2*SDEV* are greater than the respective maximum values, suggesting that threshold values based on the mean+2*SDEV* can be misleading. The boxplot-defined threshold for As in samples underlain by dacitic/andesitic rocks is equivalent to the maximum data value (82), suggesting that there are no As anomalies in the data subset.

The problems with the threshold values for As defined by each of three methods for the two data subsets according to rock type at sample location are caused by the censored values. By excluding the censored As values, the threshold values for As

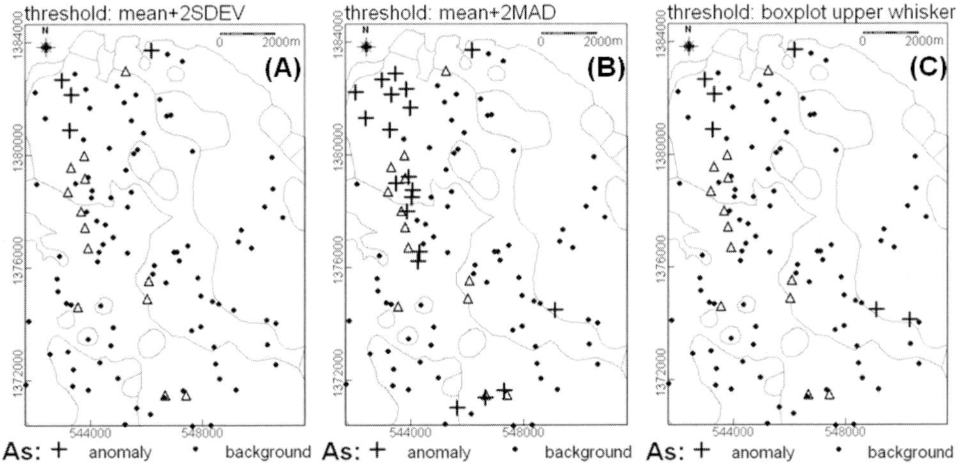

Fig. 3-17. Anomalies in the $\log_e$-transformed As data subsets according to rock type at sample points and exclusive of As censored values (see Table 3-IV), Aroroy district (Philippines). Anomalies are based on threshold defined as (**A**) mean+2*SDEV*, (**B**) median+2*MAD* and (**C**) boxplot *UW*. Triangles represent locations of epithermal Au deposit occurrences. Light-grey lines represent lithologic contacts (see Fig. 3-9).

defined by each of the three methods are apparently non-problematic (Table 3-IV), so they are mapped to study the anomalies and compare the performance of the three methods (Fig. 3-17). It is obvious that the As anomalies based on threshold defined as median+2*MAD* show the best spatial associations with the known epithermal Au deposit occurrences in the case study area (Fig. 3-17B). The northwestern most cluster of As anomalies are associated with hydrothermally altered volcano-sedimentary rocks. The As anomaly map in Fig. 3-17B is even better than the As anomaly map in Fig. 3-16B. In the former, there are 18 anomalous samples (or at least 13%) of the 135 samples, which is probably why the median+2*MAD* threshold outperforms the boxplot *UW* threshold as well as the mean+2S*DEV* threshold.

Analysis of inter-element relationships

In the preceding analysis of the uni-element data distributions, it can be perceived that probably the most dominant inter-element relationships in the study area is due to lithology (see Table 3-II and Fig. 3-13). That is, parts of the study area underlain by diorite have relatively lower concentrations of the elements under study compared to parts of the study area underlain by dacitic/andesitic volcano-sedimentary rocks. It is important to further unravel other inter-element relationships in the data, which may be useful in the interpretation of significant geochemical anomalies. For example, in this case study it is instructive to determine further (a) whether or not anomalies of As are plausibly due to scavenging by Mn-oxides and (b) whether there are inter-element

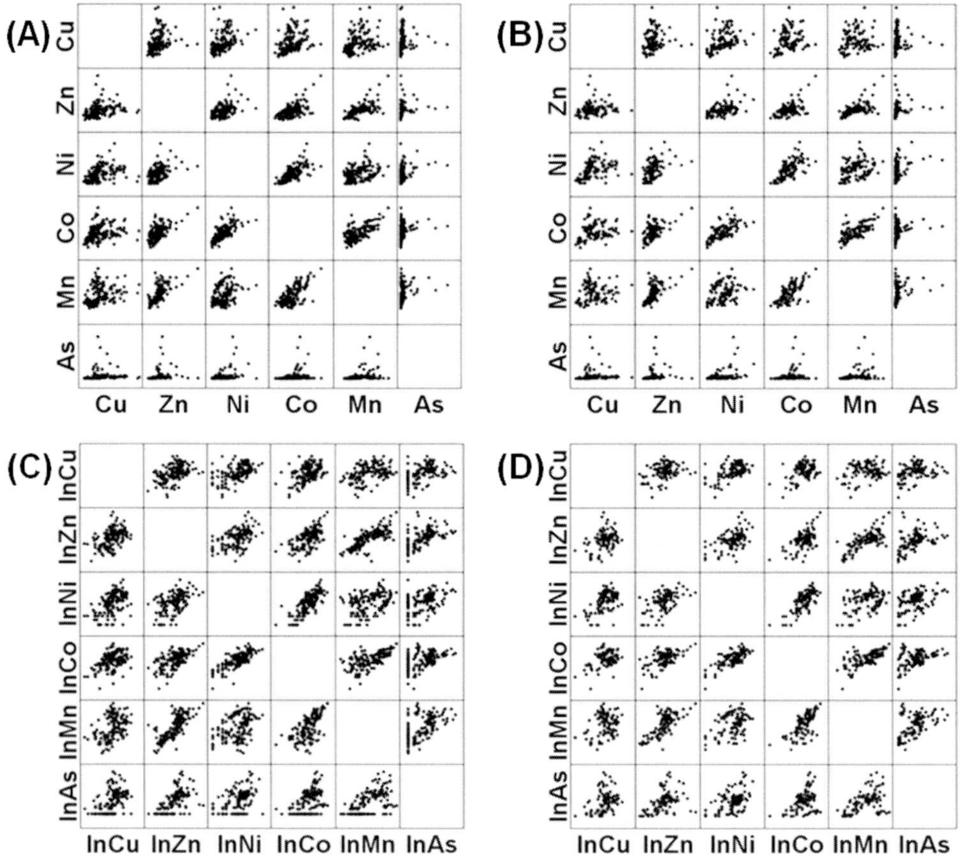

Fig. 3-18. Scatterplots of the uni-element data sets, Aroroy district (Philippines). (**A**) Raw data. (**B**) Raw data exclusive of samples with censored As values. (**C**) Log_e-transformed (ln) data. (**D**) Log_e-transformed (ln) data exclusive of samples with censored As values.

associations reflecting presence of epithermal Au deposits. Such questions can be answered by application of bivariate analytical techniques.

Scatterplots are useful for visual exploration of inter-element relationships. Scatterplots of raw or transformed data inclusive of samples with censored values (Figs. 3-18A and 3-18C) lead to misguided interpretations about inter-element relationships. Scatterplots of raw data exclusive of samples with censored values (Fig. 3-18B) also lead to misguided interpretations about inter-element relationships. Obvious data outliers (say, based on boxplots) must also be removed in creating scatterplots. The As outliers recognised in the uni-element data analyses are obvious in the scatterplots of the raw data (Figs. 3-18A and 3-18B) but not in the scatterplots of the log_e-transformed data (Figs. 3-18C and 3-18D). Thus, transformation of data values (so that they approach symmetrical distributions) and removal of samples with censored values result in

TABLE 3-V

Correlations coefficients of the $\log_e$-transformed uni-element data exclusive of samples with censored As values (n=95), Aroroy district (Philippines).

	Cu	Zn	Ni	Co	Mn
Zn	0.342				
Ni	0.560	0.537			
Co	0.396	0.589	0.754		
Mn	0.268	0.730	0.440	0.661	
As	0.341	0.414	0.496	0.391	0.462

All coefficients are significant at the 0.01 probability level (2-tailed test).

scatterplots that are optimal for visual analysis of inter-element relationships. The scatterplots in Fig. 3-18D indicate that all the elements under study have positive relationships with each other and there is no obvious presence of more than one population, except for a small cluster of high Mn values and low As values in the Mn-As plot. This small cluster in the Mn-As plot pertains to 13 samples in areas underlain by diorite with As concentrations above detection limit (see Table 3-II or 3-IV)).

Visual interpretation of a scatterplot can be aided by estimation of correlation coefficients and of covariance values between two uni-element data sets to obtain impressions about, respectively, inter-relation of data values and mutual variability of values. Estimates of correlation coefficients and covariance values are affected by data form and presence of censored values, outliers and multiple populations. The correlation coefficients and covariance values of element pairs corresponding to the data scatterplots in Fig. 3-18D are shown in Tables 3-V and 3-VI, respectively. Statistically significant positive correlations exist between all pairs of the $\log_e$-transformed uni-element data sets exclusive of samples with censored As values. The strongest correlation is between Ni and Co, followed by inter-element correlations with Mn. These correlations suggest controls by either lithology or scavenging by Mn-oxides. From the estimated correlation coefficients (Table 3-V), there are no obvious inter-element relationships reflecting presence of mineralisation. Estimates of covariance values for each pair of the uni-

TABLE 3-VI

Covariance matrix of the $\log_e$-transformed uni-element data exclusive of samples with censored As values (n=95), Aroroy district (Philippines).

	Cu	Zn	Ni	Co	Mn
Zn	0.091				
Ni	0.188	0.135			
Co	0.104	0.115	0.187		
Mn	0.066	0.134	0.103	0.120	
As	0.237	0.214	0.326	0.200	0.222

element data sets (Table 3-VI), nevertheless, indicate a Cu-Ni-Co association reflecting lithologic control, a Mn-Zn-Co association reflecting metal scavenging chemical control by Mn-oxides and an As-Ni-Cu association reflecting metallic mineralisation related to certain lithologies (i.e., andesitic rather than dacitic). The presence of obvious and subtle inter-element relationships in the case study data sets requires further application of appropriate multivariate methods that allow quantification and mapping of such inter-element relationships.

Analysis and mapping of multi-element associations

The multivariate methods most commonly employed in studying and quantifying multi-element associations in exploration geochemical data include principal components analysis (PCA), factor analysis (FA), cluster analysis (CA), regression analysis (RA) and discriminant analysis (DA). PCA and FA are useful in studying inter-element relationships hidden in multiple uni-element data sets. CA is useful for studying inter-sample relationships, whilst RA and DA are useful for studying inter-element as well as inter-sample associations. RA and DA require training data, i.e., samples representative of processes of interest (e.g., from mineralised zones). Authoritative explanations of multivariate methods applied to geochemical and geological data analysis can be found in Howarth and Sinding-Larsen (1983) and Davis (2002). In this case study, either PCA or FA is favourable for revealing inter-element relationships, a few of which may reflect presence of mineralisation.

PCA and FA are very similar techniques so that they are often confused with each other, but they have significant mathematical and conceptual differences. Howarth and Sinding-Larsen (1983) and Reimann et al. (2002) provide clear discussions about the similarities and dissimilarities between PCA and FA, which are summarised here. Both methods start with either the correlation matrix or the covariance matrix of data for a number (n) of variables. Both of them require transformation and/or standardisation of the input data. The main difference between PCA and FA is related to the proportions of the total variance of data for n variables accounted for in the analysis. The total variance is composed of the common variance in all n variables and the specific variances of each of the nth variable. In PCA, *principal components* (or PCs) are determined, without any statistical assumptions, to account for the maximum total variance of all input variables. In FA, a number of *common factors* are defined, with assumption of a statistical model with certain prerequisites, to account maximally for the common inter-correlation between the input variables. Thus, on the one hand, PCA is variance-oriented and results in a number of uncorrelated PCs (equal to n input variables) that altogether account for the total variance of all input variables. The 1st PC accounts for the highest proportion of the total variance (and thus represents the 'most common' variance) of the multivariate data, whereas the nth (or last) PC accounts for the least proportion of the total variance (and thus represents the 'most specific' variance) of the multivariate data. On the other hand, FA is correlation-oriented and results in a number (k) of uncorrelated common factors (less than the n input variables) that together do not account for the total variance of all input variables but altogether account for maximum common variance in all the input variables. Thus, the first factor accounts for the highest proportion of the 'total' common variance in the input multivariate data, whereas the kth (or last) factor accounts for the least proportion of the 'total' common variance in the input multivariate data. Because the 'total' common variance in n input multivariate data is unknown, the 'optimum' k common factors must be determined by following a number of statistical tests (Basilevsky, 1994) or 'rule-of-thumb' criteria, e.g., factors that cumulatively account for at least 70% of the total variance (Reimann et al., 2002). From the foregoing discussion, the following can be said about the applicability of either PCA or FA in geochemical data analysis (cf. Howarth and Sinding-Larsen, 1983). On the one hand, PCA is favourable in cases of geochemical data analysis in which the range of PCs representing the 'most common' variance to the 'most specific' variance in the input multi-element data sets is of interest to allow recognition of latent inter-element variations that reflect the various geochemical processes in a study area. On the other hand, FA is favourable in cases of geochemical data analysis in which the factors representing the 'most common' variance in the input multi-element data sets are of interest to allow recognition of latent inter-element relationships that describe the different geochemical processes in a study area.

Therefore, based on the preceding discussion about the difference between PCA and FA, the former is considered more appropriate to apply in the case study than the latter because of its 'exploratory' rather than 'confirmatory' nature. In PCA, it is essential to use standardised data if the correlation matrix is used to derive the PCs or to use unstandardised data if the covariance matrix is used to derive the PCs (Trochimczyk and Chayes, 1978). In addition, because estimates of either the correlation coefficient or the covariance are influenced by data form, presence of censored values, outliers and more than one population, it is also essential to 'clean' and transform the data so that they approach a (nearly) symmetrical distribution. For the same $\log_e$-transformed uni-element data sets that were used to create the scatterplots in Fig. 3-18D, the correlation matrix in Table 3-V and the covariance matrix in Table 3-VI, the derived PCs are shown in Table 3-VII. The correlation matrix (Table 3-V) was used to derive the PCs, so the uni-element data sets were first standardised using equation (3.11).

In Table 3-VII, the first two PCs (PC1 and PC2) together explain the 'most common' variance in the multivariate data and thus represent multi-element associations that reflect the major geochemical processes in the study area. PC1 accounts for at least 58% of the total variance and represents a Co-Ni-Zn-Mn-As-Cu association, which reflects a plausible combination (or overprinting) of lithologic and chemical controls. PC2 explains about 15% of the total variance and represents two antipathetic associations – a Cu-Ni association reflecting lithologic control and a Mn-Zn association reflecting metal scavenging control by Mn-oxides. Each of the last four PCs (PC3-PC6) explains the specific variances in the multivariate data and represents multi-element associations that reflect either the minor (or subtle) geochemical processes in the study area or errors in the multivariate data. PC3 accounts for at least 11% of the total variance and represents two antipathetic associations – an As-dominated multi-element association reflecting mineralisation control and a Co-Ni association reflecting lithologic control. The last

TABLE 3-VII

Principal components of the $\log_e$-transformed uni-element data sets (Aroroy district, Philippines) exclusive of samples with censored As values (n=95) and standardised according to equation (3.11).

	Cu	Zn	Ni	Co	Mn	As	% of Variance	Cum. % of variance
PC1	0.601	0.803	0.834	0.849	0.793	0.663	58.167	58.167
PC2	0.670	-0.346	0.304	-0.071	-0.473	0.085	14.942	73.109
PC3	-0.099	-0.083	-0.139	-0.295	0.008	0.733	11.011	84.120
PC4	0.405	0.293	-0.326	-0.327	0.180	-0.107	8.446	92.566
PC5	0.124	-0.367	-0.203	0.184	0.288	-0.181	5.133	97.699
PC6	0.032	0.095	-0.212	0.215	-0.181	0.063	2.301	100.000

three PCs (PC4-PC6), each explaining less than 10% of the total variance, represent multi-element associations that are somewhat difficult to interpret. Thus, PC3 can be hypothesised as an anomalous As-dominated multi-element association reflecting presence of the epithermal Au deposit occurrences. A map of the spatial distributions of PC3 scores based on boxplot-defined classes (Fig. 3-19A), shows a northwest-trending zone of high background to outlying PC3 scores following closely the trend of the epithermal Au deposit occurrences. Classification of the PC3 scores into either background or anomaly, using a threshold defined as median+2MAD of PC3 scores,

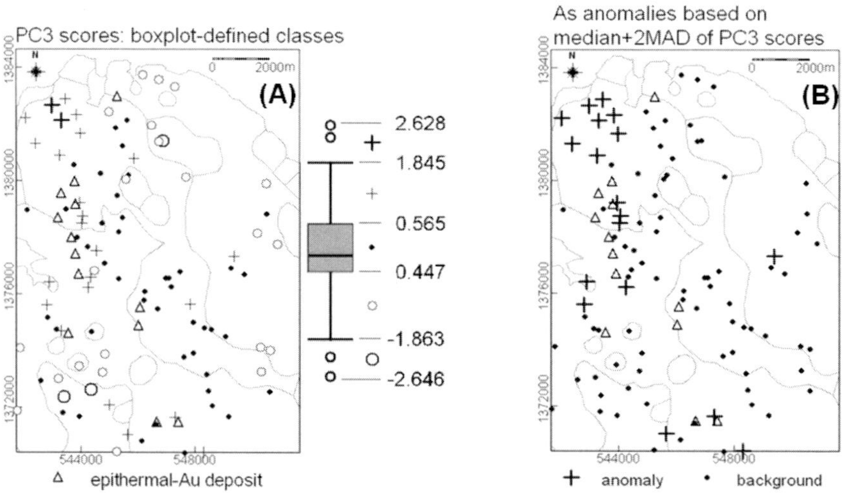

Fig. 3-19. Maps derived from PC3 in the results of PCA (shown in Table 3-VII) of stream sediment geochemical data, Aroroy district (Philippines). (**A**) Spatial distributions of boxplot-defined classes of PC3 scores. (**B**) Anomalies of an As-dominated multi-element association, based on threshold defined as median+2MAD of the PC3 scores.

TABLE 3-VIII

Principal components of the $\log_e$-transformed uni-element data subsets according to rock type at sample points (Aroroy district, Philippines), exclusive of samples with censored As values (n=95) and standardised according to equation (3.11).

	Cu	Zn	Ni	Co	Mn	As	% of Variance	Cum. % of variance
PC1	0.528	0.767	0.846	0.853	0.743	0.551	52.763	52.763
PC2	0.621	-0.403	0.275	-0.136	-0.488	0.413	17.528	70.292
PC3	-0.470	-0.120	-0.060	-0.074	0.086	0.707	12.515	82.806
PC4	0.319	0.184	-0.336	-0.370	0.286	0.141	8.114	90.920
PC5	0.104	-0.433	-0.168	0.249	0.301	-0.030	6.329	97.249
PC6	0.049	0.113	-0.254	0.223	-0.173	0.074	2.751	100.000

results in anomalies showing strong spatial association with the known epithermal Au deposits (Fig. 3-19B).

As answers to the two questions posed earlier in this section, (a) PC2 suggests that anomalies of As are not likely due to metal scavenging by Mn-oxides and, thus, are significant, whilst (b) PC3 suggests that there is an As-dominated multi-element association reflecting the presence of epithermal Au deposits. The answer to the second question requires further verification because the small cluster in the Mn-As plot (Fig. 3-18D), which pertains to 13 samples in areas underlain by diorite with As values above detection limit, certainly has an effect in the PCA. Thus, a second PCA was performed on the $\log_e$-transformed uni-element data subsets according to rock type at sample points, exclusive of samples with censored As values and standardised according to equation (3.11). The results of the second PCA (Table 3-VIII) are very similar to the results of the first PCA (Table 3-VII), but there are two main differences between them in terms PC2 and PC3. Firstly, the second PCA shows a PC2 representing a Cu-As-Ni association, whereas the first PCA shows a PC2 representing a Cu-Ni association. The Cu-As-Ni association, which is antipathetic to a Mn-Zn-Co association reflecting metal scavenging control by Mn-oxides, plausibly reflects mineralisation. Mapping and classification of PC2 scores indicate that anomalies of the Cu-As-Ni association are minor but significant because they show spatial associations with the two northernmost-lying epithermal Au deposit occurrences (Figs. 3-20A and 3-20B). So, the second PCA provides an additional answer to the second question posed earlier in this section by indicating a minor Cu-As-Ni association reflecting a few of the epithermal Au deposit occurrences. Secondly, the first PCA shows a PC3 representing weaker antipathetic relationship between As and Cu (Table 3-VII), whereas the second PCA shows a PC3 representing stronger antipathetic relationship between As and Cu (Table 3-VIII). In any case, the antipathetic relationship between As and Cu is likely due to their differences in mobility in the surficial environment. The difference between the PC3 in the first PCA and the PC3 in the second PCA can be visualised by comparing the maps in Fig. 3-19 with the maps in Figs. 3-20C and 3-20D. In Fig. 3-19 there are (stronger) As anomalies

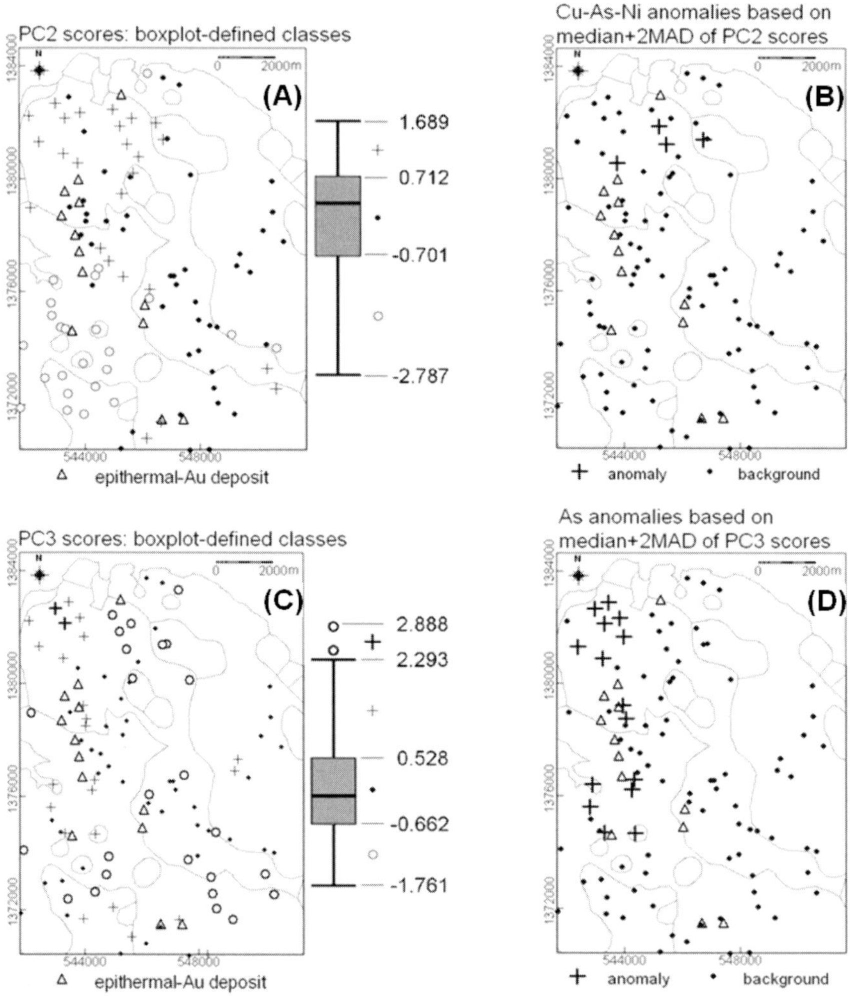

Fig. 3-20. Maps derived from PC2 and PC3 in the results of PCA (shown in Table 3-VIII) of stream sediment geochemical data, Aroroy district (Philippines). (**A**) Spatial distribution of boxplot-defined classes of PC2 scores. (**B**) Anomalies characterised by a Cu-As-Ni multi-element association, based on threshold defined as median+2*MAD* of the PC2 scores. (**C**) Spatial distribution of boxplot-defined classes of PC3 scores. (**D**) Anomalies of an As-dominated multi-element association, based on threshold defined as median+2*MAD* of the PC3 scores.

associated with the two southernmost-lying epithermal Au deposits, whereas in Figs. 3-20C and 3-20D there are no (or weaker) As anomalies associated with the two southernmost-lying epithermal Au deposits. There can be a number of plausible explanations for these results, such as differences in ore mineralogy of the individual epithermal Au deposits. Nevertheless, the As anomalies represented by the PC3 in the

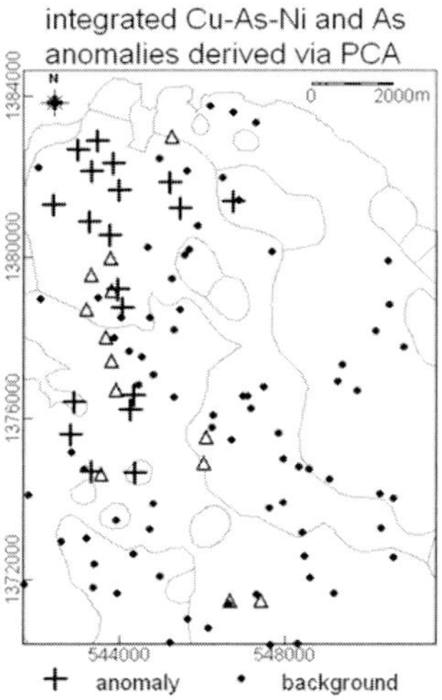

Fig. 3-21. Integrated anomalies characterised by Cu-As-Ni and As associations represented, respectively, by PC2 and PC3 from results of PCA (shown in Table 3-VIII) of stream sediment geochemical data, Aroroy district (Philippines). Triangles represent locations of epithermal Au deposit occurrences. Light-grey lines represent lithologic contacts (see Fig. 3-9).

second PCA show strong spatial association with most of the epithermal Au deposit occurrences and, thus, are significant.

Finally, if the application of multivariate methods in the analysis of multi-element geochemical data sets results in recognition and mapping of significant anomalies with different but similar multivariate signatures (i.e., they reflect the same deposit-type), such anomalies can be combined for the purpose of deriving a map of stronger evidence that can be used in mineral prospectivity mapping. For example, the individual binary anomaly maps for the Cu-As-Ni and As signatures (Figs. 3-20B and 3-20D) can be combined via a simple Boolean OR operation resulting in a new integrated binary anomaly map (Fig. 3-21). This topic will be revisited and discussed in detail later on in this book.

SUMMARY

Exploration geochemical data very seldom, if ever, show a normal distribution. Application of methods for geochemical anomaly recognition based on classical statistics can thus be misguided and potentially results in spurious anomalies. In contrast, the statistical and graphical tools of exploratory data are not based on the assumption of normal distribution of data. Careful examination of statistics and graphics in exploratory data analysis allows insight to geochemical data structure and behaviour, which must be considered in the analysis of uni-element as well as multi-element geochemical anomalies. Exploratory data analysis should be favoured over confirmatory data analysis in mapping of significant geochemical anomalies. GIS supplements exploratory data analysis with tools for data manipulation, integration (of spatial data that influence spatial variability of the geochemical landscape) and visualisation in examining and mapping of significant geochemical anomalies.

Geochemical Anomaly and Mineral Prospectivity Mapping in GIS
by E.J.M. Carranza
Handbook of Exploration and Environmental Geochemistry, Vol. 11 (M. Hale, Editor)

Chapter 4

FRACTAL ANALYSIS OF GEOCHEMICAL ANOMALIES

INTRODUCTION

The exploratory analysis of geochemical anomalies presented in the preceding chapter considers mainly the empirical density distributions of data. This may be inadequate in modeling geochemical anomalies because exploration data sets are incomplete samples of the geochemical landscape of a study area. It is possible to improve models of geochemical anomalies by considering also (a) the spatial correlation and variability of geochemical data and (b) the geometry and scale-independent properties of geochemical landscapes.

On the one hand, the consideration of spatial correlation and variability of geochemical data aims to accurately portray background and anomaly patterns (i.e., geochemical landscapes) that reflect controls by geological processes and thus facilitate recognition of significant geochemical anomalies. This objective can be achieved by application of a variety of techniques that employ weighted moving average of point data within a zone of influence, which is usually but not always circular, in order to interpolate and portray spatial distributions of a whole set of point data (e.g., uni-element concentrations geochemical samples). The applications of weighted moving average techniques, especially geostatistical techniques that rest on variogram analysis, to model geochemical landscapes have been discussed extensively in the literature (e.g., Wackernagel and Butenuth, 1989; Goovaerts, 1992; Bellehumeur et al., 1994; Jimenez-Espinosa and Chica-Olmo, 1999; Reis et al., 2003; Pardo-Igúzquiza and Chica-Olmo, 2005) are therefore not discussed in this chapter.

On the other hand, the consideration of geometry and scale-independent properties of geochemical landscapes aims to accurately separate background and anomalies in such landscapes. This has been demonstrated in the seminal works of Cheng et al. (1994) and Cheng (1999b) dealing with the *fractal* geometry of geochemical landscapes. This chapter reviews briefly the concept that geochemical landscapes have fractal properties and then demonstrates GIS-based applications of the concentration-area method for fractal analysis of geochemical anomalies.

GEOCHEMICAL LANDSCAPES AS FRACTALS

What are fractals?

Mandelbrot (1982, 1983) introduced the term fractal to describe an object or a pattern consisting of parts (i.e., fractions) that have geometries (e.g., shape or form), except scale or size, that are more or less similar to the whole object or pattern. Thus, fractals have the property of being *self-similar* or *self-affine* at various scales, meaning they are scale-invariant or scale-independent entities. As a consequence of this property, it is not possible to determine the scale of a fractal based on its shape or form alone. A fractal is strictly self-similar if it can be expressed as a union of objects or patterns, each of which is a reduced copy of (i.e., geometrically similar to) the full object or pattern. The most fractal-looking natural objects are not, however, precisely self-similar but are self-affine. On the one hand, a statistically self-similar fractal is isotropic (Turcotte, 1997), meaning that patterns with different orientations appear to have similar orientations at the same scale. On the other hand, a statistically self-affine fractal is anisotropic, meaning that patterns with different orientations appear to have similar orientations at different scales. The property of either (statistical) self-similarity or (statistical) self-affinity is an attribute that can be used to characterise seemingly-disordered natural objects or phenomena. That is to say, natural systems or patterns resulting from stochastic processes at various scales are plausibly fractals. For, example, Bölviken et al. (1992) suggested that geochemical distribution patterns (or geochemical landscapes) plausibly consist of fractals (background and anomalous patterns), because such patterns were formed by processes that have occurred throughout geological time at various rates and at various scales. They tested their hypothesis by applying various methods for measuring fractal geometry.

Fractal geometry

As originally defined by Mandelbrot (1982, 1983), a fractal has a dimension D_f, known as the Hausdorff-Besicovitch dimension, which exceeds its topological (or Euclidean) dimension D. A fractal linear feature does not have D=1 as expected from Euclidean geometry, but has a D_f between 1 (D for a line) and 2 (D for an area). An in-depth review of methods to measure fractal dimension of linear features is given by Klinkenberg (1994). Similarly, fractal areas or surfaces (e.g., geochemical landscapes) have values of D_f between 2 and 3 (D for volumes), fractal volumes have values of D_f greater than 3, and so on. An in-depth review of methods to measure fractal dimension of surfaces or landscapes is given by Xu et al. (1993). Further authoritative explanations of methods for measuring fractal dimensions of geological objects can be found in Carr (1995, 1997).

To test the fractal dimensions of geochemical landscapes, Bölviken et al. (1992) used four of the available methods for measuring fractal dimensions of areas or surfaces: (1) variography; (b) length of contour versus measuring yardstick; (3) the number-area

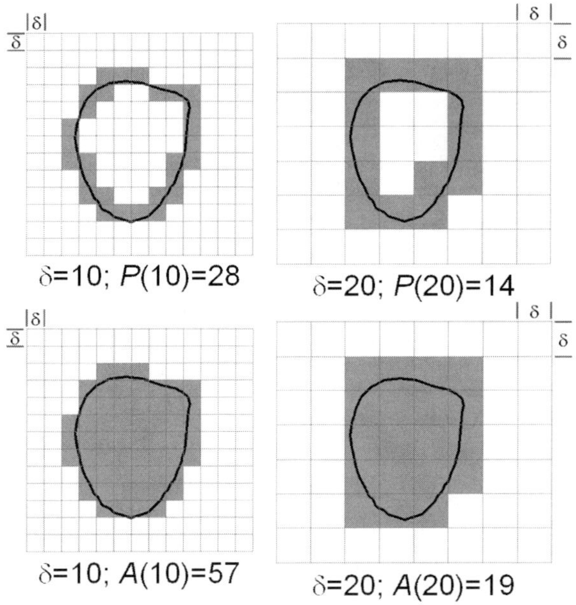

Fig. 4-1. Schematic diagram of the box-counting method for measuring perimeter (P) and area (A) of a pattern (enclosed in thick black line) in two-dimensional space. The sides of the boxes have widths of a yardstick (δ). The numbers of boxes in grey, which indicate presence of outline and parts of a pattern, are estimates of the perimeter $P(\delta)$ and area $A(\delta)$ of the pattern according to a certain yardstick (δ).

relation; and (4) the perimeter-area relation. The first method derives from geostatistical theory; it is not discussed here, but readers are referred to Journel and Huijbregts (1978), Isaaks and Srivastaa (1989) and Wackernagel (1995). The second to fourth methods derive from fractal theory. The second method derives from Mandelbrot's (1967) study of fractal dimensions of Britain's coastline. The third method derives from Mandelbrot's (1975) study of topographic landscapes. The fourth method was introduced originally by Mandelbrot (1982, 1983) and then expounded by Cheng (1994, 1995) for its application to geochemical data analysis. Of the last three methods for analysis of fractal dimensions, the perimeter-area relation is chosen for demonstration here because of its intuitive application to analysis of geochemical anomalies.

To measure the perimeter and area of a geometrical pattern in two-dimensional space, a grid of square boxes is overlaid on the geometrical pattern (Fig. 4-1). The sides of the boxes have widths of a yardstick (δ), which denotes the spatial resolution or scale at which the geometry (perimeter, area) of a pattern is measured. The perimeter of a pattern is measured by counting the boxes in which the outline of the pattern is present. Similarly, the area of a pattern is measured by counting the boxes in which any part of the pattern is present. This technique of estimating fractal geometry of patterns in two-

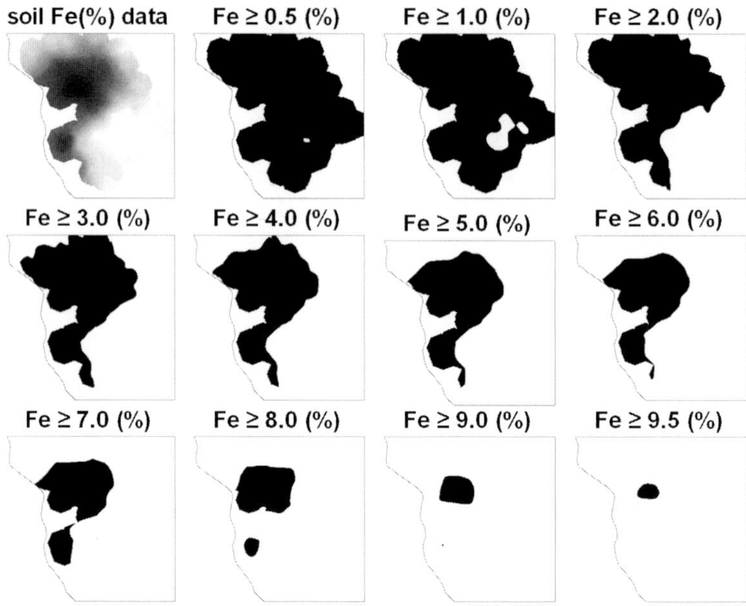

Fig. 4-2. Interpolated soil Fe data (uppermost left), which vary from lowest (white) to highest (black) values, and successive binary patterns of increasing concentration levels of Fe (from top to bottom and from left to right of the figure).

dimensional space is known as the *box-counting* method. As the yardstick δ increases, estimates of perimeter and area of a pattern in two-dimensional space decrease (Fig. 4-1). Hence, an infinitesimally small yardstick is necessary to measure accurately the perimeter and area of a pattern in two-dimensional space.

Perimeter-area relationship in geochemical patterns

Returning to the soil Fe data (see Fig. 1-1 and Chapter 3), a geochemical landscape of soil Fe values (Fig. 4-2, uppermost left map) was created by interpolating the point data via a simple weighted moving average method using a circular window. The parameters used for the interpolation are an inverse distance weight exponent of [1] and a limiting distance (or search radius) of 300 m. Interpolated/extrapolated values beyond 300 m of the outermost samples are masked out and thus excluded from the analysis. Binary maps of values equal to or greater than a contour of Fe concentration are then created (Fig. 4-2). The areas enclosed by successive Fe concentration contours change gradually in shape and the area within a concentration contour decreases as the value of the concentration contour increases.

To illustrate fractal dimension of a geochemical pattern, the concentration contours of interpolated soil Fe values are used to firstly define a threshold value to distinguish

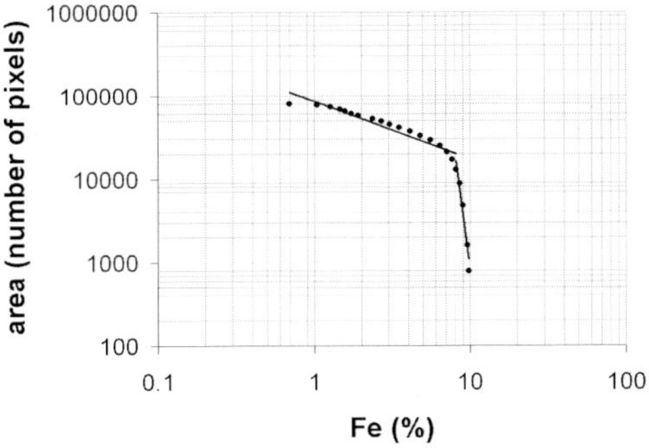

Fig. 4-3. Relationship between area and concentration levels of Fe plotted on a log-log graph. The two line segments are obtained by least squares (LS) fitting through the linear parts of the plot.

between areas of soil with "low" and "high" Fe values by means of a log-log plot of the concentration-area relationship (Fig. 4-3). The threshold value is defined by the sudden change (or inflection) in the rate of decrease of area enclosed by Fe concentration levels in the log-log plot. The threshold value in this case is 8.14% Fe, which is indicated by the intersection of two line segments fitted (by least squares method) through the linear parts of the plot. Compared to the results of the EDA analysis in the previous chapter, the threshold defined here is slightly lower than the 3rd quartile (upper hinge in the boxplot) of 8.6% Fe in the soil Fe data (see Fig. 3-3). The "high" Fe values, which are greater than the threshold value of 8.14% Fe defined here, pertain mostly to areas underlain by basalt and to areas immediately around the basalt (Fig. 4-4; see also Fig. 3-6A), whereas the "low" Fe values pertain to areas underlain by phyllite and quartzite. The concentration-area relationship is thus appropriate for classification of geochemical data.

Either the areas of "high" Fe values or the areas of "low" Fe values can be subjected to analysis of fractal dimensions by the perimeter-area method but, in keeping with the interest in geochemical exploration, the analysis is illustrated using the areas of "high" Fe values. If the areas of "high" Fe values are discretised into a series of binary patterns of values equal to or greater than a contour of Fe concentration, then one would expect a series of similarly shaped patterns. For a group of similarly shaped patterns, there exists a power-law relation between their perimeters and areas (Mandelbrot, 1982, 1983). Thus, two areas A_m and A_n, respectively enclosed by contours m and n, are related to their perimeters P_m and P_n as follows (Cheng et al., 1994):

$$\frac{P_m(\delta)}{P_n(\delta)} = \left[\frac{A_m(\delta)}{A_n(\delta)}\right]^{D_{AP}/2} . \tag{4.1}$$

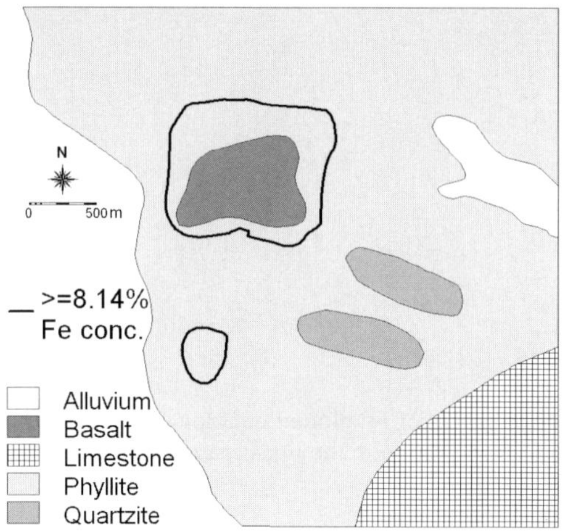

Fig. 4-4. Lithologic map and outline of areas with "high" Fe values (>8.14%) defined from the concentration-area plot (Fig. 4-3).

where δ denotes that the perimeters and areas are measured using a common yardstick in the box-counting method (Fig. 4-1). The exponent *DAP* is a fractal dimension associated with both perimeter and area and it satisfies the relation (Cheng et al., 1994):

$$D_{AP}=\frac{2D_P}{D_A}. \tag{4.2}$$

where *DP* and *DA* are fractal dimensions of perimeter and area, respectively. The relations in equations (4.1) and (4.2) are independent of length of the yardstick (δ).

The log-log plot of the perimeter-area relationship for a series of contours of "high" Fe values is shown in Fig. 4-5A. The perimeters and areas for different contours of "high" Fe values are measured using the same yardstick of 10 m, which is equivalent to the side of a pixel used in raster-based GIS operations for this example. Perimeter length was determined by first converting into polygons the series of raster binary maps defined by different contours of "high" Fe values. The straight line in the log-log plot (Fig. 4-5A) is obtained by a least squares fit through the points. The value of D_{AP} estimated by equation (4.1) is 1.4904.

For a geometrical pattern in two-dimensional space with "true" perimeter P_0, and "true" area A_0, the estimated perimeter $P(\delta)$ and estimated area $A(\delta)$ when the box-counting method with yardstick δ is used can be expressed, respectively, as (Cheng et al., 1994):

$$P(\delta)=P_0\delta^{(1-D_P)}\text{, and} \tag{4.3}$$

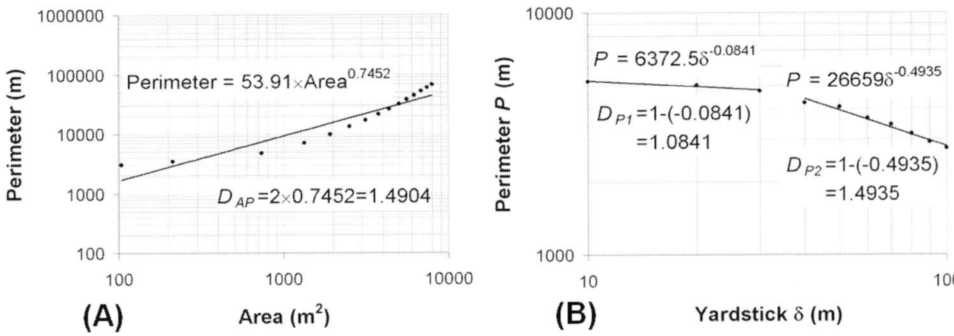

Fig. 4-5. (**A**) Perimeter-area relationship, plotted on a log-log graph, for a series of contours of "high" Fe values greater than the threshold of 8.14%. (**B**) Plots, on a log-log graph, of estimates perimeters of areas of "high" Fe values using variable yardstick.

$$A(\delta)=A_0\delta^{(1-D_A)}. \quad (4.4)$$

If *DP* can be determined from equation (4.3), then *DA* can be estimated from the relation in equation (4.2). Similarly, if *DA* can be determined from equation (4.4), then *DP* can be estimated from the relation in equation (4.2).

For our example areas of "high" Fe values, Fig. 4-5B shows the estimated perimeters using different yardsticks in the box-counting method. The straight lines in the log-log plot (Fig. 4-5B) are obtained by least squares fit through the linear parts of the plot. The estimated perimeters seem to have two fractal dimensions. The first one (D_{P1}=1.0841) for yardsticks less than 35 m is equivalent to the so-called "textural" fractal dimension (cf. Kaye, 1989, p.27; Cheng et al., 1994), which is close to 1 for surfaces assumed and explainable as a result of smoothing during interpolation. The second one (D_{P2}=1.4935) obtained by using yardsticks equal to or greater than 35 m, which is equivalent to the so-called "structural" fractal dimension (cf. Kaye, 1989, p.27; Cheng et al., 1994), is probably a more realistic geometry of the areas of "high" Fe values.

The D_{P2} (=1.4935) and the D_{AP} (=1.4904) may be used to estimate D_A using the relation in equation (4.2). In that case, D_A = 2.0042. This estimate of the fractal dimension of the areas of "high" Fe values is in agreement with the proposition of Cheng et al. (1994) that, theoretically, D_A cannot be greater than 2. In addition, Cheng (1995) found that if, and only if, D_A = 2 can D_{AP} (equation (4.2)) be used to obtain an unbiased estimate of D_P. Inasmuch as our estimate of D_A is very close to 2, then D_{P2} is a very good estimate of the fractal dimension of the perimeters of the areas of "high" Fe values based on the perimeter-yardstick relation (Fig. 4-5B).

Multifractality of geochemical landscapes

The two line segments fitted to the log-log plot of the concentration-area relationship (Fig. 4-3) indicate that there are at least two sets of fractal patterns in the soil Fe data. This suggests that geochemical landscapes can be *multifractals*. A multifractal is considered to be spatially intertwined sets of *monofractals* (Feder, 1988; Stanley and Meakin, 1988). Whereas monofractals are characterised by one fractal dimension and thus apply to binary patterns, multifractals have different fractal dimensions and thus apply to patterns with continuous spatial variability (Agterberg, 1994, 2001). As the subset of "high" Fe value can be characterised by a fractal dimension, so can the subset of "low" Fe values, although this is not to say that either subset is a monofractal because they both have continuous spatial variability. Thus, the multifractality of geochemical landscapes can be related to the probability density distributions and spatial distributions of geochemical data (Cheng and Agterberg, 1996; Gonçalves, 2001; Wei and Pengda, 2002; Panahi and Cheng, 2004; Xie and Bao, 2004; Shen and Cohen, 2005), which are influenced by various processes that have occurred throughout geological time at various rates and at various scales (e.g., Rantitsch, 2001). If that is the case, then the concentration-area relation introduced earlier is appropriately a multifractal model, which can be used to separate geochemical anomalies from background as proposed originally by Cheng et al. (1994). The concentration-area fractal method has been demonstrated by several workers to map significant anomalies using various geochemical sampling media (e.g., Cheng et al., 1996, 1997, 2000; Cheng, 1999b; Sim et al., 1999; Gonçalves et al., 2001; Panahi et al., 2004) and is reviewed and further demonstrated here.

THE CONCENTRATION-AREA METHOD FOR THRESHOLD RECOGNITION

The following discussion of the concentration-area method for separation of geochemical anomalies from background is adapted from Cheng et al. (1994). For a series of contours of uni-element concentrations, the concentration contours v and the areas of uni-element concentrations equal to or greater than v or the areas enclosed by each contour [i.e., $A(\geq v)$] satisfy the following power-law relation if they have multifractal properties:

$$A(\geq v) \propto v^{-\alpha} \tag{4.5}$$

where $\propto$ denotes proportionality and the exponent α represents the slope of a straight line fitted by least squares through a log-log plot of the relation. If, on the one hand, the concentration-area relation represents a fractal model, then the log-log plot can be fitted by one straight line and thus by one value of α corresponding to the whole range of v, representing a group of similarly-shaped concentration contours. If, on the other hand, the concentration-area relation represents a multifractal model, then the log-log plot can be depicted by at least two straight lines and thus by different values of α corresponding to different ranges of v, reflecting groups of similarly-shaped concentration contours. Consequently, the breaks in slopes of the straight lines fitted through the log-log plot of the relation can be used to distinguish different ranges of v, which intuitively represent different populations in the probability density distributions and spatial distributions of a data set of uni-element concentrations.

Suppose a bifractal a geochemical landscape, meaning there is a threshold value (v_t) separating background and anomalous uni-element concentrations. The background uni-element concentrations v_b and the areas occupied by such uni-element concentrations [i.e., $A(v_b \leq v_t)$] satisfy the following power-law relation:

$$A(v_b \leq v_t) = C_b v_b^{-\alpha_b} \tag{4.6}$$

where C_b is a constant and α_b is an exponent associated with the background component. The anomalous uni-element concentrations v_a and the areas occupied by such uni-element concentrations [i.e., $A(v_a > v_t)$] satisfy the following power-law relation:

$$A(v_a > v_t) = C_a v_a^{-\alpha_a} \tag{4.7}$$

where C_a is a constant and α_a is an exponent associated with the anomalous component. Thus, if the soil Fe data are considered to represent a bifractal geochemical landscape and an inflection point (or threshold) at 8.14% Fe is selected for the purpose of illustration, two straight lines can be fitted through the concentration-area plots for the soil Fe values have power-law equations shown in Fig. 4-6. The concentration-area relation for the "low" Fe values (≤8.14% Fe) has α_{low}=0.6939 and C_{low}= 85536, whilst the concentration-area relation for the "high" Fe values (>8.14% Fe) has α_{high}=14.574 and C_{high}= 3×10^{17}.

The interpolated soil Fe data do not portray, however, a bifractal geochemical landscape because by careful inspection of the concentration-area plots shown in either Fig. 4-3 or 4-6, three inflection points or thresholds (1.6% Fe, 7.2% Fe, 8.6% Fe) can be distinguished (Fig. 4-7). Accordingly, four straight lines can be fitted through the concentration-area plots. Each of the three straight lines that fit the plots to the left of any threshold satisfies the power-law relation in equation (4.6), whilst the straight line that fits the plots to the right of the rightmost (or highest) threshold satisfies the power-law relation in equation (4.7). Each of the lines represents a population in the soil Fe data, which, from lowest to highest, can be classified as "low-background" (or *LB*), "moderate-background" (or *MB*), "high-background" (or *HB*) and "anomaly" (or *A*). The concentration-area relation for the "low-background" class (≤1.6% Fe) has α_{LB}=0.2391 and C_{LB}= 74877. The concentration-area relation for the "moderate-background" class (1.6-7.2% Fe) has α_{MB}=0.695 and C_{MB}= 93259. The concentration-area relation for the "high-background" class (7.2-8.6% Fe) has α_{HB}=4.6729 and C_{MB}= 2×10^{8}. Finally, the

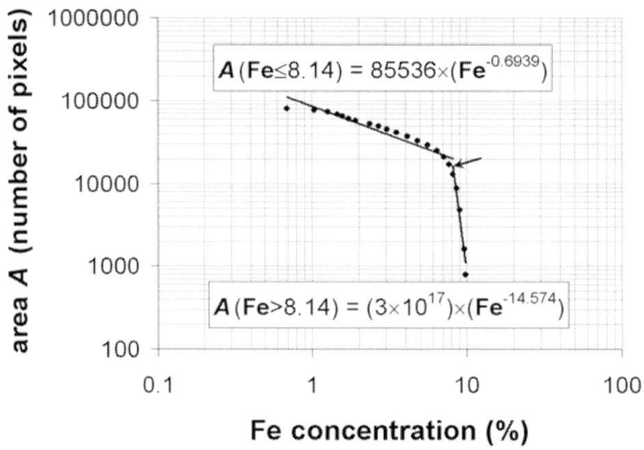

Fig. 4-6. Log-log plot of concentration-area relationship for the soil Fe data. By assuming that the soil Fe data represent a bifractal geochemical landscape and considering an inflection point (or threshold) corresponding to 8.14% Fe (pointed by short arrow), two straight lines can be fitted by least squares through the concentration-area plots. The straight line fitted through the plots for concentrations ≤8.14% Fe satisfies the power-law relation in equation (4.6), whilst the straight line fitted through the plots for concentrations >8.14% Fe satisfies the power-law relation in equation (4.7).

concentration-area relation for the "anomaly" class (7.2-8.6% Fe) has α_A=17.477 and C_A= 2×10^{20}. This sort of analysis has been demonstrated by Cheng et al. (1997, 2000), Cheng (1999b) and Panahi et al. (2004).

Discretisation of the interpolated soil Fe values based on the three threshold values (Fig. 4-8) indicates that the three threshold values are related to variations in lithology (see Fig. 4-4). Soil Fe values equal to or less than 1.6% pertain to areas underlain by the quartzites and the adjoining phyllites. Soil Fe values ranging from 1.61% to 7.2% pertain mainly to areas underlain by the phyllites. Soil Fe values ranging from 7.21% to 8.6% pertain to areas underlain by phyllites surrounding the basalt. Soil Fe values greater than 8.6% pertain mainly to areas underlain by the basalt. Thus, geochemical data classification based on threshold values determined from analysis of log-log plots representing concentration-area relationships has strong ability to portray physically meaningful spatial distributions of uni-element data.

APPLICATION OF GIS IN THE CONCENTRATION-AREA FRACTAL METHOD

Fig. 4-9 summarises the basic operations on exploration geochemical data in order to implement the concentration-area fractal method for separation of background and anomaly. Basically, a GIS can support implementation of the concentration-area fractal method in terms of (a) generation and discretisation of geochemical surfaces and (b) attribute table operation for area calculations.

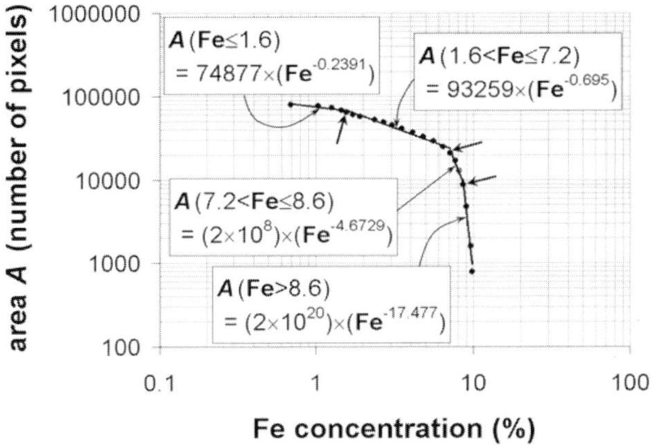

Fig. 4-7. Log-log plot of concentration-area relationship for the soil Fe data. By careful inspection of the plots, three inflection points (or thresholds) corresponding to 1.6%, 7.2% and 8.6% Fe (pointed by short arrows) can be defined and, thus, four straight lines can be fitted by least squares through the concentration-area plots. The individual straight lines fitted through the plots to the left of any threshold satisfy the power-law relation in equation (4.6), whilst the straight line fitted through the plots to the right of the rightmost threshold satisfies the power-law relation in equation (4.7).

Generation and discretisation of geochemical surfaces

Because most exploration geochemical data are recorded as attributes of sampling points, point-to-surface transformations through spatial interpolation (see Chapter 2) are essential in analysis of geochemical thresholds via the concentration-area fractal method. Spatial interpolation entails analysis of spatial correlation and variability of point geochemical data in order to determine the precise way in which to generate a geochemical surface for a certain element. This topic is, however, beyond the scope of this volume. The references to this topic cited earlier in this chapter and many other relevant publications can be consulted for further details.

Most GIS software packages support spatial interpolation of point data via either triangulation or gridding techniques. The former techniques are not appropriate but the latter techniques are appropriate for the application of the concentration-area fractal method. That is because interpolation via gridding techniques provide surfaces represented as a raster of pixels (see Figs. 2-5 and 2-14), which are amenable to the box-counting method for estimation of areas enclosed by certain uni-element concentration levels (Fig. 4-1). A geochemical surface generated from point data set must then be discretised or classified according to some intervals of the data. Classification is, of course, a basic functionality of a GIS (see Chapter 2).

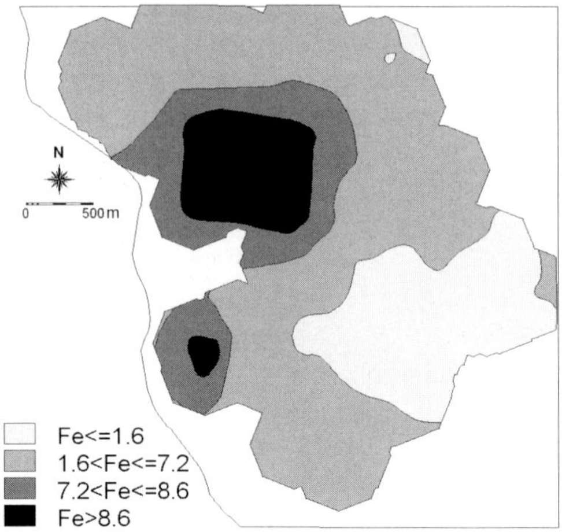

Fig. 4-8. Spatial distributions of the classified interpolated soil Fe data (Fig. 4-2, uppermost left) based on three threshold values (1.6% Fe, 7.2% Fe, 8.6% Fe) defined in the log-log plot representing relationships between areas and soil Fe concentrations (Fig. 4-7).

Attribute table operation for area calculations

For every raster map of geo-objects (see Chapter 2), such as a map of a discretised geochemical surface, there is an associated table. The areas (usually expressed in number of pixels or boxes) of polygonal geo-objects, such as areas enclosed within two contours (i.e., class limits) of uni-element concentrations, are recorded automatically in the associated attribute table during the discretisation process. For example in Fig. 4-10, the second column (`NPIX_CL`) contains the number of pixels (or boxes) representing each class of uni-element concentrations. The other columns in the table are created or calculated. The third column (`cl_min`), containing the minimum values (or lower limits) of each class, is created based on the first column. The fourth to last columns are calculated via arithmetic operations using the values in the `NPIX_CL` column. The values in the fourth column (`npix_cum`) are derived by calculating the cumulative sum of the number of pixels in the `NPIX_CL` column successively from the lowest to the highest class. The values in fifth column (`npix_tot`) are derived by aggregating the total number of pixels in the `NPIX_CL` column. The total number of pixels is recorded repeatedly in all cells of the `npix_tot` column because arithmetic operations on values in at least two columns are executed by rows. The values in the sixth column (`npix_above_cl`), which refer to cumulative sums of pixels in classes higher than a certain class, are derived by subtracting the values in the fourth column from the corresponding values in the fifth column [i.e., (`npix_above_cl`)=(`npix_tot`)–(`npix_cum`)]. The values in the last column (`npix_equal_above_cl_min`), which refer to cumulative sums of pixels of uni-element concentrations equal to or greater than

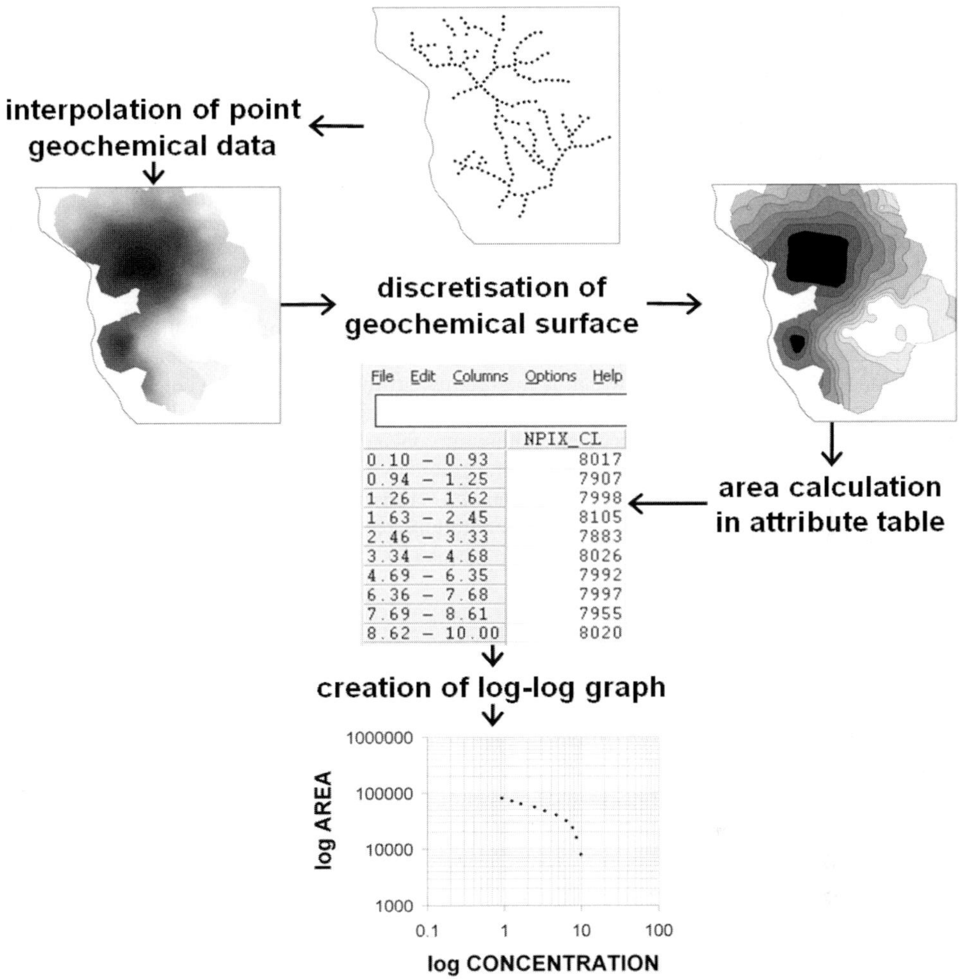

Fig. 4-9. Schematic diagram of GIS-based operations that support implementation of the concentration-area fractal method for recognition of threshold(s) separating background and anomaly in geochemical data. See text and Fig. 4-10 for further explanation.

a class minimum, are derived by adding the values in the `NPIX_CL` column to the corresponding values in the `npix_above_cl` column. The values in the `cl_min` column and the `npix_equal_above_cl_min` column are then used to create a log-log plot of the concentration-area relation.

File Edit Columns Options Help

	NPIX_CL	cl_min	npix_cum	npix_tot	npix_above_cl	npix_equal_above_cl_min
0.10 - 0.93	8017	0.10	8017	79900	71883	79900
0.94 - 1.25	7907	0.94	15924	79900	63976	71883
1.26 - 1.62	7998	1.26	23922	79900	55978	63976
1.63 - 2.45	8105	1.63	32027	79900	47873	55978
2.46 - 3.33	7883	2.46	39910	79900	39990	47873
3.34 - 4.68	8026	3.34	47936	79900	31964	39990
4.69 - 6.35	7992	4.69	55928	79900	23972	31964
6.36 - 7.68	7997	6.36	63925	79900	15975	23972
7.69 - 8.61	7955	7.69	71880	79900	8020	15975
8.62 - 10.00	8020	8.62	79900	79900	0	8020

Fig. 4-10. An example of an attribute table associated with a map of a discretised geochemical surface. The first (leftmost) column contains the names of classes of uni-element concentrations. The second column (`NPIX_CL`), which is the original attribute (or variable) column in the table, contains the number of pixels (or boxes) of each class of uni-element concentrations. The third column (`cl_min`) is created to indicate minimum concentration of each class. By performing arithmetic operations using the values in the first column, the remaining columns are derived (see text for explanations). The values in the columns `cl_min` and `npix_equal_above_cl_min` are then used to create log-log plots of the concentration-area relation.

CASE STUDY

Among the previously cited workers who demonstrated the application of the concentration-area method in mineral exploration, Cheng et al. (1996) and Cheng (1999b) applied the method using stream sediment geochemical data in different study areas. The case study here demonstrates further the concentration-area method by using the stream sediment geochemical data in the Aroroy district (Philippines). Details of the geology, mineralisation and stream sediment geochemical data of the case study area are given in Chapter 3.

Creation and classification of uni-element geochemical surface maps

Creating a geochemical surface based on stream sediment element concentrations is not a trivial procedure. Firstly, unlike uni-element concentrations in soils or rocks, uni-element concentrations in stream sediments actually do not represent spatially continuous fields or variables (i.e., they are not everywhere). Secondly, stream sediments and associated uni-element contents pertain only to a zone of influence – drainage catchment basin. Nevertheless, there are many case studies in the geochemical exploration literature wherein point data of stream sediment uni-element concentrations have been transformed, usually via 'weighted moving average' interpolation techniques, into a continuous surface (e.g., Ludington et al., 2006). Of the different 'weighted moving average' interpolation methods, inverse distance weighting and kriging are the most commonly used methods. Inverse distance weighting requires some knowledge of

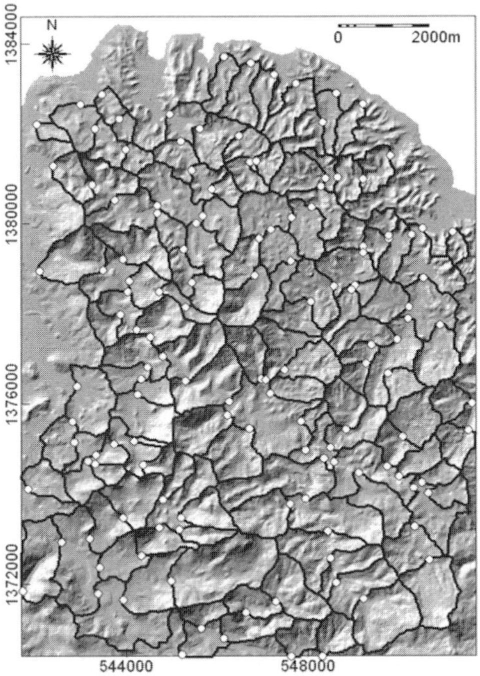

Fig. 4-11. Catchment basins (polygons in black outlines) of stream sediment sample locations (white dots), Aroroy district (Philippines). Interpolated uni-element concentrations outside any of these stream sediment sample catchment basins are excluded in the analysis of anomalies via the concentration-area fractal method.

the point uni-element concentration data, which allows for the determination of the appropriate size of the 'moving average' window (or kernel) and distance-decay parameters (i.e., limiting distance and weight exponent). Kriging assumes that spatial variability is too complex to be modeled mathematically such that it must be treated as a stochastic process and the interpolation parameters (form of variation, magnitude and spatial scale) are analyzed via variography. For simplicity, geochemical surfaces for the stream sediment uni-element concentrations in this case study are generated via the inverse distance moving average method using a limiting distance of 1700 m (which is consistent with the sampling density of about one sample per 1-2 km^2) and a weight exponent of [1].

In respect of the zone of influence of each stream sediment sample location, interpolated uni-element concentrations outside any of the stream sediment sample catchment basins (Fig. 4-11) are masked out in the analysis of uni-element threshold via the concentration-area fractal method. In addition to the continuous uni-element geochemical surfaces, discrete geochemical surfaces are created by assuming that uni-element concentrations within a stream sediment sample catchment basin are equal to uni-element concentrations in the associated sample. A case example of fractal modeling of geochemical anomalies using discrete surfaces based on stream sediment sample catchment basins is demonstrated by Shen and Cohen (2005). So, one purpose of this case study is to compare and contrast results of using continuous and discrete surfaces of stream sediment uni-element concentrations in the concentration-area fractal method for separation of background and anomaly. To ensure proper comparison and analysis of the results, the continuous and discrete geochemical surfaces are classified uniformly using 5-percentile intervals of data distributions. For classification of the discrete geochemical surfaces, the 5-percentile intervals are determined from the original point data values rather than from the map (pixel) values in order to respect the empirical density distributions of the data and because the variations in the sizes of the stream sediment sample catchment basins (Fig. 4-11) can introduce artificial data distributions and thus artifacts in the concentration-area relations.

Analysis of uni-element threshold values and anomalies

Based on the continuous geochemical surfaces, the log-log plots of the area-concentration relations for all elements, except As, can be fitted with more than two straight lines indicating the presence of at least three populations in the uni-element data sets (Fig. 4-12). Based on the discrete geochemical surfaces, the log-log plots of the area-concentration relations for all elements can be fitted with more than two straight lines indicating the presence of at least three populations in the uni-element data sets (Fig. 4-13). For each element, the log-log curves of the concentration-area plots based on the continuous geochemical surfaces and on the discrete geochemical surfaces have very similar shapes (Figs. 4-12 and 4-13). This suggests that, in this case study, either the continuous geochemical surfaces or discrete geochemical surfaces can be used in the concentration-area fractal analysis of geochemical anomalies.

The breaks in slopes of the straight lines fitted to the log-log plots of the concentration-area relations represent thresholds that can be used to classify the uni-element data sets into background and anomalous populations. The numbers of thresholds defined per element based on the continuous and discrete geochemical surfaces are equal except for Cu, Mn and As. For each of these three elements, the number of thresholds obtained from the discrete geochemical surfaces is greater than the number of thresholds obtained from the continuous geochemical surfaces. Nevertheless, the values of the thresholds defined using the continuous and discrete geochemical surfaces are closely similar (Table 4-I), especially for Zn, Ni, Mn and As, although for each of the last two elements a third threshold and a second threshold was defined, respectively, using the discrete geochemical surfaces. For Cu, the higher threshold based on the continuous geochemical surface is closely similar to the highest threshold based on the discrete geochemical surface. In addition, for Cu, the lower threshold based on the continuous geochemical surface is roughly equivalent to the average of the lowest and intermediate thresholds based on the discrete geochemical surface. For Co, the two higher thresholds based on the continuous geochemical surface are respectively

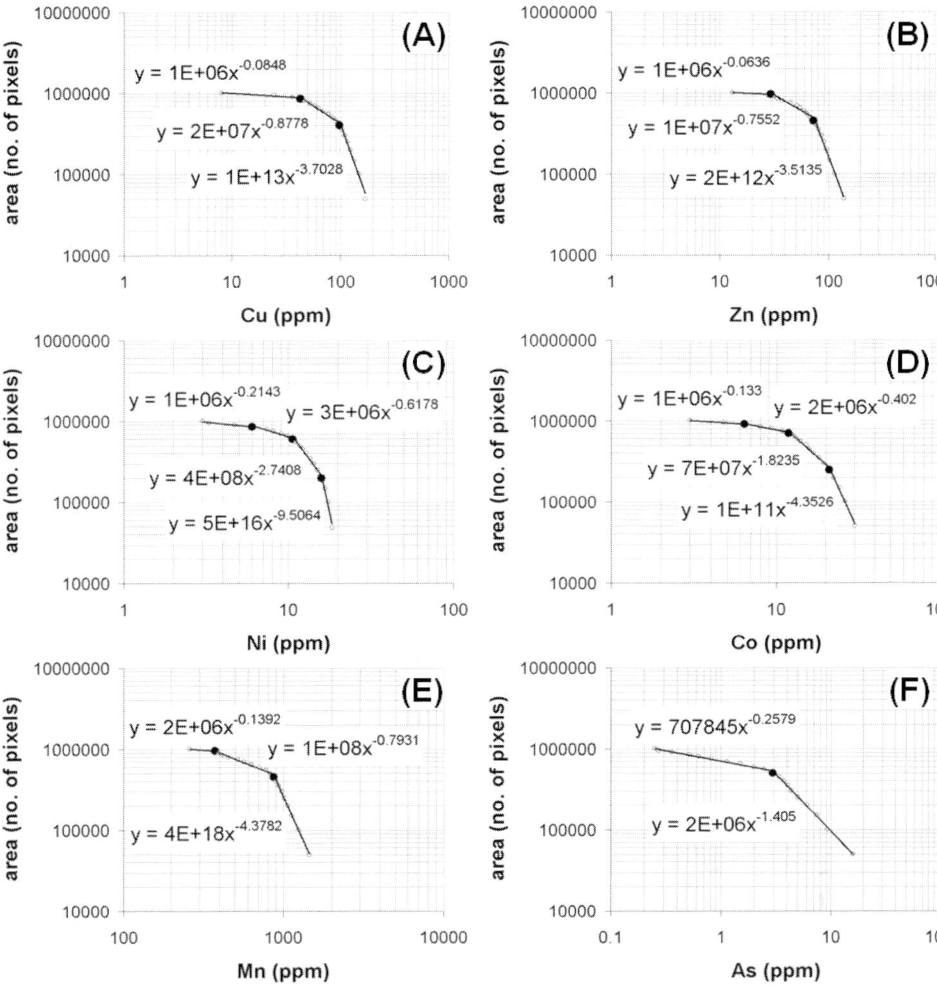

Fig. 4-12. Log-log plots of relation between areas bounded by concentration contours $A(\geq v)$ and uni-element concentration values, Aroroy district (Philippines). Solid lines are obtained by least squares fitting through linear parts of the plots. Dots at breaks in slopes of the lines represent threshold concentrations (v_t). Lines to the left of any threshold follow a power-law relation represented by equation (4.6), whereas lines to the right of rightmost thresholds follow a power-law relation represented by equation (4.7).

equivalents of the two lower thresholds based on the discrete geochemical surface. The differences in the results obtained are plausibly due to the fact that the continuous geochemical surfaces are smoother than the discrete geochemical surfaces.

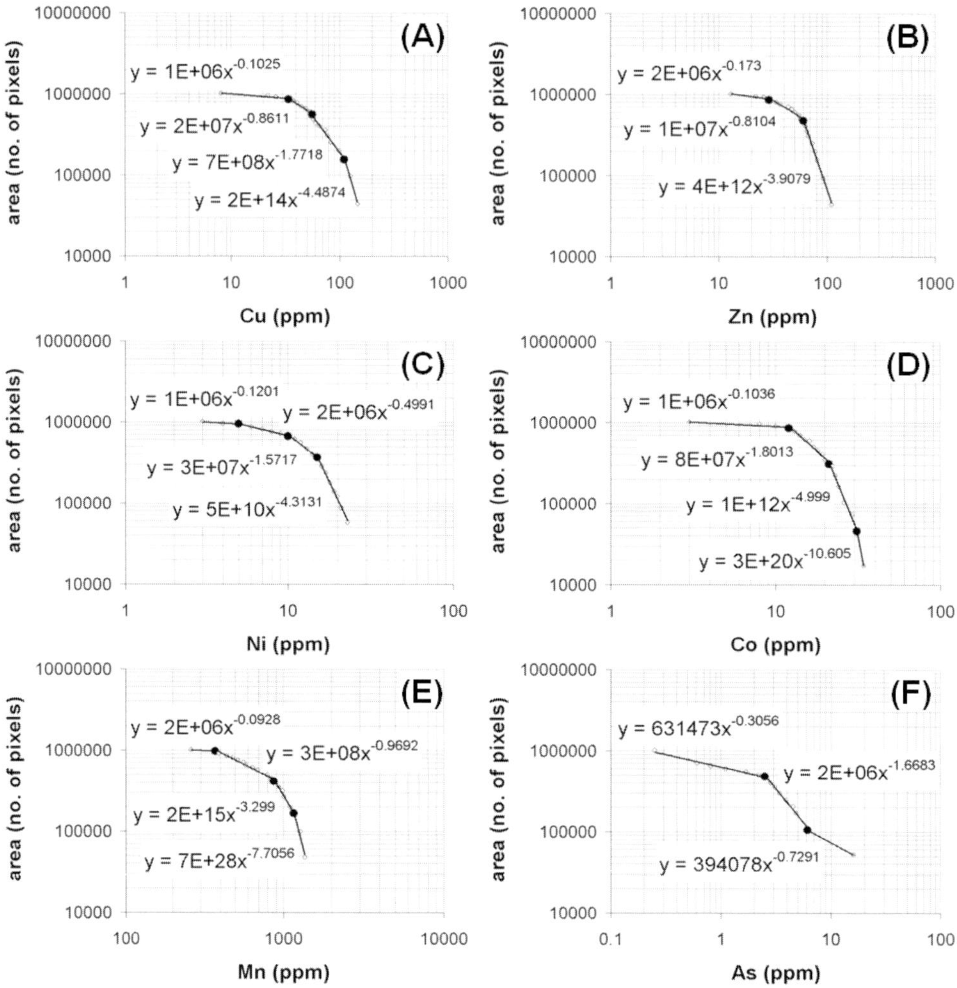

Fig. 4-13. Log-log plots of relation between areas of stream sediment sample catchment basins with uni-element concentrations above certain levels $A(\geq v)$ and uni-element concentration values, Aroroy district (Philippines). Solid lines are obtained by least squares fitting through linear parts of the plots. Dots at breaks in slopes of the lines represent threshold concentrations (v_t). Lines to the left of any threshold follow a power-law relation represented by equation (4.6), whereas lines to the right of rightmost thresholds follow a power-law relation represented by equation (4.7).

Most of the ranges of uni-element concentrations based on the thresholds defined from the analyses of the continuous and discrete geochemical surfaces are interpreted as different levels background populations (Table 4-I). The populations in the Ni, Co and Mn data are interpreted as different levels of background because the whole range of

TABLE 4-I

Interpretations of ranges of uni-element concentrations based on thresholds* defined by concentration-area fractal analysis of continuous surfaces and discrete sample catchment basin surfaces of stream sediment geochemical data, Aroroy district (Philippines).

Element	Analysis based on continuous surfaces		Analysis based on discrete surfaces	
	Concentration (in ppm) range*	Interpretation	Concentration (in ppm) range*	Interpretation
Cu			≤**34**	Low background
	≤**42.8**	Low background	35 – **56**	High background
	42.9 – **97.4**	High background	57 – **99**	Low anomaly
	>97.4	Anomaly	>99	High anomaly
Zn	≤**29.3**	Low background	≤**29**	Low background
	29.4 – **72.4**	High background	30 – **60**	High background
	>72.4	Anomaly	>**60**	Anomaly
Ni	≤**6.0**	Very low background	≤**5**	Very low background
	6.1 – **10.6**	Low background	6 – **10**	Low background
	10.7 – **15.8**	Background	11 – **15**	Background
	>15.8	High background	>**15**	High background
Co	≤**6.4**	Very low background		
	6.5 – **11.8**	Low background	≤**12**	Low background
	11.8 – **20.8**	Background	13 – **21**	Background
	>20.8	High background	21 – **31**	High background
			>**31**	Very high background
Mn	≤**375.9**	Very low background	≤**371**	Very low background
	376.0 – **874.5**	Low background	372 – **871**	Low background
	>875.4	Background	872 – **1161**	Background
			>1161	High background
As	≤**2.95**	Background	≤**2.51**	Background
	>2.95	Anomaly	2.52 – **6.11**	Low anomaly
			>6.11	High anomaly

*Values in bold are thresholds defined in the log-log plots of concentration-area relations in geochemical surfaces of the data (see Figs. 4-12 and 4-13).

concentrations for each of these elements is comparable to their respective average normal abundances (or Clarke values) in the Earth's crust (see Levinson, 1974; Rose et al., 1979). For example, Fig. 4-14 shows spatial distributions of different background levels of the Co data based on thresholds defined from the analyses of the continuous and discrete geochemical surfaces. Very low to low background Co values are mostly distributed in areas underlain by the Aroroy Diorite (see Fig. 3-9). Background Co values are mostly distributed in areas underlain by the Mandaon and Lanang Formations. High to very high background Co values are mostly distributed in areas underlain by the

the total variance of the stream sediment uni-element data explained by PC3 compared to that explained by PC4 suggests that Cu-As anomalies are slightly more widely distributed in the area than As-Ni anomalies. In addition, the magnitude of the loadings of Cu and As on PC3 and PC4 suggests that the former is slightly mobile (thus more dispersed) than the latter in the surficial environments of the study area. Based on these arguments, it can hypothesised that, in terms of indicating presence of epithermal Au mineralisation in the area, (a) the Cu-As association represented by PC3 are distal anomalies whilst the As-Ni association represented by PC4 are proximal anomalies and, thus, (b) the latter multi-element association (or PC4) is more important than the former multi-element association (or PC3). Thus, the scores of PC3 and PC4 are further subjected to the concentration-area fractal method for recognition of anomalies, although results of analysis based on PC4 scores are explained first followed by results of analysis based on PC3.

The scores of PC3 and PC4 for the point geochemical data are interpolated via inverse distance moving average method to derive continuous geochemical surfaces. In addition, the scores of PC3 and PC4 obtained here for the point geochemical data are attributed to pixels in the associated stream sediment sample catchment basins to derive discrete geochemical surfaces. The multi-element geochemical surfaces are discretised in the same way the uni-element geochemical surfaces are discretised (see above). The plots of concentration-area relations for the multi-element geochemical surfaces are shown in Fig. 4-17. Note that the 'concentration' variables represented by the PC scores do not have the normal concentration units because the PCs are linear combinations of the $\log_e$-transformed uni-element data. For this reason and because negative PC scores cannot be transformed to logarithms, the 'concentration' axes of the concentration-area plots are not in the logarithmic scale. The PC scores at the breaks in slopes of the straight lines fitted to the concentration-area relations represent thresholds that can be used to classify the multi-element association scores into background and anomalous populations. The very similar shapes of the concentration-area curves and the equal numbers of thresholds defined per set of PC scores represented as continuous and discrete geochemical surfaces (Fig. 4-17) suggest that, in this case study, either continuous geochemical surfaces or discrete geochemical surfaces can be used in the concentration-area fractal analysis of geochemical anomalies.

For the PC4 scores, the thresholds based on analysis of either continuous or discrete geochemical surfaces indicate five populations, which are interpreted, from lowest to highest, as (a) low background, (b) background, (c) high background, (d) low anomaly and (e) high anomaly. The spatial distributions of the background and anomalous populations of PC4 scores, representing As-Ni association in stream sediments, show some degree of similarity (Fig. 4-18). For the PC3 scores, the thresholds based on analysis of either continuous or discrete geochemical surfaces indicate four populations, which are interpreted, from lowest to highest, as (a) low background, (b) high background, (c) low anomaly and (d) high anomaly. The spatial distributions of the background and anomalous populations of PC3 scores represented as continuous and

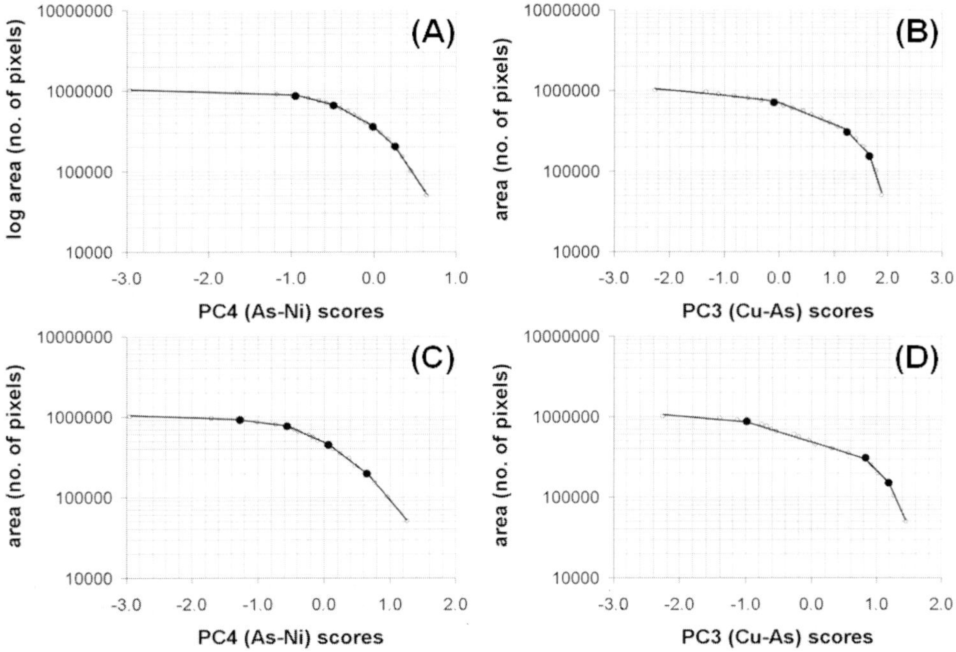

Fig. 4-17. Concentration-area models for PC4 and PC3 scores of the stream sediment geochemical data (Table 4-II), Aroroy district (Philippines), based on their [(**A**) and (**B**), respectively] transformations into continuous geochemical surfaces and [(**C**) and (**D**), respectively] representations as discrete geochemical surfaces. Solid lines are obtained by least squares fitting through linear parts of the plots. Dots at breaks in slopes of the lines represent threshold PC scores (v_t). Lines to the left of any threshold follow a power-law relation represented by equation (4.6), whereas lines to the right of rightmost thresholds follow a power-law relation represented by equation (4.7).

discrete surfaces show some degree of similarity (Fig. 4-19). The low and high anomalies based on analysis of the continuous surface of PC3 scores show weaker spatial association with the epithermal Au deposit occurrences compared to the low and high anomalies based on analysis of the discrete surface of PC3 scores. This indicates that spatial interpolation of stream sediment geochemical data potentially results in loss of important geo-information due to smoothing of the data.

The low and high anomalies of the As-Ni association based on analysis of the continuous and discrete surfaces of PC4 scores (Fig. 4-18) show stronger spatial associations with the epithermal Au deposit occurrences compared to the low and high anomalies of the Cu-As association based on analysis of either continuous or discrete surface of PC3 scores (Fig. 4-19). This indicates that most of the low and high anomalies of PC4 scores are significant, whereas most of the low and high anomalies of PC3 scores are not. Nevertheless, some low and high anomalies of PC3 scores (Fig. 4-19) are

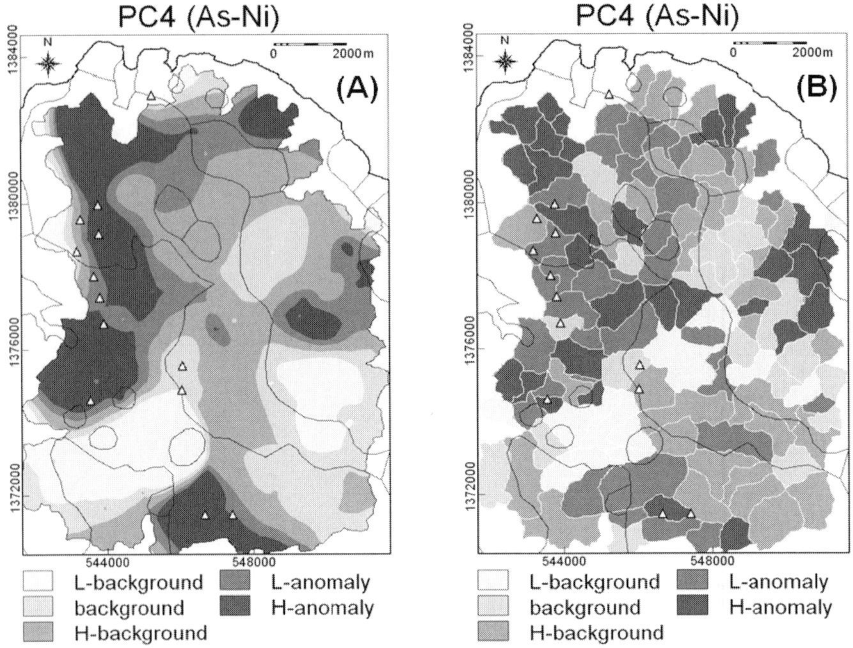

Fig. 4-18. Spatial distributions of background and anomalous populations of PC4 scores representing an As-Ni association in stream sediments, Aroroy district (Philippines), based on thresholds defined from the (**A**) continuous surface of PC4 scores (Fig. 4-17A) and (**B**) discrete catchment basin surface of PC4 scores (Fig. 4-17C). L = low; H = high. Triangles represent locations of epithermal Au deposit occurrences, whilst thin black lines represent lithologic contacts (see Fig. 3-9).

situated along the intrusive contact of the Aroroy Diorite with the Mandaon Formation (see Fig. 3-9), suggesting that such anomalies are possibly associated with hydrothermally altered rocks.

Synthesis of multi-element associations

It is appealing to integrate the PC3 and PC4 scores of the point stream sediment geochemical data because there are some parts of the area where anomalies of PC3 scores and anomalies of PC4 scores are spatially coincident (Fig. 4-18 and 4-19). For example, the southern parts of the area where two epithermal Au deposits occur and the northern part of the area below one epithermal Au deposit occurrence. In addition, in the northwestern parts of the area where many epithermal Au deposits occur, there is spatial coincidence between high background PC3 scores and anomalous PC4 scores. A simple multiplication can be applied to integrate the PC3 and PC4 scores, although this will create false anomalies from PC3 and PC4 scores that are both negative. This problem

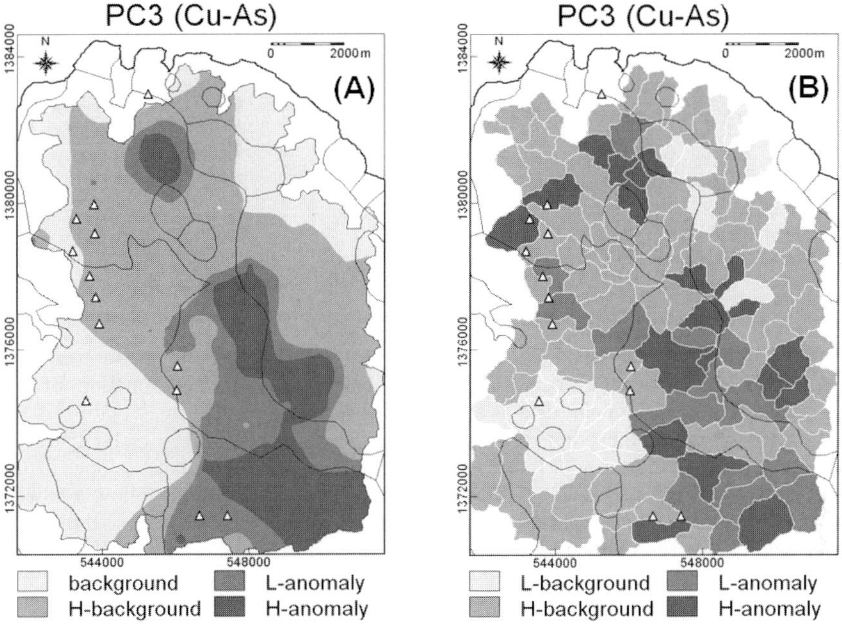

Fig. 4-19. Spatial distributions of background and anomalous populations of PC3 scores representing Cu-As association in stream sediments, Aroroy district (Philippines), based on thresholds defined from the (**A**) continuous surface of PC3 scores (Fig. 4-17B) and (**B**) discrete catchment basin surface of PC3 scores (Fig. 4-17D). L = low; H = high. Triangles represent locations of epithermal Au deposit occurrences, whilst thin black lines represent lithologic contacts (see Fig. 3-9).

can be overcome by first re-scaling the PC3 and PC4 scores linearly to the range [0,1] and thereafter performing multiplication on the re-scaled variables. The product 'integrated As-Ni-Cu' scores can then be transformed into either a continuous or discrete surface for the application of the concentration-area fractal method of modeling anomalies.

The concentration-area models for the continuous and discrete surfaces of the integrated As-Ni-Cu scores are very similar (Fig. 4-20). The three thresholds recognisable from the concentration-area curves indicate that there are four populations in the integrated As-Ni-Cu scores, which are interpreted, from lowest to highest, as (a) low background, (b) high background, (c) low anomaly and (d) high anomaly.

The spatial distributions of low and high anomalies of the integrated As-Ni-Cu scores show strong spatial associations with several of the epithermal Au deposit occurrences (Fig. 4-21). Based on the discrete surface of the integrated As-Ni-Cu scores (Fig. 4-21B), there seem to be two parallel north-northwest trending anomalies, one following the trend of epithermal Au deposit occurrence and the other following the intrusive

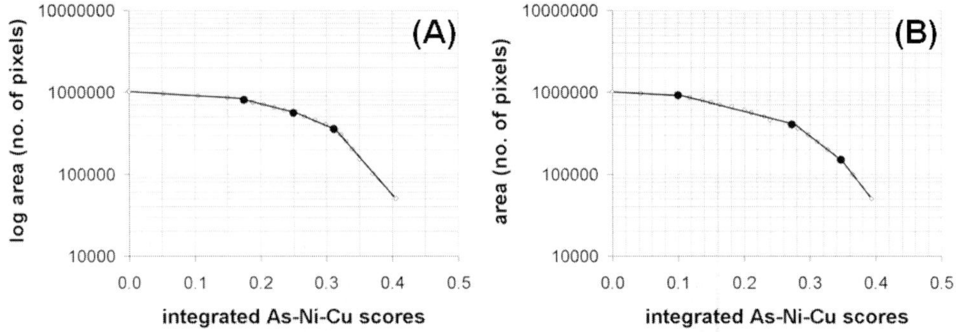

Fig. 4-20. Concentration-area models for the integrated As-Ni-Cu scores the stream sediment geochemical data in the Aroroy district (Philippines), based on (**A**) transformation of the integrated As-Ni-Cu scores of the point data as a continuous surface and (**B**) representation of the integrated As-Ni-Cu scores of the point data as discrete geochemical surfaces. Solid lines are obtained by least squares fitting through linear parts of the plots. Dots at breaks in slopes of the lines represent threshold integrated scores (v_t). Lines to the left of any threshold follow a power-law relation represented by equation (4.6), whereas lines to the right of rightmost thresholds follow a power-law relation represented by equation (4.7).

contact of the Aroroy Diorite with the Mandaon Formation (see Fig. 3-9). The latter could be related to feeble hydrothermal alteration in the Mandaon Formation due to the intrusion of the Aroroy Diorite. The continuous surface of the integrated As-Ni-Cu scores suggests that the anomalies related to the epithermal Au deposit occurrences and the anomalies associated to possible hydrothermally altered rocks along the contact between Mandaon Formation and the Aroroy Diorite seem to overlap rather than run parallel. This could be a result, however, of the smoothing effect during interpolation of the point data of integrated As-Ni-Cu scores.

Combining the PC3 (Cu-As) and PC4 (As-Ni) scores as integrated As-Ni-Cu scores results in downgrading of importance of anomalies (e.g., from anomalous PC3 or PC4 scores to high background integrated scores) in four localities indicated by Roman numerals in Fig. 4-21. In localities I, II and III, which are underlain by the Aroroy Diorite, anomalies of PC3 scores (Fig. 4-18), which are mostly due to Cu rather than As (see Figs. 4-14 and 4-15), and anomalies of PC4 scores (Fig. 4-19), which are mostly due to Ni rather than As (see Fig. 4-15) are downgraded in importance (i.e., they now map as high background) after deriving and analysing the integrated As-Ni-Cu scores (Fig. 4-19). These observations imply that the anomalies of PC3 and PC4 scores in localities I, II and III are probably non-significant but due to variations in mineralogical composition of the Aroroy Diorite (see Fig. 3-9). Thus, combining the PC3 (Cu-As) and PC4 (As-Ni) scores into the integrated As-Ni-Cu scores has positive effects in these cases. In locality I, which is underlain by the Sambulawan Formation and some alluvial deposits (see Fig. 3-9), anomalies of PC4 scores (mainly due to As rather than Ni; see Fig. 4-15) are downgraded in importance (i.e., they now map as high background) after

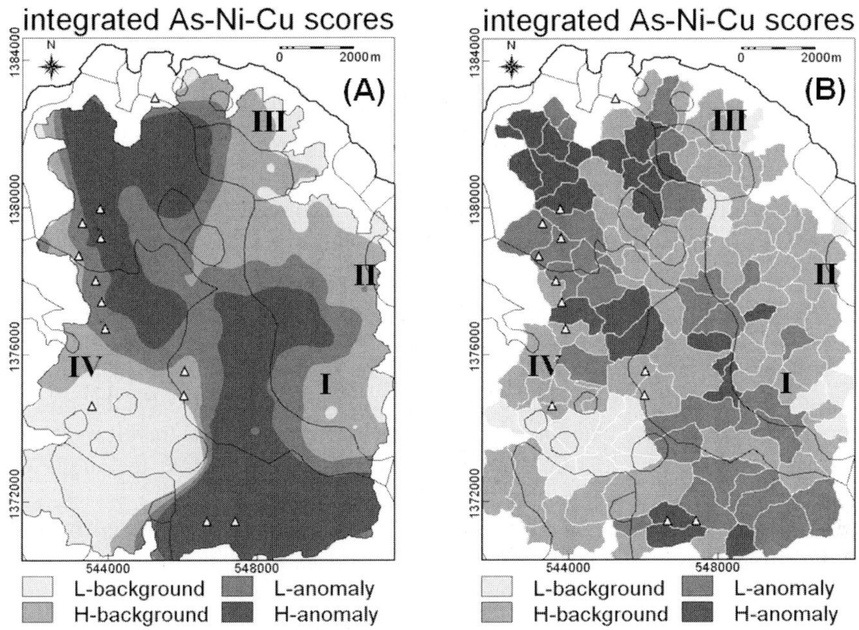

Fig. 4-21. Spatial distributions of background and anomalous populations of integrated As-Ni-Cu scores derived from the stream sediment geochemical data, Aroroy district (Philippines) based on thresholds defined from the (**A**) continuous surface of integrated As-Ni-Cu scores (Fig. 4-20A) and (**B**) discrete catchment basin surface of integrated As-Ni-Cu scores (Fig. 4-20B). L = low; H = high. Triangles represent locations of epithermal Au deposit occurrences, whilst thin black lines represent lithologic contacts (see Fig. 3-9). Roman numbers are locations referred to in the text discussing the effect of combining the PC3 and PC4 scores into the integrated As-Ni-Cu scores.

deriving and analysing the integrated As-Ni-Cu scores (Fig. 4-19). The anomalies of PC4 scores in locality IV are probably significant as they occur downstream of epithermal Au deposit occurrences, so combining the PC3 (Cu-As) and PC4 (As-Ni) scores has a negative effect in this case. Combining the PC3 (Cu-As) and PC4 (As-Ni) scores into integrated As-Ni-Cu scores has, nevertheless, an overall positive effect in this case study and is therefore defensible.

CONCLUSIONS

In the analysis of exploration geochemical data to recognise significant anomalies, it can be useful to consider that geochemical landscapes have spatial variability, geometrical properties and scale-invariant characteristics. The concentration-area fractal method, developed originally by Cheng et al. (1994), takes into account such attributes of geochemical landscapes. The concentration-area method is not, however, the only method for fractal analysis of geochemical anomalies. There are variants of the concentration-area method, e.g., a concentration-distance method (Li et al., 2003) and a summation method (Shen and Cohen, 2005). The concentration-area method and its variants are appropriate for geochemical data analysis in the spatial domain, in which the scale-invariant characteristics of geochemical landscapes are related to the empirical density distributions, spatial variability and geometrical patterns of geochemical data sets.

Other methods for fractal analysis of geochemical anomalies involve converting geochemical data into a function of 'wave numbers' in the frequency domain, in which the scale-invariant characteristics of geochemical landscapes are represented by means of a power spectrum. For details of fractal analysis of geochemical anomalies in the frequency domain, readers are referred to authoritative explanations by Cheng et al. (2000). Nevertheless, Cheng et al. (2000) aver that "because the spatial distribution of power spectrum is determined not only by the wave numbers but also by the power-spectrum function, it should be characterised by the concentration-area fractal method". This means that the concentration-area fractal method is a fundamental technique for modeling of geochemical anomalies.

The case study on GIS-based application of the concentration-area fractal method shows the multifractal nature of the geochemical landscape in the Aroroy district (Philippines) based on stream sediment uni-element data. The results of the case study illustrate that significant anomalies related to the epithermal Au deposit occurrences are modeled better when the geochemical landscape is represented as discrete surfaces rather than as continuous surfaces. Perhaps the application of inverse distance moving average to derive the continuous geochemical surfaces from the point stream sediment uni-element data is not optimal in this case, particularly because stream sediments do not represent continuous geo-objects. The case study demonstrates, nonetheless, that either continuous or discrete geochemical surfaces can be used in GIS-based concentration-area multifractal modeling of significant anomalies in stream sediment uni-element data.

As shown in the case study, application of traditional methods of spatial interpolation to derive continuous geochemical surfaces can undermine multifractal analysis of geochemical anomalies. However, methods for multifractal interpolation (Cheng, 1999a) and fractal filtering (Xu and Cheng, 2001) have been developed to take into account the concentration-area relationship in deriving, from point geochemical data, 'ready-to-use' interpolated maps in which background and anomalies are already separated. These methods have not been demonstrated here, not only because extensive discussion spatial interpolation and filtering is beyond the scope of this volume but also because they require specialised software, which is dedicated to such purposes and which is not available in most commercial or shareware GIS software packages.

Chapter 5

CATCHMENT BASIN ANALYSIS OF STREAM SEDIMENT ANOMALIES

INTRODUCTION

Stream sediments are commonly used as the chief sampling medium in exploration geochemical surveys. They represent, at any point along drainage systems, composite materials derived from the weathering and erosion of one or more sources upstream. It follows that uni-element concentrations in stream sediments are derived from multiple (mostly background and rarely anomalous) sources. In most cases, a major proportion of variation in stream sediment uni-element concentrations is due to lithologic units underlying the areas upstream of sample points. Rose et al. (1970) have demonstrated that certain chemical contents of stream sediments have (a) positive relationships with areas of the lithologic units in a catchment basin and (b) negative relationships with the total area of a catchment basin. Hawkes (1976) considered such relationships as due to downstream of dilution of chemical contents of stream sediments and suggested an idealised formula that relates the measured uni-element concentrations (Y_i) in stream sediment sample i and the area of its sample catchment basin (A_i) to the uni-element concentrations in and surface areas of source materials in the sample catchment basin:

$$Y_i A_i = Y_a A_a + Y_i'(A_i - A_a) \quad (5.1)$$

where Y_a represents the uni-element concentrations due to anomalous sources occupying a cumulative area A_a, Y_i' represents the uni-element concentrations due to background sources occupying a cumulative area equal to A_i–A_a. Hence, if a catchment basin contains only background sources, then Y_i is equal to Y_i'; whereas if a catchment basin contains anomalous sources (e.g., mineral deposits), then Y_i is greater than Y_i'. As stream sediment geochemical exploration aims to identify catchment basins containing anomalous sources, equation (5.1) can be re-arranged as:

$$Y_a A_a = A_i (Y_i - Y_i') + Y_i' A_a \quad (5.2)$$

and the term $Y_a A_a$ can be considered an 'anomaly rating'. The term $A_i(Y_i - Y_i')$ in equation (5.2) is equivalent to the 'productivity' of a catchment basin (Polikarpochkin, 1971), which was demonstrated by Moon (1999) to be useful in identifying anomalous catchment basins.

In another study, Pan and Harris (1990) proposed a generalised equation to predict uni-element concentration (Y_s) at a source as a function of uni-element concentration (Y_i) in a stream sediment sample i and the distance (D_i) of that sample from the source:

$$Y_s = Y_i D_i^{\theta/\sin\alpha} \tag{5.3}$$

where θ is a uni-element coefficient and α is the angle of topographic slope. The generalised equation proposed by Pan and Harris (1990) worked well in their case study with discrete anomalous sources of Au, Ag and Cu in soil. It can be noted, however, that equation (5.3) does not take into account the area and uni-element background in a sample catchment basin. Although background concentrations of Au can be relatively insignificant compared to anomalous concentrations of Au, it is important to consider background of other elements in order to recognise anomalies.

Analysis of stream sediment geochemical data alone can therefore be insufficient for recognition of significant anomalies. Effective interpretation of stream sediment geochemical data requires integration of every available piece of spatial information pertinent to the zone of influence of every stream sediment sample location – its catchment basin. The relation in equation (5.2) indicates that, in order to recognise stream sediment anomalies, (a) background uni-element concentrations Y_i' must first be estimated for each sample catchment basin, (b) background uni-element concentrations Y_i' must then be removed from measured uni-element concentrations Y_i (i.e., $Y_i - Y_i'$) leaving geochemical residuals, which may include significant anomalies (i.e., derived from mineral deposits) and (c) geochemical residuals must be corrected for downstream dilution by taking into account area of sample catchment basin (i.e., $A_i(Y_i - Y_i')$) to enhance anomalies. By considering areal proportions of lithologic units in every sample catchment basin, it is possible to estimate local background uni-element concentrations due to lithology in every sample catchment basin (Bonham-Carter and Goodfellow, 1984, 1986; Bonham-Carter et al., 1987; Carranza and Hale, 1997). Instead of areal proportions of lithologic units in every sample catchment basin, Peh et al. (2006) demonstrated that linear proportions of lithologic units along perennial streams in every sample catchment are also useful in estimating local background uni-element concentrations.

The objective of this chapter is to explain techniques, which can be implemented in a GIS, for catchment basin analysis of uni-element anomalies in stream sediments in order to (a) estimate local uni-element background concentrations due to lithology and (b) derive and correct uni-element residuals for downstream dilution. Dilution-corrected uni-element residuals are then used in the analysis of uni-element and multi-element anomalies. These techniques are then demonstrated in a case study of the same stream sediment geochemical data used to demonstrate the EDA and fractal analysis explained, respectively, in Chapters 3 and 4.

ESTIMATION OF LOCAL UNI-ELEMENT BACKGROUND DUE TO LITHOLOGY

Two techniques are explained here: (1) multiple regression analysis; and (2) analysis of weighted mean uni-element concentrations. The former is demonstrated by Bonham-Carter and Goodfellow (1984, 1986), Bonham-Carter et al. (1987) and Carranza and Hale (1997), whilst the latter is demonstrated by Bonham-Carter et al. (1987).

Multiple regression analysis

In multiple regression analysis, measured stream sediment uni-element concentrations (Y_i) and areal proportions (X_{ij}) of j (=1,2,...,m) lithologic units in sample catchment basin i (=1,2,...,n) are used, respectively, as dependent and independent variables in order to estimate for every sample catchment basin local background uni-element concentrations (Y_i') due to lithology in sample catchment basin i, thus:

$$Y_i' = b_o + \sum_{j=1}^{m} b_j X_{ij} , \quad (5.4)$$

where $\sum_{j=1}^{m} X_{ij} = 1$, b_o and b_j are the regression coefficients determined by the least-squares method to minimise the quantity $\sum_{i=1}^{n} (Y_i - Y_i')^2$. The multiple regression equation implies that estimates of background uni-element concentrations are a result of additive mixing of weathering products of lithologic units in sample catchment basins.

The coefficient b_o can be interpreted as regional average uni-element content, whereas the coefficient b_j can be interpreted as average uni-element content of lithologic unit j (=1,2,...,m) in any sample catchment basin i (=1,2,...,n). However, by inclusion of b_o, equation (5.4) is indeterminate because the regression matrix is singular, unless one independent variable is discarded (Bonham-Carter et al., 1987). This problem can be overcome by allowing round-off errors (e.g., using two decimals) in calculating areal proportions of lithologic units so that $\sum_{j=1}^{m} X_{ij} \cong 1.00$. The multiple regression modeling can also be forced through origin (i.e., setting b_o=0) so that the singularity problem is avoided and equation (5.4) is determinate (Bonham-Carter and Goodfellow, 1984, 1986).

In order to determine relative contributions of the independent variables and their ability to account for total variation in Y_i, the multiple regression analysis is performed via forward and forced simultaneous inclusion of independent variables. That is to say, the most significant independent variables are not searched and included in the final regression equation according to a statistical criterion; rather, all independent variables are included in the final regression model.

The ability of the independent variables to account for the variation of the dependent variables can be characterised using R^2 (usually expressed as percentage), the ratio of sum of squares explained by regression to the total sum of squares, which indicates goodness-of-fit of the multiple regression model. Invariably, regression models have poor fit to uni-element concentration data that are significantly positively skewed. Logarithmic transformation of uni-element concentration data invariably results in

TABLE 5-I

Results of multiple regression analysis (forced simultaneous inclusion of independent variables and forced through the origin) using $\log_e$-transformed stream sediment Cu and Zn data (n=102; see Figs. 2-7A and 2-9A, respectively) as dependent variables and areal proportions of lithologic units (see Fig. 1-1) in sample catchment basins as independent variables.

Dependent variable	Anti-$\log_e$ of regression coefficients of independent variables				R^2
	Basalt	Limestone	Phyllite	Quartzite	
Cu	49.2	50.6	48.7	29.8	99.58
Zn	25.2	84.9	41.9	11.1	96.21

multiple regression models with high R^2, although the models become multiplicative rather than additive. Logarithmic transformation of uni-element concentration data results, however, in regression coefficients that are usually positive and easy to interpret. Whether raw or log-transformed uni-element concentration data are used, statistical tests of significance of the regression are inappropriate if the stream sediment sampling density is high and/or if the stream sediment uni-element concentration data exhibit some degree of spatial autocorrelation and thus are not independent (Bonham-Carter et al., 1987).

For the stream sediment Cu and Zn data shown in Figs. 2-7A and 2-9A, respectively, Table 5-I shows the regression coefficients for the lithologic units (see Fig. 1-1) occurring in the sample catchment basins. The regression analysis was performed after $\log_e$-transformation of the data to reduce asymmetry in their empirical density distributions. The regression coefficients of the independent variables given in Table 5-I are transformed back to normal values to illustrate that they represent average uni-element concentrations in individual lithologic units. The average Cu concentrations (in ppm) in each lithologic unit, except quartzite, are more or less uniform at about 50 ppm, while the average Zn concentrations (in ppm) in each lithologic unit are different. The values of R^2 indicate that the Cu and Zn concentrations in stream sediments in the area are mostly entirely due to lithology, although about 4% of variability in Zn is not accounted for by lithology. Consequently, the spatial distributions of local background Cu and Zn concentrations in stream sediments per sample catchment basin, as estimated according to equation (5.4), reflect mostly the patterns of the lithologic units (Fig. 5-1).

Analysis of weighted mean uni-element concentrations due to lithology

Local background uni-element concentrations due to lithology in every sample catchment basin can also be estimated by first calculating a weighted mean uni-element concentration M_j in each of the j (=1,2,…,m) lithologic units in i (=1,2,…,n) sample catchment basins, thus:

$$M_j = \sum_{i=1}^{n} Y_i \hat{X}_{ij} \Big/ \sum_{i=1}^{n} \hat{X}_{ij} \,, \tag{5.5}$$

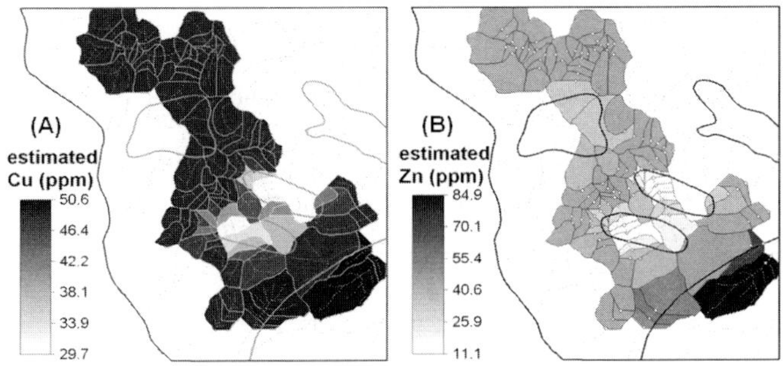

Fig. 5-1. Spatial distributions of local background element concentrations [(**A**) Cu, (**B**) Zn] in stream sediments per sample catchment basin estimated via multiple regression analysis (see text; Table 5-I). Polygons in thick grey or black outlines in the maps are lithologic units (see Fig. 1-1).

where Yi represents uni-element concentrations in stream sediment sample i (1,2,…,n) and $\hat{X}_{ij}$ is the area of each of the j (=1,2,…,m) lithologic units, but not their areal proportions, in sample catchment basin i (=1,2,…,n) as in equation (5.4). Prior to application of equation (5.5), raw uni-element concentration data showing asymmetric empirical density distributions must be appropriately transformed to values whose empirical density distributions are approximately symmetrical. The statistical significance of the value of M_j for each of the j (=1,2,…,m) lithologic units can be inferred from its weighted standard deviation (SD_j), which is estimated as:

$$SD_j=\sqrt{\left(\sum_{i=1}^{n}(Y_i-M_j)^2\,\hat{X}_{ij}\right)\Big/\sum_{i=1}^{n}\hat{X}_{ij}}\ . \tag{5.6}$$

Then, the local background uni-element concentrations (Y_i') due to j (=1,2,…,m) lithologic units in each sample catchment basin i (=1,2,…,n) can be estimated as:

$$Y_i'=\sum_{j=1}^{m}M_j\hat{X}_{ij}\Big/\sum_{j=1}^{m}\hat{X}_{ij}\ . \tag{5.7}$$

Bonham-Carter et al. (1987) have shown that M_j in equation (5.5) and $b_o+\sum b_j$ in equation (5.4) usually have good agreement; that means, there are usually small differences between estimates of Y_i' by using either equation (5.4) or equation (5.7). Only with lithologic units having small areas (about 1% of total area covered by sample catchment basins) do large differences between the two methods of estimation occur. Nevertheless, the method of estimating weighted uni-element concentrations in lithologic units has an advantage over the multiple regression method because the former can be implemented readily in a GIS whilst the latter is usually handled outside a GIS.

TABLE 5-II

Mean element concentrations (M_j) in individual lithologic units, in the area shown in Figs. 1-1 and 5-1, as estimated according to equation (5.5) (see text).

Element	Estimated mean element concentrations (M_j; in ppm) in individual lithologic units			
	Basalt	Limestone	Phyllite	Quartzite
Cu	50.4	56.8	47.5	34.1
Zn	33.8	73.0	43.8	16.9

In our example data of Cu and Zn in stream sediment samples, the mean uni-element concentrations (M_j) in the individual lithologic units (Table 5-II) as estimated according to equation (5.5) are more or less similar to the results of the multiple regression analysis (Table 5-I), depending on the element examined. These results demonstrate good agreement between b_j in equation (5.4) and M_j in equation (5.5). Consequently, the spatial distributions of local background Cu and Zn concentrations in stream sediments per sample catchment basin as estimated according to equation (5.7) (Fig. 5-2) are similar to the spatial distributions of the local background Cu and Zn concentrations in stream sediments per sample catchment basin as estimated according to equation (5.4) (Fig. 5-1). The agreement between b_j in equation (5.4) and M_j in equation (5.5) is further verified in the case study below.

DILUTION CORRECTION OF UNI-ELEMENT RESIDUALS

Subtraction of estimated local background uni-element concentrations from the corresponding measured uni-element concentrations results in geochemical residuals (i.e., $Y_i - Y_i'$), which are either positive or negative. On the one hand, a positive residual can be interpreted as enrichment of uni-element concentrations in stream sediments due to anomalous sources (e.g., mineral deposits). On the other hand, a negative residual can

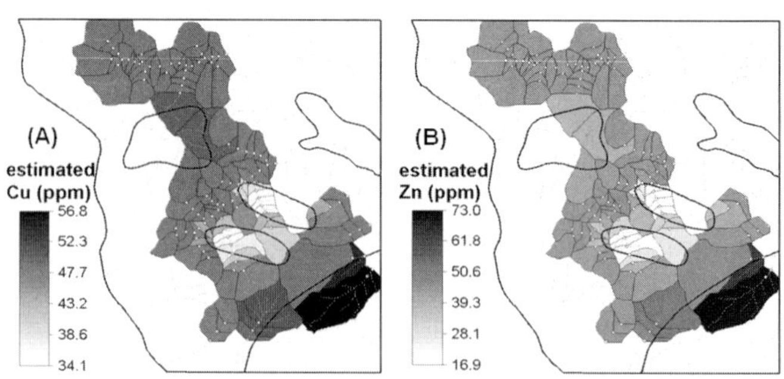

Fig. 5-2. Spatial distributions of local background element concentrations [(**A**) Cu, (**B**) Zn] in stream sediments per sample catchment basin estimated according to equation (5.5) (see text; Table 5-II). Polygons in black outlines in the maps are lithologic units (see Fig. 1-1).

be interpreted as depletion of uni-element concentrations in stream sediments due to certain intrinsic or anthropogenic factors or processes. Negative residuals could also arise, however, if some values in the uni-element data used in the analysis are 'anomalously' high values or outliers, which cause upward bias in multiple regression modeling or in calculation of weighted means and thus result in estimates of local background uni-element concentrations that may be, for some catchment basins, rather artificially high. Recognition and removal of such outliers prior to the regression analysis or the analysis of weighted means may yield unbiased estimates of local background uni-element concentrations, but only for samples retained in the analysis. Nevertheless, as mineralisation is partially or completely unknown, absolute values of estimates of local background uni-element concentrations are trivial but the magnitude of uni-element residuals is useful in ranking of anomalies.

The magnitude of uni-element residuals (i.e., $Y_i - Y_i'$) is controlled by downstream dilution due to mixing of stream sediments from various and mostly non-anomalous sources in a sample catchment basin, thereby obscuring contributions of anomalous sources. The relation proposed by Hawkes (1976) re-arranged in equation (5.2) indicates that uni-element residuals can be corrected for downstream dilution by considering the sample catchment basin area to enhance uni-element anomalies based on positive uni-element residuals, which might indicate presence of mineralisation.

To correct uni-element residuals for downstream dilution, Bonham-Carter and Goodfellow (1984, 1986), assumed a unit area of 1 km^2 (i.e., A_a = 1 km^2) for exposed mineral deposits of interest contributing to stream sediments and defined, by slightly modifying equation (5.2), a dilution-corrected 'mineralisation rating' variable R_i as:

$$R_i = A_i (Y_i - Y_i') + Y_i' . \tag{5.8}$$

However, by using equation (5.8) to correct residuals for downstream dilution, it can be argued that estimates of local background uni-element concentrations (i.e., Y_i') are added back. Rose et al. (1979, pp. 399) point out, nonetheless, that the term $Y_i'A_a$ in equation (5.2) can be neglected if A_i is much larger than A_a. Carranza and Hale (1997) assumed a small unit area of 1 ha (i.e., A_a = 0.01 km^2) of exposed anomalous sources, which is 10× to 200× smaller than the sample catchment basins in their study area. They then neglected the term $Y_i'A_a$ in equation (5.2) and considered Y_a to represent dilution-corrected residuals of uni-element concentrations, thus:

$$Y_a = 100 \times A_i (Y_i - Y_i') . \tag{5.9}$$

Equation (5.9) is adopted here to derive dilution-corrected residuals of uni-element concentrations in stream sediments.

For our example stream sediment Cu and Zn data, Fig. 5-3 displays the similarity between the spatial distributions of dilution-corrected uni-element residuals based on either local background uni-element concentrations estimated via regression analysis

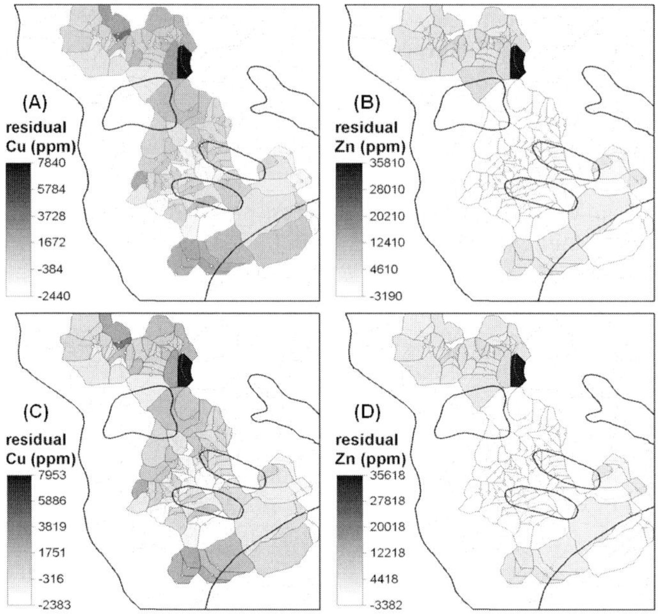

Fig. 5-3. Spatial distributions of dilution-corrected residuals of Cu and Zn contents of stream sediments based on local background uni-element contents estimated [(**A**), (**B**)] via regression analysis (equation (5.4)) and [(**C**), (**D**)] from weighted uni-element contents of lithologic units (equation (5.7)). Polygons in black outlines in the maps are lithologic units (see Fig. 1-1).

(equation (5.4)) or local background uni-element concentrations estimated from weighted mean uni-element concentrations in lithologic units (equation (5.7)). This goes to show further the agreement between dilution-corrected uni-element residuals based on local background uni-element concentrations derived by application of each of the two above-explained techniques.

To recognise anomalous sample catchment basins for certain elements, dilution-corrected uni-element residuals must be subjected to analytical techniques for distinguishing between background and anomaly, such as those explained in Chapter 3 and Chapter 4. This is demonstrated further below in a case study.

ANALYSIS OF ANOMALOUS MULTI-ELEMENT SIGNATURES

Modeling of uni-element anomalies is useful in analysis of specific pathfinder elements for certain mineral deposits. Certain types of mineral deposits are characterised, however, by a suite or an association of one or more 'ore' elements, so it is instructive to analyse relationships among dilution-corrected uni-element residuals in order to determine multi-element geochemical signatures that reflect the presence of mineralisation in a study area. Because the derivation of dilution-corrected uni-element residuals involves removal of background due to lithology and because positive dilution-corrected uni-element residuals suggest enrichment due to anomalous sources, it is intuitive to constrain the analysis of multi-element geochemical signatures by using only a subset of samples with positive dilution-corrected residuals for at least one of the elements under study (e.g., Carranza and Hale, 1997; Carranza, 2004a).

Analysis of multi-element geochemical signatures could be realised through a variety of mathematical multivariate techniques, such as cluster analysis, correspondence analysis, discriminant analysis, factor analysis, regression analysis, principal components analysis (PCA), etc. Explanations of the fundamentals of such multivariate analytical techniques can be found in textbooks (e.g., Davis, 2002) and explanations of applications of such methods to analysis of multivariate geochemical data can be found in Howarth and Sinding-Larsen (1983). For cases where there are few or no known occurrences of mineral deposits of the type sought in a study area, PCA is a useful multivariate analytical technique because it serves as an exploratory approach to discriminate between background and anomalous multi-element signatures.

A brief explanation about PCA is given in Chapter 3. Because results of PCA tend to be dominated by non-anomalous populations, recognition of anomalous multi-element associations can be enhanced by using a subset of samples consisting of anomalous dilution-corrected residuals of at least one of the elements under study. So, classification of anomalous dilution-corrected uni-element residuals must be performed prior to PCA. For the purpose of illustration using the stream sediment Cu and Zn data, an arbitrary threshold representing the 70th percentile of positive dilution-corrected uni-element residuals is used for uni-element anomaly classification. However, after the classification of anomalous dilution-corrected uni-element residuals, there are still some problems that must be overcome. Firstly, inspection of histograms or boxplots of data for a subset of samples consisting of anomalous dilution-corrected residuals of at least one of the elements under study can reveal the presence of multiple populations, outliers and asymmetric empirical density distributions in the data. These factors undermine reliable estimation of a covariance matrix or a correlation matrix, either of which is used as a starting point of PCA. Secondly, logarithmic transformation of the data to alleviate the effects of these factors is not feasible because negative dilution-corrected uni-element residuals can be present in the data subset. A remedy to such problems is to perform a simple rank-ordering approach. Thus, by considering 1 as lowest rank, descending ranks (i.e., n to 1) are assigned to descending n values of dilution-corrected uni-element residuals and averaging ranks in case of ties. A Spearman rank correlation matrix can then be computed for the rank-transformed dilution-corrected uni-element residuals, which can be used in PCA (e.g., George and Bonham-Carter, 1989; Carranza and Hale, 1997).

Table 5-III shows the results of PCA using the whole set of dilution-corrected Cu and Zn residuals derived from multiple regression analysis of the stream sediment Cu and Zn data and the results of PCA using a subset of samples with anomalous dilution-corrected of either Cu or Zn residuals derived from multiple regression analysis of the stream sediment Cu and Zn data. If all samples are used, then PC1 can be interpreted to represent an anomalous inter-element association because the loadings on Cu and Zn are

TABLE 5-III

Principal components of rank-transformed dilution-corrected Cu and Zn residuals derived from results of multiple regression analysis of the stream sediment Cu and Zn data.

		Cu	Zn	% of Variance	Cum. % of variance
For all samples (n=102)	PC1	0.805	0.805	64.7	64.7
	PC2	–0.594	0.594	35.3	100.0
For samples with anomalous dilution-corrected residuals (n=30)	PC1	0.826	–0.826	68.2	68.2
	PC2	0.564	0.564	31.8	100.0

both positive and high. For the same reason, the PC2 obtained by PCA of a subset of samples with anomalous dilution-corrected residuals of either Cu or Zn can be interpreted to represent an anomalous inter-element association. Such interpretations can be verified by mapping of PC scores, which can be calculated according to the following formula (George and Bonham-Carter, 1989):

$$P_{ci}=\sum_{j=1}^{k} L_{cj} r_{ij} \tag{5.10}$$

where P_{ci} is score for sample i (=1,2,...,n) on principal component c, L_{cj} is loading on variable j (=1,2,...,k) and r_{ij} is rank of sample i for variable j.

High PC1 scores based on PCA of rank-transformed dilution-corrected residuals of Cu and Zn for all samples (Fig. 5-4A) coincide with low (mostly negative) dilution-corrected residuals of Cu and/or Zn (Fig. 5-3), meaning that the analysis is be dominated by non-anomalous inter-element associations. By contrast, high PC2 scores based on PCA of samples with rank-transformed anomalous dilution-corrected residuals of either Cu or Zn (Fig. 5-4B) coincide with high positive dilution-corrected residuals of Cu and/or Zn (Fig. 5-3), meaning that the analysis results in enhancement of anomalous inter-element associations. These results demonstrate that PCA of dilution-corrected uni-element residuals for all samples can be dominated by non-anomalous populations. So, it is useful to first classify background and anomalous populations in dilution-corrected uni-element residuals and then to perform PCA on a subset of samples consisting of anomalous dilution-corrected residuals of at least one of the elements under study in order to enhance recognition of anomalous multi-element associations.

The next section digresses to the discussion of application of GIS in catchment basin analysis of stream sediment geochemical anomalies. Then, the foregoing concepts and explanations of methods pertinent to catchment basin analysis of stream sediment geochemical anomalies are demonstrated further in a case study.

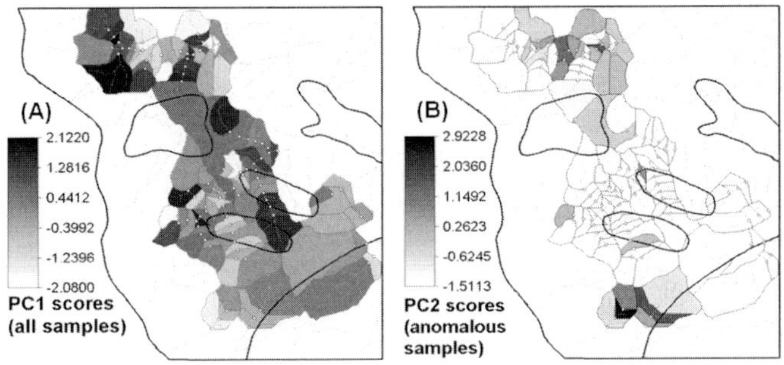

Fig. 5-4. Spatial distributions of PC scores derived by PCA of rank-transformed dilution-corrected Cu and Zn residuals obtained from results of multiple regression analysis of the stream sediment Cu and Zn data (Table 5-III). (**A**) PC1 scores based on PCA of all samples. (**B**) PC2 scores based on PCA of a subset of samples with anomalous dilution-corrected residuals of either Cu or Zn. Polygons in black outlines in the maps are lithologic units (see Fig. 1-1).

APPLICATION OF GIS IN CATCHMENT BASIN ANALYSIS

A GIS supports implementation of catchment basin analysis of stream sediment anomalies in terms of (a) creating polygons representing sample catchment basins, (b) estimating areal proportions of lithologic units in sample catchment basins, (c) estimating local background uni-element concentrations attributable to lithologic units, (d) correcting uni-element residuals for downstream dilution and (e) classifying geochemical anomalies based on dilution-corrected uni-element residuals.

Creation of sample catchment basins

Sample catchment basins may be hand-drawn or hand-digitised by tracing drainage divides on a map of sample points overlaid on a topographic map. The difficulty of this procedure lies in deciding the position of a catchment basin boundary laterally upstream/upslope away from a sample point. Alternatively, sample catchment basins may be created automatically using a digital elevation model (DEM), digitised streams and digitised sample points (Fig. 5-5).

There are several ways by which a DEM can be obtained. One way is to digitise elevation contours on a topographic map and then perform interpolation of elevation using digitised elevation contours. A DEM can also be created automatically from some air- or space-borne images (e.g., Baldi et al., 2002), such as stereo images of bands 3B and 3N of ASTER (Advanced Spaceborne Thermal Emission Reflection Radiometer) data (Hirano et al., 2003). The SRTM (Shuttle Radar Topography Mission) also provides ready-made DEMs (see http://www2.jpl.nasa.gov/srtm/). Whatever method is used, the spatial resolution of a DEM can be crucial in generation of accurate stream sediment sample catchment basins.

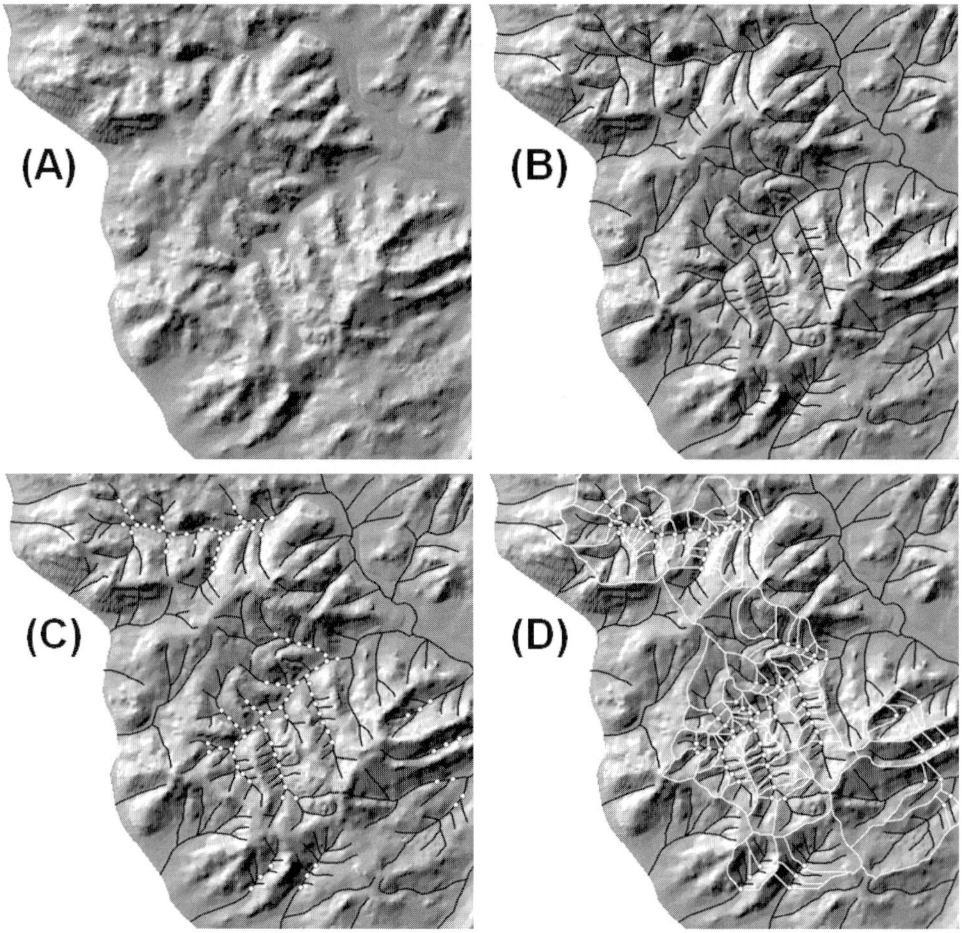

Fig. 5-5. Spatial data and sequential steps involved in creation of stream sediment sample catchment basins. (**A**) Digital elevation model (DEM) created either manually from analogue topographic elevation data or automatically from digital topographic elevation data. (**B**) Stream networks (black lines) digitised from analogue topographic maps or delineated automatically from a DEM. (**C**) Stream sediment sample locations (small white dots) are used as references in further processing of digital stream networks to indicate points to which overland water flows. (**D**) DEM and processed digital streams are used as input data in spatial neighbourhood algorithms for automatic delineation of sample catchment basins (in white outlines).

A DEM can be used for automatic determination of stream networks (e.g., Martz and Garbrecht, 1993) instead of digitising streams from a topographic map. Digitised streams are processed further to indicate stream sediment sample locations as points toward which water flows from different points on topographic surfaces represented by a DEM. The DEM and the processed digital streams are then used via certain spatial neighbourhood algorithms to create automatically sample catchment basins. Detailed explanations of such algorithms can be found in Jenson and Domingue (1988), Martz and Garbrecht (1993) and Jones (2002). Sample catchment basins are labeled with unique IDs corresponding to the stream sediment samples they represent. The area of each sample catchment basin can then be incorporated in the stream sediment geochemical database.

Measuring areas and areal proportions of lithologic units in catchment basins

Calculation of areas and areal proportions of lithologic units in stream sediment sample catchment basins involve map (cross or intersect) overlay and table (join) operations as illustrated and explained schematically in Fig. 5-6. Maps used in the map overlay operation can be either in vector or in raster format and the measured areas can be either in units of, usually, square metres or in number of pixels. The table operation uses simple arithmetic calculations. The operations may vary slightly from one GIS software package to another; however, the overall procedure is straightforward.

Estimation of local background uni-element concentrations

Estimation of weighted mean uni-element concentrations due to lithology (see above) can be readily performed in the attribute table of the stream sediment geochemical data, in which areas of individual lithologic units per sample catchment basin have been appended to the stream sediment geochemical database. Estimation of local background uni-element concentrations in every sample catchment basin via the multiple regression analysis explained above may not be readily performed in a GIS. That is because many GIS software packages do not have algorithms for multivariate data analysis, such as regression analysis. However, an attribute table of the stream sediment geochemical data, which contains the areal proportions of lithologic units in sample catchment basins, can be exported to file formats supported by statistical software packages in which the multiple regression analysis can be performed. Estimates of local background uni-element concentrations per stream sediment sample catchment basin can then be exported back to the GIS for further analysis.

Correction for downstream dilution

Application of either equation (5.8) or equation (5.9) to correct for downstream dilution can be readily implemented in an attribute table of stream sediment geochemical data containing areas of sample catchment basins and estimates of local background uni-element concentrations per sample catchment basin. It is also possible to calculate productivity per sample catchment basin (Polikarpochkin, 1971; Moon, 1999), which also represents dilution-corrected geochemical residuals. However, unlike in equation (5.8) or (5.9), where A_i is incremental catchment area between a sample site and the next sample site upstream, the A_i used for calculation of productivity is total catchment area upstream of a sample location. Explanations for automatic calculation of total catchment areas from points along drainage networks can be found in Martz and Garbrecht (1993). Alternatively, it has been shown that, to estimate productivity, stream order (based on

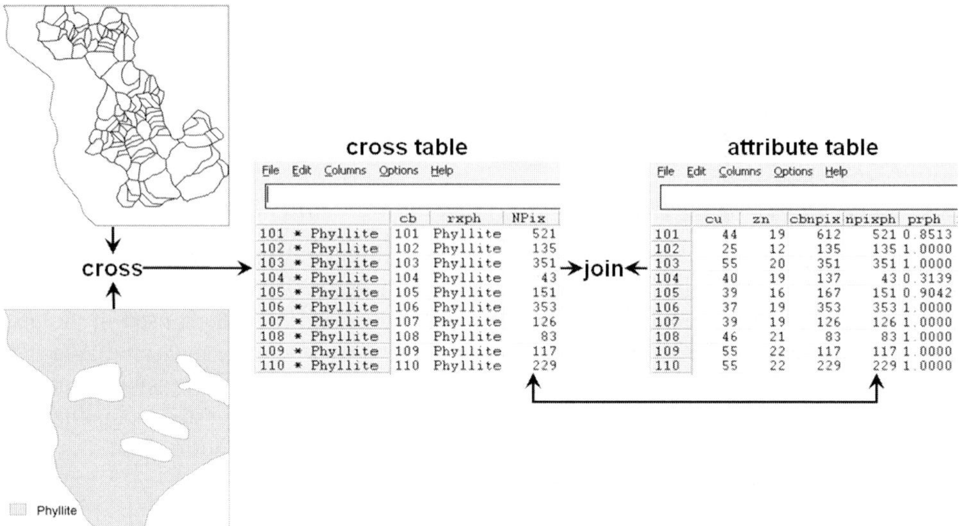

Fig. 5-6. GIS operations to measure area and areal proportions of lithologic units in stream sediment sample catchment basins. Map of sample catchment basins and map of a lithologic unit are crossed (i.e., intersect operation), resulting in a cross table, which records the area, as number of pixels (Npix), of a lithologic unit (rxph) in individual sample catchment basins (cb). Values in the column Npix of the cross table are appended, via table join operation, to a new column (npixph) in the attribute table of the stream sediment geochemical data. Areal proportions of that lithologic unit (in column prph) in individual sample catchment basins are obtained by dividing the values in column npixph by the values in column cbpix (areas, as number of pixels, of sample catchment basins).

Strahler (1957)) at sample sites is a useful surrogate for total catchment area of individual sample sites (Carranza, 2004a). Determining order of stream can be performed automatically in a GIS (e.g., Garbrecht and Martz, 1997).

Analysis and classification of geochemical anomalies

The GIS-aided techniques for application of EDA and fractal analysis, as explained in Chapters 3 and 4, respectively, can be applied to the analysis and classification of geochemical anomalies based on dilution-corrected uni-element residuals. Such methods will be demonstrated further in the case study below.

CASE STUDY

This case study demonstrates further catchment basin analysis of stream sediment anomalies by using the data in the Aroroy district (Philippines). Details of the geology, mineralisation and stream sediment geochemical data of the case study area are given in Chapter 3. The catchment basins of the stream sediment samples are shown in Fig. 4-11.

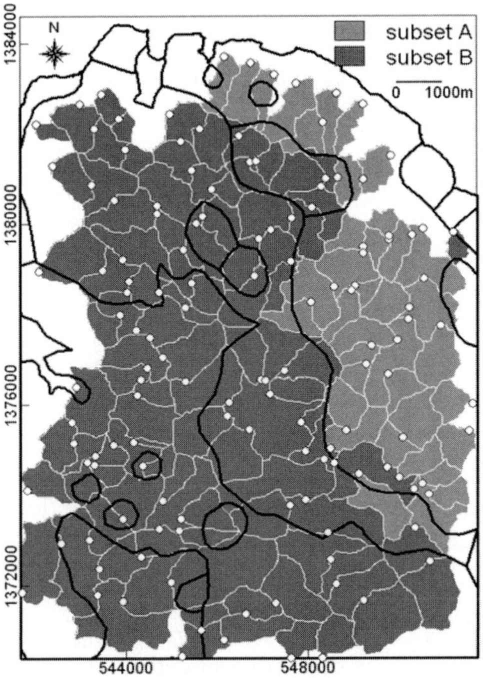

Fig. 5-7. Subsets of stream sediment sample catchment basins (in white outlines) according to rock type at sampling locations (white dots), Aroroy district (Philippines). Polygons in black outlines are lithologic units (see Fig. 3-9).

Estimation of local uni-element background

Estimation of local background uni-element concentrations per stream sediment sample catchment basin via either the multiple regression analysis or the analysis of weighted mean uni-element concentrations in lithologic units described above can be undermined by the presence of multiple populations in the geochemical data. Thus, prior to analysis, the uni-element geochemical data must be examined for presence of multiple populations, for example, by analysis of cumulative probability graphs (Sinclair, 1974), or Q-Q plots or Normal Q-Q plots. The analysis must be performed on 'homogenous' subsets of uni-element geochemical data. In addition, uni-element geochemical data may require some form of transformation so that the values approach a symmetrical empirical density distribution.

The multiple regression analysis and the analysis of weighted mean uni-element concentrations in lithologic units are performed on two subsets of $\log_e$-transformed stream sediment geochemical data in the case study area. The two subsets (A and B) of the data are based on rock type at sample points (Fig. 5-7), following the exploratory data analysis demonstrated in Chapter 3 (see Tables 3-II and 3-IV), although here the

TABLE 5-IV

Regression coefficients of lithologic units (represented as areal proportions in sample catchment basins) for subset A of $\log_e$-transformed uni-element stream sediment data (n=38) at sampling sites underlain by Aroroy Diorite (see Fig. 5-7), Aroroy district (Philippines).

	Regression coefficients of lithologic units (data subset A; n=38)		
	Mandaon Formation	Aroroy Diorite	Sambulawan Formation
Cu	4.882	3.623	4.447
Zn	4.394	3.888	3.614
Ni	3.024	1.951	3.730
Co	2.628	2.422	3.775
Mn	6.983	6.185	6.066
As	1.319	-1.115	0.084

samples with censored values for As are not excluded as in the exploratory data analysis in Chapter 3. That is because the censored values are needed to estimate local background uni-element concentrations for samples with uni-element concentrations below detection limits. Data for subset A pertain to stream sediment sample sites (n=38) underlain by Aroroy Diorite, whereas data for subset B pertain to stream sediment sample sites (n=97) underlain by dacitic/andesitic volcano-sedimentary rocks (Fig. 5-7).

The multiple regression analysis performed on each of the two subsets of the data is forced through the origin (i.e., setting b_o=0) so that the problem of matrix singularity is avoided and equation (5.4) is determinate. In addition, because the uni-element data are $\log_e$-transformed, the regression coefficients (b_j) of j (=1,2,…,m) independent variables (lithologic units represented as areal proportions in individual sample catchment basins) are usually positive and readily interpretable as the geometric mean of uni-element concentrations in individual lithologic units. Likewise, because the uni-element data are $\log_e$-transformed, the M_j values for the j (=1,2,…,m) lithologic units (represented by their areas in individual sample catchment basins) are interpreted as the weighted geometric mean uni-element concentrations in individual lithologic units.

The estimates of regression coefficients and M_j values for the lithologic units in data subset A are given in Tables 5-IV and 5-V, respectively. Note that the sample catchment basins for stream sediment data subset A are underlain by at least one of the three lithologic units listed in either Table 5-IV or 5-V. The regression coefficients of lithologic units in data subset A show that the Aroroy Diorite has the lowest geometric means of concentrations of most elements compared to the Mandaon and Sambulawan Formations, whilst the Mandaon Formation has the highest geometric means of concentrations of most elements, particularly As, compared to Aroroy Diorite and Sambulawan Formation. Likewise, the M_j values of lithologic units in data subset A show that the Aroroy Diorite has the lowest geometric means of concentrations of most elements compared to the Mandaon and Sambulawan Formations, whilst the Mandaon Formation has the highest geometric means of concentrations of most elements, particularly As, compared to Aroroy Diorite and Sambulawan Formation. However, for

TABLE 5-V

Values of M_j (calculated according to equation (5.5)) for lithologic units (represented by their areas in sample catchment basins) based on subset A of $\log_e$-transformed uni-element stream sediment data (n=38) at sampling sites underlain by Aroroy Diorite (see Fig. 5-7), Aroroy district (Philippines).

	M_j values of lithologic units (data subset A; n=38)		
	Mandaon Formation	Aroroy Diorite	Sambulawan Formation
Cu	4.387	3.711	3.787
Zn	3.992	3.452	3.436
Ni	2.617	2.065	3.436
Co	2.635	2.512	2.833
Mn	6.659	6.210	6.107
As	0.441	-0.983	-0.669

the Mandaon and Sambulawan Formations, most of the regression coefficients are higher than the corresponding M_j values; whilst for the Aroroy Diorite, the regression coefficients and M_j values are more or less similar.

The estimates of regression coefficients and the M_j values for the lithologic units in data subset B are given in Tables 5-VI and 5-VII, respectively. Note that the sample catchment basins for stream sediment data subset B are underlain by at least one of the six lithologic units listed in either Table 5-VI or 5-VII. The regression coefficients of lithologic units in data subset B show that the alluvial deposits have the lowest geometric means of concentrations of most elements, followed by the Nabongsoran Andesite, compared to the other lithologic units. In contrast, the estimated M_j values of lithologic units in data subset B suggest that the Nabongsoran Andesite and the alluvial deposits have more or less similar geometric mean uni-element concentrations as the other

TABLE 5-VI

Regression coefficients of lithologic units (represented as areal proportions in sample catchment basins) for subset B of $\log_e$-transformed uni-element stream sediment data (n=97) at sampling sites underlain by dacitic/andesitic volcano-sedimentary rocks (see Fig. 5-7), Aroroy district (Philippines).

	Regression coefficients of lithologic units (data subset B; n=97)					
	Mandaon Formation	Aroroy Diorite	Sambulawan Formation	Lanang Formation	Nabongsoran Andesite	Alluvial deposits
Cu	4.262	4.176	4.089	4.028	2.203	-4.911
Zn	4.051	3.699	3.593	4.218	4.218	2.140
Ni	2.537	2.684	2.619	2.466	1.497	-5.320
Co	2.905	2.607	3.065	2.798	2.629	0.678
Mn	6.663	6.407	6.093	6.794	6.785	7.267
As	1.511	-2.209	-0.303	0.864	-3.604	2.994

TABLE 5-VII

Values of M_j (calculated according to equation (5.5)) for lithologic units (represented by their areas in sample catchment basins) based on subset B of $\log_e$-transformed uni-element stream sediment data (n=97) at sampling sites underlain by dacitic/andesitic volcano-sedimentary rocks (see Fig. 5-7), Aroroy district (Philippines).

	M_j values of lithologic units (data subset B; n=97)					
	Mandaon Formation	Aroroy Diorite	Sambulawan Formation	Lanang Formation	Nabongsoran Andesite	Alluvial deposits
Cu	4.248	4.175	4.074	4.103	3.534	3.045
Zn	4.054	3.722	3.677	4.230	4.217	3.989
Ni	2.498	2.452	2.523	2.454	2.140	1.609
Co	2.873	2.655	3.003	2.838	2.834	2.565
Mn	6.669	6.376	6.267	6.816	6.834	6.846
As	1.127	-0.405	-0.050	0.937	-0.456	1.099

lithologic units. The regression coefficients and M_j values are more or less similar for the Aroroy Diorite and Mandaon, Sambulawan and Lanang Formations. Among the lithologic units, excluding the alluvial deposits, the Mandaon Formation seems to be the most enriched in As.

There is either agreement or disagreement between the regression coefficients and the M_j values, depending on the element and lithologic unit examined (Tables 5-IV to 5-VII). In view of the inconsistencies between the regression coefficients and the M_j values, it is instructive to analyse the results further in order to determine whether to use the regression coefficients or the M_j values for reliable estimates of local uni-element background per sample catchment basin. A plausible explanation for the inconsistencies between the regression coefficients and M_j values can be deduced by plotting $(b_j - M_j)$ against the percentage of individual lithologic units in the total area covered by the sample catchment basins (Fig. 5-8) (cf. Bonham-Carter et al., 1987). Based on data subset A, there are large differences (i.e., $|b_j - M_j| > 0.5$) between the regression coefficients and M_j values if lithologic units occupy less than 7% of the total area covered by the sample catchment basins (Fig. 5-8A). Based on data subset B, there are large differences (i.e., $|b_j - M_j| > 0.5$) between the regression coefficients and M_j values if lithologic units occupy less than 5% of the total area occupied by the sample catchment basins (Fig. 5-8B). From Figs. 5-8A and 5-8B, it is evident that the results for data subset A show that the regression coefficients are mostly overestimated and that the results for data subset B show that the regression coefficients are mostly underestimated if lithologic units occupy, on average, less than 6% of the total area covered by the sample catchment basins. In particular, regression coefficients of the independent variables with respect to As – the pathfinder element for mineral deposits of interest in this case study – are usually either overestimated or underestimated. The plots in Fig. 5-8 indicate that estimates of the regression coefficients are highly sensitive to variations in areal proportions of lithologic units in sample catchment basins, whilst the M_j values

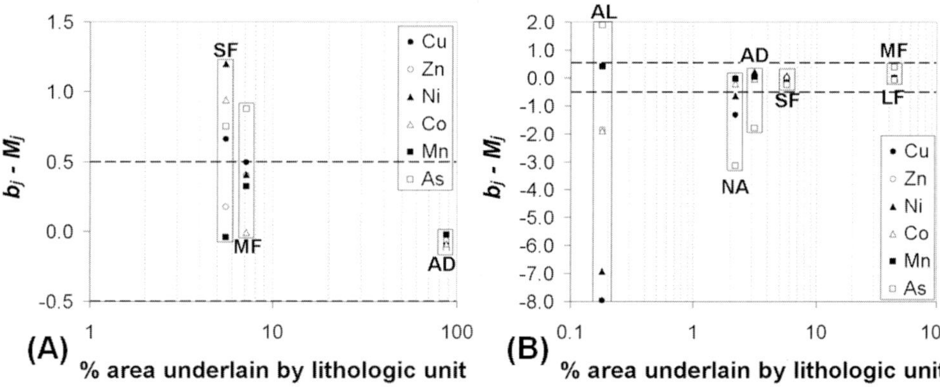

Fig. 5-8. Plots of $b_j - M_j$, the difference between regression estimates (b_j) of geometric means of uni-element contents of stream sediment and weighted estimates (M_j) of geometric means of uni-element contents of stream sediment, versus percent area occupied by individual lithologic units in sample catchment basins, Aroroy district (Philippines). Results from analysis of (**A**) data subset A and (**B**) data subset B. Abbreviations in the plots refer to lithologic units (see Fig. 3-9): MF = Mandaon Formation; AD = Aroroy Diorite; SF = Sambulawan Formation; LF = Lanang Formation; NA = Nabongsoran Andesite; AL = alluvial deposits. Lengths of rectangular boxes indicate for which lithologic units the differences between values of b_j and M_j are large (i.e., $|b_j - M_j| > 0.5$, as indicated by the dashed horizontal lines in the plots).

seem to represent more stable estimates of average uni-element concentrations in individual lithologic units. Therefore, the M_j values and equation (5.7) are used further in this case study to estimate local uni-element background concentrations due to lithology in individual sample catchment basins.

Derivation and standardisation of uni-element residuals

In order to recognise uni-element anomalies, the estimated values of Y_i', which represent local uni-element background concentrations due to lithology in individual sample catchment basins, are subtracted from the corresponding measured uni-element concentrations in stream sediment samples. This means empirically-derived uni-element concentrations due to background sources (e.g., lithology) must be removed from measured uni-element concentrations in order to leave geochemical residuals, from which the contributions or effects of anomalous factors (e.g., mineralisation) may be recognised. Because there are two subsets of data from which local uni-element background concentrations due to lithology were estimated, then there are also two subsets of geochemical residuals. It is important that uni-element residuals per subset are standardised prior to analysis of anomalies, via downstream dilution correction, so that dilution-corrected uni-element residuals and thus uni-element anomalies from both subsets are comparable and can be analysed as one data set.

To standardise the dilution-corrected uni-element residuals per subset, the median and inter-quantile range (*IQR*) are determined and equation (3.10) (see Chapter 3) is

TABLE 5-VIII

Changes in numbers (and percentages) of positive and negative uni-element residuals in data subsets A and B, Aroroy district (Philippines), after standardisation by using robust statistics in equation (3.10) for exploratory data analysis (EDA) and using classical statistics in equation (3.9) for confirmatory data analysis (CDA).

	No. of residuals		No. of EDA-standardised residuals*		No. of CDA-standardised residuals*	
	Positive	Negative	Positive	Negative	Positive	Negative
Data subset A (n=38)						
Cu	19	19	19 (0)	19 (0)	**15** (–21)	*23* (+21)
Zn	16	22	*20* (+25)	**18** (–18)	**14** (–13)	*24* (+9)
Ni	21	17	19 (0)	19 (0)	**18** (–13)	*20* (+18)
Co	19	19	19 (0)	19 (0)	**17** (–11)	*21* (+11)
Mn	17	21	*19* (+12)	**19** (–10)	**14** (–18)	*24* (+14)
As	13	25	13 (0)	25 (0)	**10** (–23)	*28* (+12)
Data subset B (n=97)						
Cu	45	52	*50* (+11)	**47** (–10)	**41** (–9)	*56* (+8)
Zn	49	48	49 (0)	48 (0)	**39** (–20)	*58* (+21)
Ni	55	42	**49** (–11)	*49* (+10)	**41** (–25)	*56* (+33)
Co	50	47	50 (0)	47 (0)	**48** (–4)	*51* (+13)
Mn	52	45	**49** (–6)	*48* (+7)	**46** (–12)	*51* (+14)
As	53	44	**49** (–8)	*48* (+10)	**17** (–68)	*80* (+82)

*Values in italics represent increase in number of either positive or negative residuals. Values in bold represent decrease in number of either positive or negative residuals. Values in parentheses indicate percentage increase (+) or decrease (–) in number of positive or negative residuals resulting from standardisation.

applied. Equation (3.10) is used in this case instead of equation (3.11) because (a) the *MAD* (median of absolute deviations of data values from the data median) is less resistant to outliers than the *IQR* and (b) it can be expected theoretically that the geochemical residuals consist mainly of outliers. For example, Bonham-Carter and Goodfellow (1984) found that uni-element residuals lack spatial autocorrelation, meaning that geochemical residuals in a data (sub)set have large deviations from their central tendency. In addition, Table 5-VIII shows the results of standardisation based on robust statistics (median, *IQR*), as used in equation (3.10) for exploratory data analysis (EDA), compared to results of standardisation based on classical statistics (mean, standard deviation), as used in equation (3.9) for confirmatory data analysis (CDA).

On the one hand, depending on the element examined in data subset A, the EDA-based standardisation results in a 0-25% increase in number of positive residuals and in a 0-18% decrease in the number of negative residuals. In addition, depending on the element examined in data subset B, the EDA-based standardisation results in either a 0-11% increase or a 0-11% decrease in the number of positive residuals and in either a 0-10% decrease or a 0-10% increase in the number of negative residuals. Moreover, depending on the element examined in the whole data set, the EDA-based

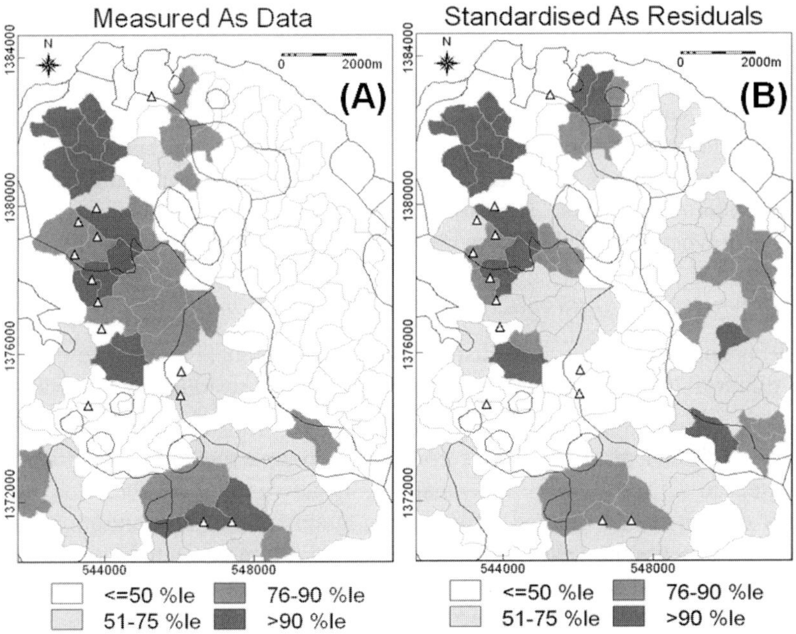

Fig. 5-9. Spatial distributions, based on percentile (%le) classification, of As in stream sediments represented as sample catchment polygons, Aroroy district (Philippines). (**A**) Measured (or raw) As data. (**B**) Standardised As residuals after removing from the measured As data the estimated uni-element background As contents due to lithology in individual sample catchment basins. Triangles represent locations of epithermal Au deposit occurrences, whilst thin black lines represent lithologic contacts (see Fig. 3-9).

standardisation results in a median 0% increase/decrease in number of positive residuals and in a median 0% increase/decrease in number of negative residuals.

On the other hand, depending on the element examined in data subset A, the CDA-based standardisation results in an 11-23% decrease in number of positive residuals and in a 9-21% increase in the number of negative residuals. In addition, depending on the element examined in data subset B, the CDA-based standardisation results in a 4-68% decrease in the number of positive residuals and in an 8-82% increase in the number of negative residuals. Moreover, depending on the element examined in the whole data set, the CDA-based standardisation results in a median 16% decrease in number of positive residuals and in a median 14% increase in number of negative residuals.

Results of the CDA-based standardisation are therefore deleterious in the synthesis of uni-element residuals derived from different 'homogenous' subsets of uni-element geochemical data aimed at performing conjunctive modeling of anomalies. Therefore, the EDA-based standardisation according to equation (3.10) is applied to uni-element residuals in subsets A and B. Fig. 5-9 shows the spatial distributions of the whole set of measured As values and the synthesised set of standardised As residuals in subsets A and

B. In contrast to the spatial distributions of the measured As values (Fig. 5-9A), the spatial distributions of the standardised As residuals show not only enrichment of As along the north-northwest trend of the epithermal Au deposit occurrences but also enrichment of As in the eastern parts of the area underlain by the Aroroy Diorite (Fig. 5-9B). These results demonstrate the value of estimating and then removing local background uni-element contents attributable to lithology from measured uni-element concentrations in stream sediments. The standardised uni-element residuals are then further subjected to downstream dilution correction to identify significant anomalies.

Correction for downstream dilution and classification of anomalies

Equation (5.9) is applied to the standardised uni-element residuals in order to correct for downstream dilution. Note that the standardised geochemical residuals are the values used for the term $Y_i - Y_i'$ in equation (5.9). The dilution-corrected uni-element residuals are then subjected to fractal analysis via the concentration-area method (Cheng et al., 1994) demonstrated in Chapter 4. In this analysis, negative dilution-corrected uni-element residuals cannot be considered in a log-log graph of the concentration-area relation. This is not a concern, however, because negative dilution-corrected uni-element residuals can already be considered to represent background populations and thus excluded in the analysis, leaving high background to anomalous populations to be recognised from the positive dilution-corrected uni-element residuals.

Fig. 5-10A shows that the log-log graph of the concentration-area model for positive dilution-corrected As residuals follows power-law relations indicating multifractal patterns in the data. Three straight lines satisfactorily fit the concentration-area plots, indicating the presence of three populations that can be separated by threshold values at the breaks in slopes of the straight lines. These three populations, from lowest to highest values, are considered to represent high background, low anomaly and high anomaly of As. The three recognised thresholds, plus a threshold for the highest negative or zero dilution-corrected As residual to represent upper limit of background, are used to display the spatial distributions of the background and anomalous populations of dilution-corrected As residuals (Fig. 5-10B).

The spatial distributions of the classified dilution-corrected As residuals (Fig. 5-10B) show a series of anomalies that follow the north-northwest trend of the epithermal Au deposit occurrences and indicate some low anomalies in the eastern sections of the area underlain by the Aroroy Diorite. These latter anomalies cannot be readily considered as significant or not without considering other uni-element anomalies. Fractal analysis via the concentration-area method was also applied to separate background and anomalies in the dilution-corrected residuals of Cu, Zn, Ni, Co and Mn, so that analysis of multi-element anomalies can be performed.

Modeling of multi-element anomalies via principal components analysis

Principal components analysis is performed on a subset of samples (n=93 out of the total 135) with anomalous dilution-corrected residuals for at least one of the elements

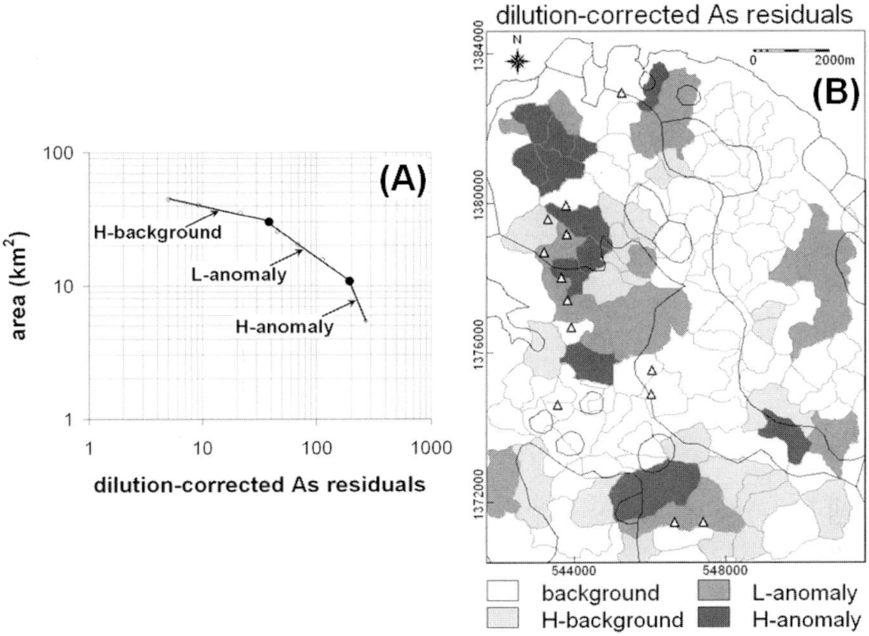

Fig. 5-10. (**A**) Log-log plot of concentration-area fractal model of positive dilution-corrected As residuals, Aroroy district (Philippines), showing two thresholds (dots) that separate high (H) anomaly, low (L) anomaly and high background populations in the data. (**B**) Spatial distributions of high background and anomalous populations of dilution-corrected As residuals based on thresholds recognised in the concentration-area plot. Background class pertains to negative dilution-corrected As residuals, which are not represented in the log-log plot of the concentration-area relation. Triangles in the map represent locations of epithermal Au deposit occurrences, whilst thin black lines represent lithologic contacts (see Fig. 3-9).

under study. Prior to PCA, descending ranks of n to 1 (i.e., considering 1 as the lowest rank) are assigned to descending n values of dilution-corrected uni-element residuals and tied ranks are averaged. A Spearman rank correlation matrix is computed for rank-transformed dilution-corrected uni-element residuals, and is used in the PCA (e.g., George and Bonham-Carter, 1989; Carranza and Hale, 1997). Table 5-IX shows the results of the PCA.

The PC1, accounting for about 34% of the total variance, plausibly represents a Co-Zn-Mn multi-element association reflecting metal-scavenging by Mn-oxides in the drainage environments in most parts of the area; Cu, Ni and As have antipathetic, weak sympathetic and very weak sympathetic relations, respectively, with this multi-element association. The PC2, accounting for about 22% of the total variance, represents a Ni-Cu-As association having an antipathetic relation with Mn and thus plausibly reflects an anomalous multi-element association. The association between Cu and As in PC2 plausibly reflects enrichment of these elements in stream sediments due to weathering and erosion of anomalous sources containing sulphide (chalcopyritic and arsenopyritic)

TABLE 5-IX

Principal components of rank-transformed dilution-corrected uni-element residuals for samples (n=93) with anomalous dilution-corrected uni-element residuals for at least one of the elements under study, Aroroy district (Philippines).

	Cu	Zn	Ni	Co	Mn	As	% of Variance	Cum. % of variance
PC1	-0.193	0.722	0.514	0.817	0.700	0.251	34.05	34.05
PC2	0.710	0.050	0.731	-0.037	-0.455	0.298	22.29	56.34
PC3	0.436	0.416	-0.048	-0.121	0.147	-0.780	16.82	73.16
PC4	0.378	0.225	-0.357	-0.365	0.387	0.483	13.95	87.11
PC5	0.345	-0.482	-0.090	0.352	0.270	-0.063	9.33	96.44
PC6	0.085	0.143	-0.254	0.243	-0.242	0.056	3.55	99.99

minerals, which generally characterise the mineralogy of epithermal Au deposits. The association of Ni with Cu and As in PC2 can be due to the dacitic/andesitic rocks that hosts the epithermal Au deposits in the area. The PC3, accounting for about 17% of the total variance, is dominated by As and is plausibly an anomalous pathfinder signature for epithermal Au deposits. The PC4, accounting for about 14% of the total variance, represents an antipathetic relation between an As-Mn-Cu association and a Co-Ni association; the former is possibly due to metal-scavenging by Mn-oxides in some parts of the area whilst the latter is possibly due to lithologies with slightly more mafic compositions that are not represented in the analysis because they are not mappable at the scale of the lithologic map shown in Fig. 3-9. The last two PCs, together accounting for about 13% of the total variance, are not easily interpretable in terms of geological significance.

The presence of two anomalous multi-element associations in the subset of samples with anomalous dilution-corrected residuals for at least one of the elements under study is possibly due to differences in mobility of As, Cu and Ni in the surficial environments. Thus, the scores of PC2 and PC3 (computed according to equation (5.10)) are analysed further to determine if they represent significant anomalies. Because loading on As in PC3 is negative, the PC3 scores are negated (i.e., multiple by -1) so that high negated PC3 scores represent As anomalies. The PC2 and negated PC3 scores are separately portrayed, using the sample catchment basins, as discrete geochemical surfaces, which are then subjected to the concentration-area fractal method for separation of background and anomaly (see also Chapter 4). On the one hand, the spatial distributions of the PC2 scores, representing a multi-element Ni-Cu-As association, show a northwest-trending zone of low to high anomalies roughly following the north-northwest trend of the epithermal Au deposit occurrences (Fig. 5-11A). The low to high anomalies of the PC2 scores seem to decay from the southeastern parts to the northwestern parts of the area. On the other hand, the spatial distributions of the negated PC3 scores, representing an As-dominated multi-element association, show a north-northwest-trending zone of low to very high anomalies closely following the north-northwest trend of the epithermal Au

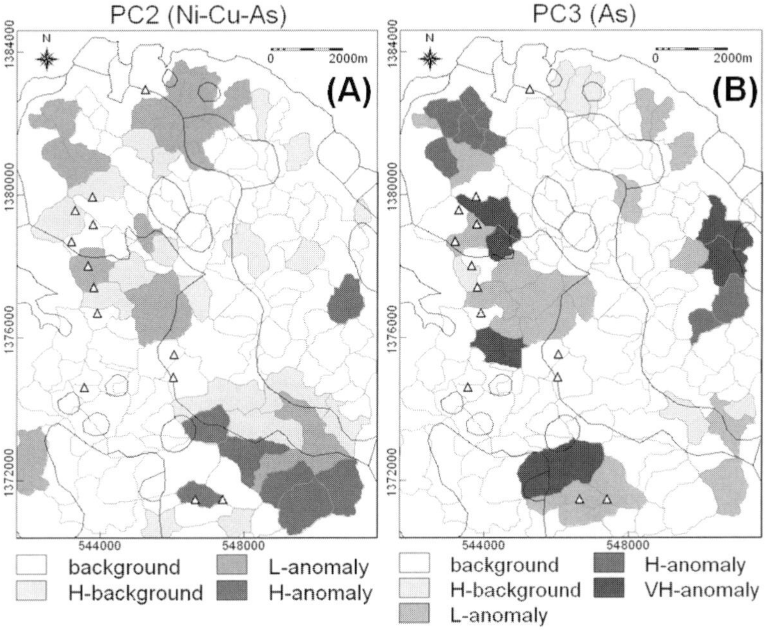

Fig. 5-11. Spatial distributions of element associations in stream sediment sample catchment basins, Aroroy district (Philippines), determined via principal components analysis (see Table 5-IX) and then classified via concentration-area fractal analysis. (**A**) Multi-element Ni-Cu-As association represented by PC2 scores. (**B**) As-dominated multi-element association represented by negated PC3 scores. L = low; H = high; VH = very high. Triangles represent locations of epithermal Au deposits, whilst thin black lines represent lithologic contacts (see Fig. 3-9).

deposit occurrences. There is also a north-trending zone of low to high anomalies of negated PC3 scores in the eastern parts of the area where epithermal Au deposits are not known to occur.

The apparent similarity and difference between the spatial distributions of anomalous PC2 and negated PC3 scores may be explained as follows. On the one hand, the low and high anomalies of PC2 scores along a zone roughly following the trend of the epithermal Au deposits plausibly represent weathered materials derived from mineralised outcrops, which have been transported downstream and farther away from the deposits. On the other hand, the low, high and very high anomalies of negated PC3 scores along a zone closely following the trend of the epithermal Au deposits plausibly represent weathered materials derived from mineralised outcrops, which have been transported downstream but not farther away from the deposits. Thus, anomalies of either PC2 scores or negated PC3 scores are both significant, although anomalies PC2 are possibly related to 'allochthonous' mineralisation whilst anomalies of negated PC3 scores are possibly related to 'autochthonous' mineralisation.

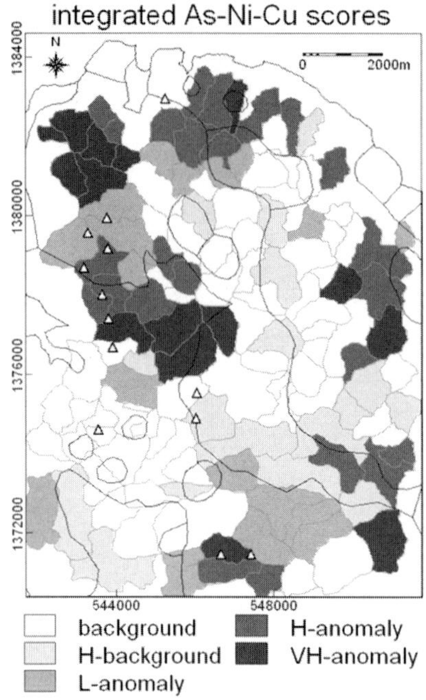

Fig. 5-12. Spatial distributions of integrated As-Ni-Cu scores obtained as products of PC2 and negated PC3 scores representing anomalous multi-element associations derived via principal components analysis (Table 5-IX; Fig. 5-11) of rank-transformed dilution-corrected uni-element residuals in stream sediment samples, Aroroy district (Philippines). The background and anomalous populations of integrated As-Ni-Cu scores were modeled via concentration-area fractal analysis. L = low; H = high; VH = very high. Triangles represent locations of epithermal Au deposit occurrences, whilst thin black lines represent lithologic contacts (see Fig. 3-9).

Because of the apparent similarity between the spatial distributions of anomalous PC2 and negated PC3 scores, it is appealing to integrate such variables into one variable representing a multi-element As-Ni-Cu association reflecting the presence of epithermal Au deposits. A simple multiplication can be applied to integrate the PC2 and negated PC3 scores, although this creates false anomalies from PC2 and negated PC3 scores that are both negative. This problem is overcome by first re-scaling the PC2 and negated PC3 scores linearly to the range [0,1] and thereafter performing multiplication on the re-scaled variables. The resulting 'integrated As-Ni-Cu' scores are then portrayed as a discrete surface, using the sample catchment basins, for the application of the concentration-area fractal method to separate background and anomaly.

The spatial distributions of integrated As-Ni-Cu scores (Fig. 5-12) show adjoining high and very high anomalies, which coincide or are proximal to most epithermal Au deposit occurrences. Most of the high anomalies of PC2 scores in the southeastern quadrant of the area (Fig. 5-11A) have been downgraded in importance (i.e., they now map as low anomalies as shown in Fig. 5-12), but many of the low and high anomalies of both PC2 and negated PC3 scores in the northwestern quadrant of the area (Fig. 5-11B) have been upgraded in importance (i.e., they now map as high anomalies as shown in Fig. 5-12). In addition, many of the low anomalies of negated PC3 scores in the eastern parts of the area (Fig. 5-11B) have been enhanced. Combining the PC2 (Ni-Cu-As) and negated PC3 (As) scores into integrated As-Ni-Cu scores has an overall positive effect in this case study and is therefore defensible.

Screening of multi-element anomalies with fault/fracture density

The presence of stream sediment uni-element or multi-element anomalies does not always mean presence of mineral deposits, so it is necessary to apply certain criteria for ranking or prioritization of anomalies prior to any follow-up work. Criteria for ranking or prioritization can be related to indicative geological features of the mineral deposit type of interest or to factors that could influence localisation of stream sediment anomalies.

In the study area, faults/fractures can influence localisation of stream sediment anomalies because (a) such geological features are common loci of epithermal Au deposits, whose element contents find their way into streams due to weathering and erosion and (b) the presence of such geological features indicates enhanced structural permeability of rocks in the subsurface, which facilitates upward migration of groundwaters that have come in contact with and have leached substances from buried deposits. These arguments suggest that the significance of multi-element stream sediment anomalies in sample catchment basins can be screened or examined further by using fault/fracture density as a factor (cf. Carranza and Hale, 1997).

Fig. 5-13A shows a map of faults/fractures in the study area, indicating that the epithermal Au deposits are localised mostly along certain north-northwest-trending faults/fractures. A fault/fracture density map can be created by calculating, per sample catchment basin, the ratio of number of pixels representing faults/fractures in a sample catchment basin to number of pixels in that sample catchment basin. Most of the epithermal Au deposit occurrences in the study area are situated in sample catchment basins with moderate to high fault/fracture density (Fig. 5-13B). In order to further screen the multi-element stream sediment anomalies (e.g., as shown in Fig. 5-12), the product of integrated As-Ni-Cu scores and fault/fracture density can be obtained and then subjected to classification via the concentration-area fractal method.

The results shown in Fig. 5-14 clearly indicate that, on the one hand, anomalies of integrated As-Ni-Cu scores in the western half of the study area (see Fig. 5-12) are mostly significant in terms of indicating localities that contain or are proximal to epithermal Au deposit occurrences. The anomalous sample catchment basins in the western half of the study area (Fig. 5-14) are aligned along the north-northwest trend of the epithermal Au deposit occurrences. On the other hand, anomalies of integrated As-Ni-Cu scores related to the Aroroy Diorite in the eastern half of the study area (see Fig. 5-12) are downgraded in importance (i.e., they now map mostly as background as shown in Fig. 5-14) after using fault/fracture density in the analysis. This latter result suggests that the Aroroy Diorite is possibly non-mineralised.

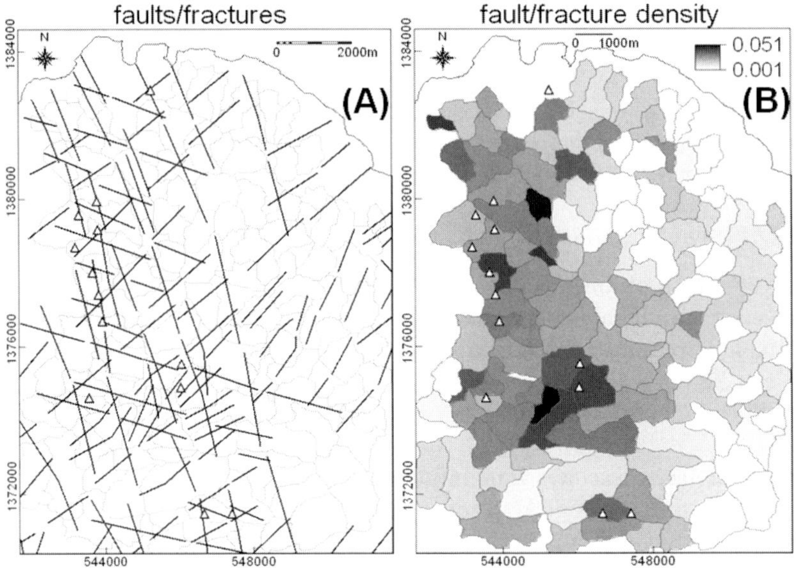

Fig. 5-13. (**A**) Map of faults/fractures in the Aroroy district (Philippines), compiled mostly from unpublished literature and interpretations of shaded-relief images of DEM illuminated from different directions (e.g., Fig. 4-11). (**B**) Fault/fracture density measured as the ratio of number of pixels representing faults/fractures in a sample catchment basin to number of pixels per sample catchment basin. Triangles represent locations of epithermal Au deposit occurrences.

DISCUSSION AND CONCLUSIONS

There are various factors that influence variation in stream sediment background uni-element concentrations. For example, it has been shown in some case studies that drainage sinuosity (Seoane and De Barros Silva, 1999), which is a geogenic factor, and selective logging (Fletcher and Muda, 1999), which is an anthropogenic factor, can influence the variability of background uni-element concentrations in stream sediments. Nevertheless, a universal factor of variation in stream sediment background uni-element concentration is lithology. Estimation and removal of local background uni-element concentrations in stream sediments due to lithology from measured stream sediment uni-element concentrations is vital to the recognition of significant geochemical anomalies.

The results of the case study demonstrate that significant uni-element and multi-element anomalies can be extracted from stream sediment geochemical data through a 5-stage GIS-based methodology involving: (1) estimation of local background uni-element concentrations due to lithology per sample catchment basin; (2) removal of estimated local background uni-element concentrations due to lithology from measured uni-element concentrations, which results in geochemical residuals; (3) dilution-correction of geochemical residuals using a modified formula from the relation proposed by Hawkes

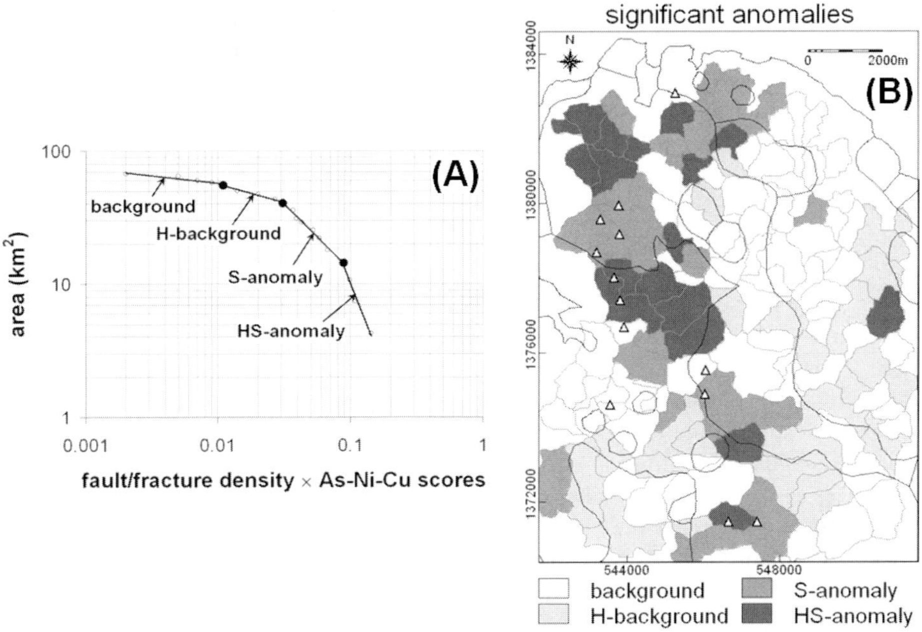

Fig. 5-14. Results of synthesis and analysis of fault/fracture density and integrated As-Ni-Cu scores, Aroroy district (Philippines). (**A**) Log-log plot of concentration-area fractal model of products of fault/fracture density and integrated As-Ni-Cu scores, showing three thresholds (dots) at breaks in slopes of straight lines representing populations of background, high (H) background, significant (S) anomalies and highly significant (HS) anomalies. (**B**) Spatial distributions of background and anomalous populations of products of fault/fracture density and integrated As-Ni-Cu scores based on thresholds recognised in the concentration-area plot. Triangles in the map represent locations of epithermal Au deposit occurrences, whilst thin black lines represent lithologic contacts (see Fig. 3-9).

(1976); (4) modeling of uni-element and/or multi-element anomalies via application of the concentration-area fractal method; (5) screening of significant anomalies by integration of factors that favour mineral deposit occurrence, such as fault/fracture density.

Consideration of the area of influence of every stream sediment sample location – its catchment basin – is the key in stages (1) and (3). In stage (1), more stable estimates of local background uni-element concentrations due to lithology can be obtained by using areas of lithologic units per sample catchment divided by the total area of sample catchment basins instead of using areal proportions of lithologic units per sample catchment basin. In stage (3), dilution-correction of uni-element residuals is based on area of sample catchment basins plus an assumption of a small unit area of exposed anomalous sources (e.g., mineral deposits). Correction for downstream dilution using either equation (5.8) or (5.9) (i.e., based on the assumption of a unit area for exposed anomalous sources contributing to uni-element concentrations in stream sediments), is appropriate for analysis of stream sediment geochemical data in regions where there are some occurrences of mineral deposits of interest. In regions where there are no known occurrences of mineral deposits of interest, it is prudent to calculate productivity (Moon, 1999) or 'stream-order-corrected' residuals (Carranza, 2004a). Note, nonetheless, that correcting for downstream dilution by using either equation (5.8) or (5.9) neglects contributions from overbank materials, assumes lack of interaction between sediment and water and that erosion is uniform in each catchment basin. Certainly, the downstream dilution-correction model based on the idealised relation proposed by Hawkes (1976), which is adopted in the case study, does not apply universally. However, the idealised formula proposed Hawkes (1976) shows reasonable agreement between theory and prediction of known porphyry copper deposits in his study area. In addition, considering that sizes of catchment basins differ and that sizes of anomalous sources (if present) could differ from one catchment basin to another, dilution-correction is warranted despite limitations of the model.

Prior to stage (4), if dilution-corrected residuals are derived from 'homogeneous' subsets of stream sediment geochemical data, then application of robust statistics for exploratory data analysis is preferred for standardisation of dilution-corrected residuals per data subset instead of the conventional application of classical statistics for standardisation. In stage (4), the results of the case study further demonstrate usefulness of fractal analysis (Cheng et al., 1994) of discrete geochemical surfaces (i.e., catchment basins polygons) of uni-element residuals and derivative scores representing multi-element data. In the past, recognition of anomalies from dilution-corrected residuals was made by visual inspection of spatial distributions of percentile-based classes of such variables (Bonham-Carter and Goodfellow, 1986; Bonham-Carter et al., 1987; Carranza and Hale, 1997). Now, recognition of anomalies from dilution-corrected residuals can be made objectively by application of the concentration-area fractal method. In stage (5), the area of influence of individual sample catchment basins is further useful in screening of anomalies, as demonstrated in the case study using fault/fracture density estimated as the ratio of number of pixels representing faults/fractures in a sample catchment basin to number of pixels in that sample catchment basin. In another GIS-based case study, Seoane and De Barros Silva (1999) prioritised sediment sample catchment basins that are anomalous for gold by using catchment basin drainage sinuosity, which is estimated as the ratio of total length of streams within a sample catchment basin to the total distance between the start and end points of the main stream and its tributaries in that sample catchment basin. Finally, it is clear that GIS supplements catchment basin analysis of stream sediment anomalies with tools for data manipulation, integration and visualisation in discriminating and mapping of significant geochemical anomalies.

PART III. MINERAL PROSPECTIVITY MAPPING

Geochemical Anomaly and Mineral Prospectivity Mapping in GIS
by E.J.M. Carranza
Handbook of Exploration and Environmental Geochemistry, Vol. 11 (M. Hale, Editor)

Chapter 6

ANALYSIS OF GEOLOGIC CONTROLS ON MINERAL OCCURRENCE

INTRODUCTION

Occurrences (or locations) of mineral deposits of the type sought are themselves significant geochemical anomalies and are samples of a mineralised landscape. The spatial distribution of such samples is invariably considered by many geoscientists, especially mineral explorationists, to be non-random because of the knowledge that an inter-play of certain geological processes has controlled their occurrence. It follows then that such samples of a mineralised landscape are associated spatially, as well as genetically, with certain geological features. It is for these reasons that mineral prospectivity for the type of deposits sought is, in many cases, modeled by way of probabilistic techniques. That is to say, the probability or likelihood that mineral deposits of the type sought are contained or can be found in a part of the Earth's crust is considered greater than would be expected due to chance.

We recall from Chapter 1 that, in any scale of target generation, modeling of mineral prospectivity usually starts with the definition of a *conceptual model of mineral prospectivity* for mineral deposits of the type sought (Fig. 1-2). Analysis of the spatial distribution of occurrences of mineral deposits of the type sought and analysis of their spatial associations with certain geological features are useful in defining a conceptual model of mineral prospectivity of the type sought in a study area. Analysis of the spatial distribution of occurrences of mineral deposits of the type sought can provide insights into which geological features have plausibly controlled their localisation at certain locations. In addition, the analysis of spatial associations between occurrences of mineral deposits of the type sought and certain geological features is instructive in defining and weighting relative importance of certain geological features to be used as spatial evidence in predictive modeling of mineral prospectivity.

This Chapter explains techniques for analysis of the spatial distribution of occurrences of mineral deposits of the type sought and techniques for the analysis of spatial associations between occurrences of mineral deposits of the type sought and certain geological features. These techniques are demonstrated by using a map of occurrences of epithermal Au deposits in the case study Aroroy district (Philippines) (Fig. 3-9) in order to define a conceptual model of prospectivity for this type of mineral deposit in that district. Models of multi-element anomalies, which are derived by applications of the methods explained and demonstrated in Chapters 3 to 5, are examined further here in terms of their spatial associations with the known occurrences of epithermal Au deposits in the case study area. Faults/fractures in the case study area (Fig. 5-13) are also used as input data in the spatial association analysis in order to define prospectivity recognition criteria representing structural controls on epithermal Au mineralisation. However, let us proceed first with the analysis of the spatial distribution of occurrences of the epithermal Au deposits in the case study area in order to gain insights into their geologic controls.

SPATIAL DISTRIBUTION OF MINERAL DEPOSITS

In most case studies of mineral prospectivity mapping, the locations of known mineral deposits of the type sought are depicted as points. Thus, a univariate point map (of mineral deposits of the type sought) is used as input data in the analysis of the spatial distribution of mineral deposits. Three methods to characterise the spatial distribution of occurrences of mineral deposits of the type sought are explained and demonstrated here: point pattern analysis, fractal analysis and Fry analysis.

Point pattern analysis

Point pattern analysis is a technique that is used to obtain information about the arrangement of point data in space to be able to make an inference about the spatial distribution of occurrences of certain geo-objects represented as points. There are three basic types of point patterns (Diggle, 1983) (Fig. 6-1).

1. A pattern of complete spatial randomness (CSR), in which points tend to lack interaction with each other. This pattern suggests geo-objects resulting from independent processes that occur by chance.
2. A clustered pattern, in which points tend to form groups compared to points in CSR. This pattern suggests geo-objects resulting from an inter-play of processes that involve '*concentration*' of groups of points to certain locations.
3. A regular pattern, in which points tend to be farther apart compared to points in CSR. This pattern suggests geo-objects resulting from an inter-play of processes that involve '*circulation*' of individual points to certain locations.

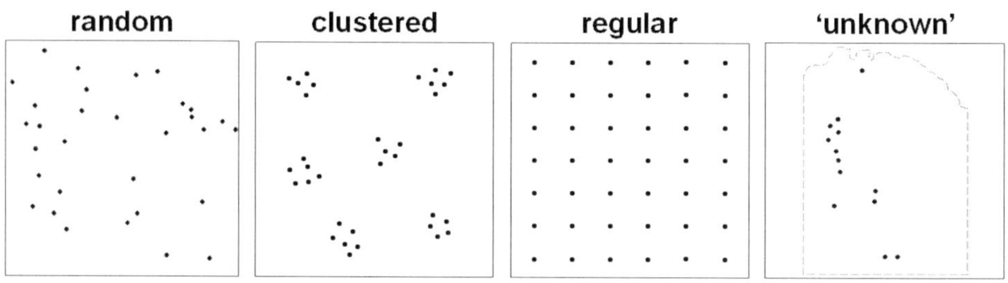

Fig. 6-1. Basic types of point patterns: random; clustered; regular. The 'unknown' point pattern represents known occurrences of epithermal Au deposits in the Aroroy district (Philippines) demarcated in light-grey dashed outline (see Fig. 3-9).

A point pattern under consideration is thus compared to a point pattern of CSR. The null hypothesis in point pattern analysis is, therefore, that the point pattern under examination assumes CSR and that the geo-objects represented by the points are independent of each other and each point is a result of a random (or Poisson) process. The plainest alternative or 'research' hypothesis in point pattern analysis is that the point pattern under investigation does not assume CSR and that the geo-objects represented by the points are associated with each other because they were generated by common processes. Thus, if occurrences of mineral deposits of the type sought are non-random, they may display a clustered distribution or a more or less regular distribution. There are various techniques by which the null hypothesis or the alternative hypothesis can be tested and they can be grouped generally into two types of measures (Boots and Getis, 1988): (1) measures of dispersion; and (2) measures of arrangement.

Measures of dispersion study the locations of points in a pattern with respect to the study area. Measures of dispersion can be further subdivided into two classes: (a) quadrat methods; and (b) distance methods. Quadrat methods make use of sampling areas of a unit size and consistent shape (e.g., a square pixel), which can be either scattered or contiguous, to measure and compare frequencies (or occurrences) of observed points to expected frequencies of points in CSR. It is preferable to make use of contiguous quadrats (e.g., a grid of square pixels) instead of scattered quadrats in the analysis of the spatial distribution of occurrences of mineral deposits of the type sought. That is because scattered quadrats are positioned at randomly selected locations, producing a bias toward CSR. Choosing a quadrat size, however, is a difficult issue in using contiguous quadrats: large quadrats tend to result in more or less equal frequencies, which generate bias toward a regular or a clustered pattern; small quadrats can break up clusters of points, resulting in a bias toward CSR. The best option is to apply distance methods, which compare measured distances between individual points under study with expected distances between points in CSR.

In a GIS, distance between two points is determined, based on the Pythagorean theorem, as the square root of the sum of the squared difference between their easting (or x) coordinates and the squared difference between their northing (or y) coordinates. In a set of n points, measured distances from one point to each of the other points are referred to as 1st-, 2nd-, 3rd- or $(n-1)^{th}$-order neighbour distances; the 1st-order neighbour distance being the nearest neighbour distance. If, on the one hand, the mean of measured n^{th}-order neighbour distances is smaller than the mean of expected n^{th}-order neighbour distances in CSR, then the set of points under examination assumes a clustered pattern. If, on the other hand, the mean of measured n^{th}-order neighbour distances is larger than the mean of expected n^{th}-order neighbour distances in CSR, then the set of points under examination assumes a regular pattern. The significance of the difference between the mean of measured n^{th}-order neighbour distances and the mean of expected n^{th}-order neighbour distances in CSR may be determined statistically based on the normal distribution; for details, readers are referred to Boots and Getis (1988).

The occurrences of epithermal Au deposits in the Aroroy district (Philippines), according to the results of analysis of up to the 6th-order neighbour distances (Table 6-I),

TABLE 6-I

Means of different orders of neighbour distances in the point pattern of occurrences of epithermal Au deposits in Aroroy district (Philippines).

Order of neighbour distances	Mean of measured distances (m)	Mean of expected distances (m) in CSR
1st	992.2	966.6
2nd	1880.3	1449.9
3rd	2249.9	1812.3
4th	2660.9	2114.3
5th	3222.0	2378.8
6th	3765.6	2616.5

assume a regular distribution. Boots and Getis (1988) aver that the choice of how many orders of neighbour distances to examine depends on the point pattern being studied. Here, the choice of examining up to the 6th-order neighbour distances is arbitrary, but is based on the assumption that unknown (or undiscovered) occurrences of mineral deposits of the type sought are located close to the known occurrences. Note that the mean of the measured 6th-order neighbour distances (about 4 km) is not an unrealistic search radius from a known mineral deposit occurrence within which to explore for undiscovered occurrences of the same type of mineral deposit. Determining the statistical significance of the results is also considered here to be inappropriate because the boundary of the study area is arbitrary (i.e., geologically non-real) and occurrences of epithermal Au deposits outside the study area are thus excluded from the analysis. Nevertheless, the results suggest that individual occurrences of epithermal Au deposits in the study area were formed by an inter-play of geological features that 'circulated' mineralising hydrothermal fluids into certain localities. This generalisation is discussed later in the synthesis of results from this analysis and the results of fractal and Fry analyses that follow below.

In contrast with measures of dispersion, measures of arrangement study the locations of points in a pattern with respect to each other. Measures of arrangement are useful when the actual boundary of a study area is unknown or difficult to define or if it is not necessary to impose an arbitrary boundary. In measures of arrangement, the observed number of reflexive (or reciprocal) nearest neighbour (RNN) points is compared with the expected number of RNNs in a situation of CSR. The CSR is simulated for the same area and the same number of points. Two points are considered 1st-order RNN if they are each other's nearest neighbour in a neighbourhood of points, whereas 2nd-order RNNs are points that are each other's 2nd-nearest neighbours in a neighbourhood of points, and so on (Boots and Getis, 1988). In the study area, there are eight 1st-order RNNs (Fig. 6-2). RNNs are always pairs of points, so that the observed number of RNNs is always an even number. The expected number of jth-order RNNs is estimated according to the probability that a point in CSR is the jth-nearest neighbour of its own j^{th}-nearest neighbour (see Cox (1981) for details). If the observed number of jth-order RNNs is

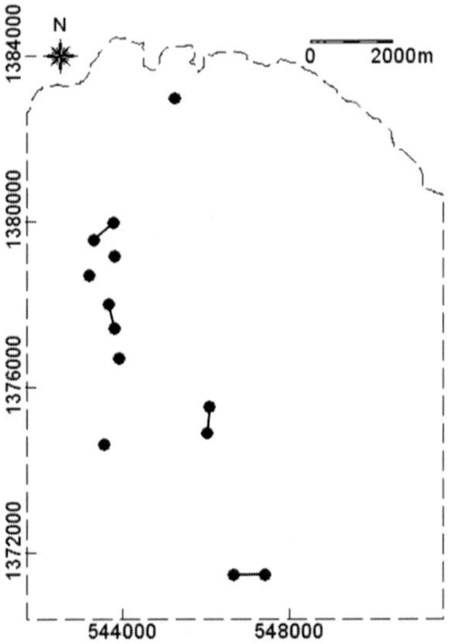

Fig. 6-2. 1st-order reflexive nearest neighbours (dots connected by a solid line) in the pattern of points (dots) representing occurrences of epithermal Au deposits in the Aroroy district (Philippines) demarcated in light-grey dashed outline.

higher than the expected number of jth-order RNNs in CSR, then the points under examination assume a clustered pattern. If the observed number of jth is lower than the expected number of jth-order RNNs in CSR expectations, then the points under examination assume a regular pattern.

In a GIS, RNNs are determined automatically by software because they can be computationally complex. The occurrences of epithermal Au deposits in the Aroroy district (Philippines), according to the results of analysis of up to the 6th-order RNNs (Table 6-II), tend to assume a regular distribution. The reason for examining up to the 6th-RNNs is simply for consistency with the analysis of dispersion discussed earlier. Table 6-II shows that the observed numbers of jth-order RNNs are mostly slightly lower than the expected numbers of jth-order RNNs. Unfortunately, according to Boots and Getis (1988), there are no statistical tests available to examine the significance of the difference between the observed and expected numbers of jth-order RNNs.

Tables 6-I and 6-II show that some discrepancy can occur between the results of measures of arrangement and the results of measures of dispersion. Analysis of dispersion by distance methods is more advantageous than analysis of arrangement because the statistical theory for analysis of dispersion in terms of distances is better developed than the statistical theory behind analysis of arrangement, so the former is less subjective than the latter. That is to say, analysis of dispersion is parametric whilst analysis of arrangement is non-parametric. The main advantage of analysis of arrangement over analysis of dispersion by distance methods is that the former is free of so-called edge-effects, because locations of points are studied with respect to each other

TABLE 6-II

Numbers of different orders of reflexive nearest neighbours (RNNs) in the point pattern of occurrences of epithermal Au deposits in Aroroy district (Philippines).

Order of RNNs	Observed number	Expected number in CSR
1st	8	8.08
2nd	4	4.28
3rd	4	3.16
4th	0	2.62
5th	2	2.29
6th	2	2.06

rather than with respect to the study area. However, edge effects can be compensated by way of a number of methods (see Boots and Getis (1988) for details).

Despite this slight discrepancy, the results of analysis of arrangement and analysis of dispersion of the occurrences of epithermal Au deposits in the case study area are coherent. Thus, a generalisation can be made from the results of point pattern analysis shown in Tables 6-I and 6-II that the spatial distribution of occurrences of epithermal Au deposits in the Aroroy district is not random but assumes a regular pattern. The results of the analysis can imply that more-or-less regularly-spaced geological features (e.g., faults/fractures) may have controlled the circulation of mineralising hydrothermal fluids and thus the localisation of epithermal Au deposits at certain locations. This implication can be examined further via applications of fractal analysis and Fry analysis.

Fractal analysis

As defined in Chapter 4, a fractal pattern has a dimension D_f, known as the Hausdorff-Besicovitch dimension, which exceeds its topological (or Euclidean) dimension D (Mandelbrot, 1982, 1983). Fractal analysis of a point pattern of occurrences of certain types of mineral deposits has been demonstrated by Carlson (1991), Cheng and Agterberg (1995), Cheng et al. (1996), Wei and Pengda (2002), Weiberg et al. (2004), Hodkiewicz et al. (2005) and Ford and Blenkinsop (2008).

The fractal dimension of a point pattern can be determined by the box-counting method (see Fig. 4-1). A square grid or raster with a cell or pixel size δ (i.e., length or width of a pixel) is overlaid on a map of points. The number of pixels $n(\delta)$ containing one or more points is counted. The procedure is repeated for different values of δ and the results are plotted in a log-log graph. If the point pattern is a fractal, the plots of $n(\delta)$ versus δ satisfy a power-law relation (Mandelbrot, 1985; Feder, 1988), thus:

$$n(\delta)=C\delta^{-D_b} \tag{6.1}$$

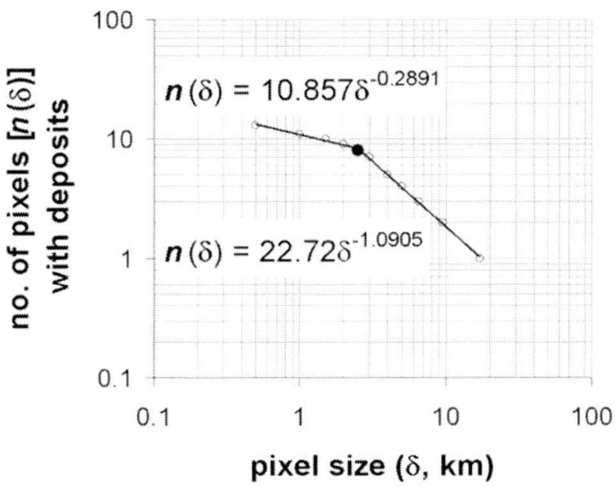

Fig. 6-3. Log-log plots of $n(\delta)$ versus δ for the pattern of points representing occurrences of epithermal Au deposits in the Aroroy district (Philippines). The individual straight lines fitted through the linear portions of plots satisfy the power-law relation in equation (6.1).

where $0 \leq D_b \leq 2$ is the box-counting fractal dimension and C is a constant. The relation in equation (6.1) can be represented as a linear function in a log-log graph:

$$\log n(\delta) = \log C - D_b \log\delta\ . \tag{6.2}$$

The slope D_b of the linear log-log plots of $n(\delta)$ versus δ is therefore a useful measure of the fractal dimension of a pattern of points. Linear log-log plots of $n(\delta)$ versus δ for point patterns with Poisson or random distributions have slopes of -2, whilst linear log-log plots of $n(\delta)$ versus δ for point patterns with fractal distributions have fractional slopes between 0 and -2 (Carlson, 1991).

The point pattern of the occurrences of epithermal Au deposits in the Aroroy district has two box-counting fractal dimensions (Fig. 6-3). The straight line fit through the plots when $\delta \leq 2.5$ km has a slope of -0.2891, whereas the straight line fit through the plots when $\delta > 2.5$ km has a slope of -1.0905. These results indicate that the spatial distribution of the occurrences of epithermal Au deposits in the Aroroy district is non-random but fractal. A plausible interpretation of the results shown in Fig. 6-3, together with the results shown in Tables 6-I and 6-II, is that, in the Aroroy district, there are fractal fracture systems that controlled the epithermal mineralisations over scales (i.e., lengths or widths) ranging from about 0.5 km to at most 2.5 km and that there are fractal hydrothermal systems that controlled the epithermal mineralisations over scales (i.e., lengths, widths or diameters) ranging from about 2.5 km to at least 10 km.

Mandelbrot (1983) also reported that fractal point patterns follow a radial-density power-law relation, thus:

$$d = Cr^{D_r - 2}\ . \tag{6.3}$$

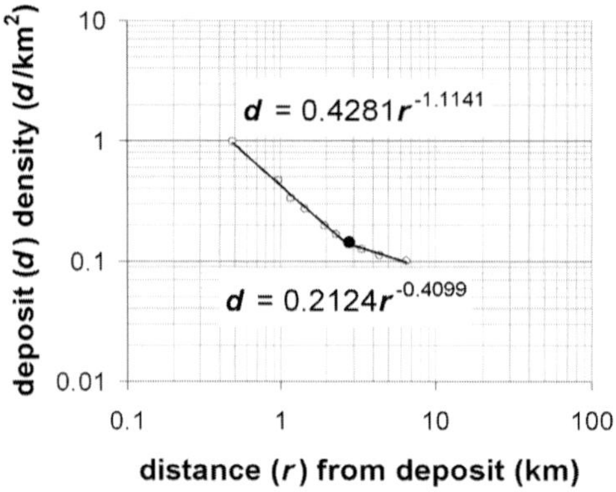

Fig. 6-4. Log-log plots of density (*d*) of points representing occurrences of epithermal Au deposits in the Aroroy district (Philippines) as a function of distance or radius *r* from each point. The individual straight lines fitted through the linear portions of plots satisfy the power-law relation in equation (6.3).

where d is the density of points in polygonal patterns defined by circles of radius r from such points, C is a constant and D_r is the radial-density fractal dimension of the pattern defined by the points. Feder (1988) calls D_r the cluster dimension of fractal point pattern. The radial-density relation has been applied by Carlson (1991), Agterberg (1993) and Wei and Pengda (2002) to characterise spatial distributions of mineral deposit occurrences. For the application of the radial-density relation in a GIS, a small pixel size is used to represent each point (i.e., mineral deposit occurrence) as just one pixel. The radial-density is then derived as the ratio of the number of points to the total number of pixels common to overlapping circles of radius r from certain points plus the number of pixels in non-overlapping circles of radius r from certain points. If the spatial distribution of the points is fractal, then the radial-density of the points should decrease, following a power-law function, with increasing radius from the points.

The point pattern of the occurrences of epithermal Au deposits in the Aroroy district has two radial-density fractal dimensions (Fig. 6-4). The straight line fit through the plots when $r \leq 2.8$ km has a slope of -1.1141, whereas the straight line fit through the plots when $r > 2.8$ km has a slope of -0.4099. These results indicate that the spatial distribution of the occurrences of epithermal Au deposits in the Aroroy district is non-random but fractal. A plausible interpretation of the results shown in Fig. 6-4, together with the results shown in Tables 6-I and 6-II, is that, in the Aroroy district, there are fractal fracture systems that controlled the epithermal mineralisations over scales (lengths or widths) ranging from about 0.5 km to at most 2.8 km and that there are fractal hydrothermal systems that controlled the epithermal mineralisations over scales (lengths, widths or diameters) ranging from about 2.8 km to about 7 km.

The estimated fractal dimensions of the pattern of epithermal Au deposits in the Aroroy district are somewhat inconsistent [0.2891 (D_b) versus 1.1141 (D_r)] at scales less 2.8 km and [1.0905 (D_b) versus 0.4099 (D_r)] at scales greater than 2.8 km, although the scales of the fractal systems interpreted from the results of the two methods are similar. Discrepancies in fractal dimensions estimated via the box-counting method and the radial-density method are difficult to explain, although, according to Carlson (1991), such inconsistencies commonly occur in measuring fractal dimensions. For example, in estimating the radial-density of points in a raster-based GIS, there is no rule-of-thumb for choosing an ideal pixel size except that one should apply sound reasoning related to the geo-objects represented by the points. The results of the box-counting and the radial-density fractal analyses of the spatial distribution of the occurrences of epithermal Au deposits in the Aroroy district commonly imply, however, that certain types of controls are operating on at least two scales (lengths, widths, or diameters). One type of control (e.g., fracture systems) possibly operates at scales of at most 2.8 km, which is plausibly at the 'deposit-to-another-deposit' scale. The other type of control (e.g., hydrothermal systems) possibly operates at scales of at least 7 km, which is plausibly at the scale of a mineralised landscape (i.e., district scale in this case). These interpretations can be investigated further via the application of Fry analysis

Fry analysis

Fry analysis (Fry, 1979), which is a geometrical method of spatial autocorrelation analysis of a type of point geo-objects, is another useful technique to study spatial distribution of points representing occurrences of mineral deposits of the type sought. The method plots translations (so-called Fry plots) of point geo-objects by using each and every point as a centre or origin for translation. Fig. 6-5 shows the basic principle in creating a Fry plot using analogue maps (e.g., tracing paper). A map of data points is marked with a series of parallel (either north-south trending or east-west trending) reference lines. On a second but empty map, a centre or origin is indicated by the intersection of a north-south trending line and an east-west trending line. The centre or origin in the second map is then placed on top of one of the data points (point 1 in Fig. 6-5), the reference lines of the same directions in both maps are kept parallel and the positions of all the data points are recorded in the second map. The centre or origin in the second map is then placed on top of a different data point (point 2 in Fig. 6-5), the reference lines of the same directions in both maps are kept parallel and the positions of all the data points are recorded again in the second map. The procedure is continued until all the data points have been used as the centre or origin in the second map. For n data points there are n^2-n translations created. These are called 'all-object-separations' plots and are more commonly known as 'Fry plots', developed originally for the investigation of strain and strain partitioning in rocks (Fry, 1979; Hanna and Fry, 1979). Fry plots have been used in the analysis of spatial distributions of occurrences of mineral deposits (Vearncombe and Vearncombe, 1999, 2002; Stubley, 2004; Kreuzer et al., 2007) and geothermal fields (Carranza et al., 2008c) in order to infer their structural controls.

It is clear in Fig. 6-5 that a Fry plot enhances subtle patterns in a spatial distribution of points and it also records distances and orientations between pairs of translated points, which can be used to construct a rose diagram as a complementary tool for visual analysis of trends reflecting controls by certain geological features. A rose diagram can

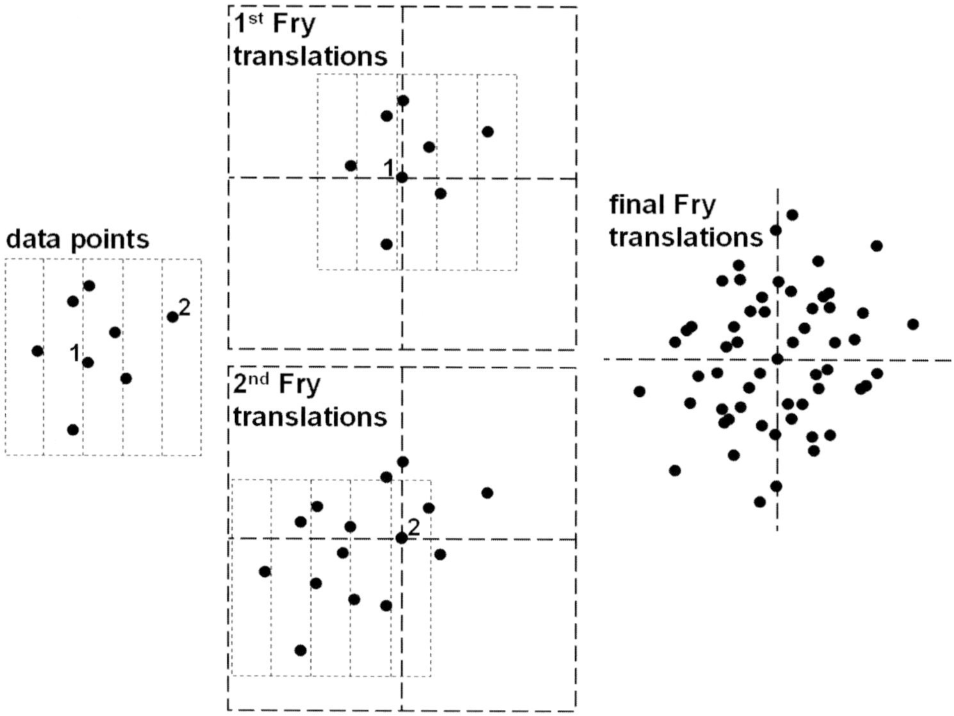

Fig. 6-5. Schematic procedure of creating Fry plots for a set of points. See text for explanation.

be created for orientations and frequency of orientations between (a) all pairs of translated points and (b) pairs of translated points within specified distances from each other. The former case may reveal trends due to processes operating at the scale of a mineralised landscape (i.e., regional or district), whereas the latter case may reveal trends due to processes operating at the 'deposit-to-another-deposit' scale. For the latter case, therefore, it is instructive to use a distance within which there is maximum probability of only two neighbouring points (i.e., analysis of trends between any two neighbouring occurrences of mineral deposits of the type sought). This distance can be determined via point pattern analysis (see Boots and Getis (1988) for details).

Creation of Fry plots is not part of routine functionalities of any GIS software package. Whilst it is possible to create Fry plots manually (Fig. 6-5), the procedure can be cumbersome especially as the number of points increases. There are software packages specialised in creating graphics, including Fry plots and rose diagrams, to support the analysis of problems in structural geology. The digitally-captured data of the locations (i.e., map coordinates) of mineral deposit occurrences stored in a GIS can be exported to formats supported by software packages for creating Fry plots. The output of Fry point coordinates can, in turn, be exported to a GIS software package in order to visualise and analyse the spatial distributions of the Fry plots and the original data points

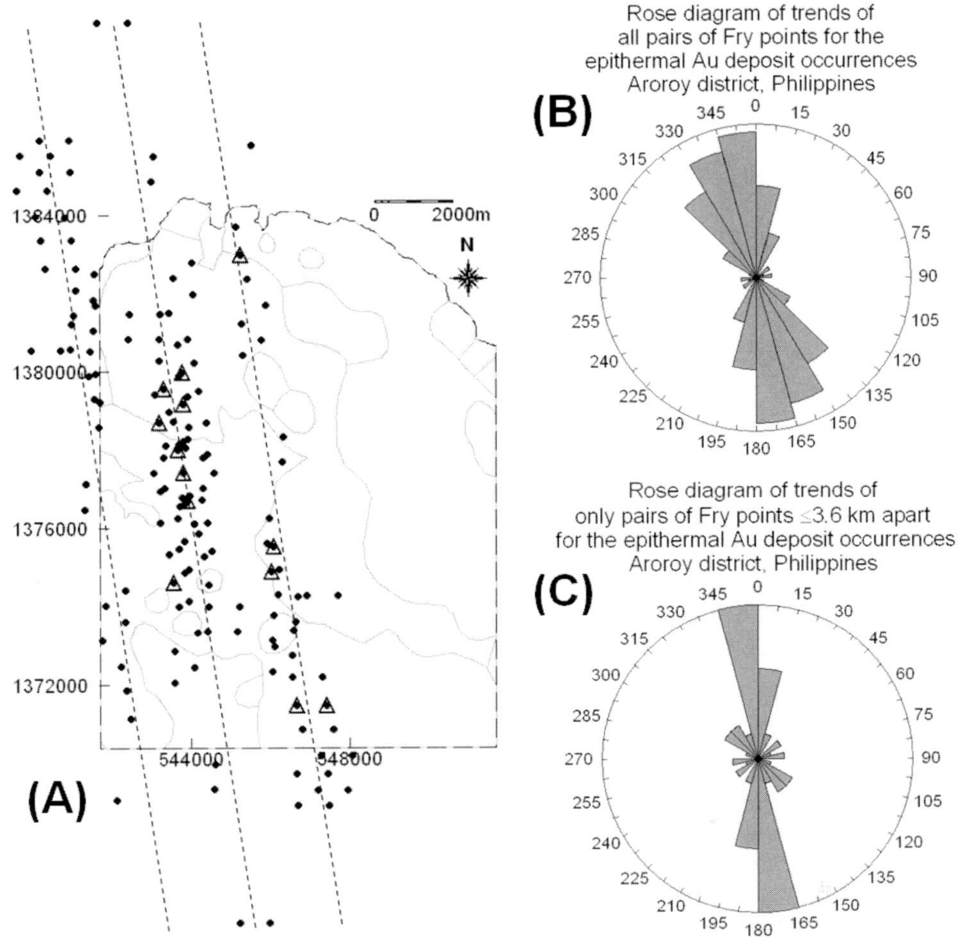

Fig. 6-6. (**A**) Fry plots (black dots) of occurrences of epithermal Au deposits (triangles) in the Aroroy district (Philippines). Light-grey lines are boundaries of lithologic units (see Fig. 3-9). Parallel north-northwest trending dashed lines are approximate axes of imaginary corridors of Fry points (see text for discussion). Rose diagrams of trends based on (**B**) all pairs of Fry points and (**C**) only pairs of Fry points within 3.6 km of each other.

together. In order to properly interpret the spatial and geological meaning of the results, it is useful to 'centre' the Fry plots on the original data points (Fig. 6-6). To do so, in accordance with procedure for creating Fry plots explained above and illustrated in Fig. 6-5, one has to first find which of the original data points is approximately at the 'centre' of the pattern. The 'centre' data point is one whose easting (or x) coordinate is closest to the median of easting coordinates of all points and whose northing (or y) coordinate is closest to the median of northing coordinates of all points. The 'centre' data point should then be made as the first record in the data file to be used in creating a Fry plot (although this might be software dependent).

The Fry plots for the 13 occurrences of epithermal Au deposits in the Aroroy district show prominent 150-180° (or 330-360°) trends (Figs. 6-6A and 6-6B), which indicates structural controls by north-northwest-trending faults/fractures (see Fig. 5-13). The overall north-northwest trend of the 13 occurrences of the epithermal Au deposits in Aroroy district is, independent of the Fry plots, perceivable from the map shown in Fig. 3-9. So, one could say that Fry plots of mineral deposit occurrences can be biased by pre-existing data. Nevertheless, the Fry plots of the 13 occurrences of epithermal Au deposits in the study area show at least two patterns that are not obvious in the map of the original data points. Firstly, the Fry plots suggest the presence of north-northwest trending corridors of epithermal Au deposits (Fig. 6-6A), which seem to be regularly-spaced at about 2 km intervals. These regularly-spaced corridors possibly represent parallel district-scale hydrothermal systems controlled by the general spacing of north-northwest trending faults/fractures (see Fig. 5-13). Secondly, the rose diagram of orientations of only pairs of Fry points within 3.6 km of each other (Fig. 6-6C) shows a subsidiary 120-150° (or 300-330°) trend, which suggests that northwest trending faults/fractures are important local-scale controls on epithermal Au mineralisations in the study area. Thus, the Fry plots of the 13 occurrences of epithermal Au deposits in the Aroroy district complement as well as supplement the results of the point pattern analysis (in terms of regularity of spatial distribution) and the fractal analysis (in terms controls at local- and district-scales).

Knowledge synthesis and results of spatial analysis

In order to develop a model of plausible geologic controls on localisation of the type of mineral deposits sought in a study area, it is instructive to synthesise the results of analyses of spatial distribution of mineral deposits of the type sought with published knowledge of geological processes relevant to the formation of mineral deposits of the type sought. There is no particular way of doing the synthesis. The following discussions provide an example of doing so for the case study area.

Based on a review of models of epithermal precious metal deposits worldwide, there is a variety of geologic settings in which such type of mineral deposits can occur (Mitchell and Garson, 1981; Sawkins, 1989; Robb, 2004). Igneous activity plays an essential role in the formation of most epithermal Au deposits in terms of providing the heat necessary to generate hydrothermal convection cells (White and Hedenquist, 1990). In most cases, epithermal Au deposits are spatially and temporally associated with subaerial volcanic rocks and/or their related subvolcanic intrusions (Sillitoe and Bonham, 1984). Subaerial volcanism may occur in a variety of tectonic settings, although it occurs mostly along volcanic arcs in subduction tectonic settings characteristic of oceanic-continental or oceanic-oceanic plate collision environments (Le Pichon et al., 1973). Strong structural control is almost universally recognised for various types of gold deposits, including epithermal gold deposits (Henley, 1990; Henley and Adams, 1992). Faults/fractures in the near-surface enhance the permeability of potential host rocks. Regional faults are also important (Mitchell and Balce, 1990), perhaps in guiding the emplacement of magmatic heat sources and influencing structural permeability and subsequent hydrothermal activity (Hedenquist, 1986; Sillitoe, 1993).

Based on a review of the general characteristics of epithermal systems in the Philippines, most epithermal Au deposits in the archipelago, including those in the Aroroy district, are deposited along volcanic arcs during the mid-Miocene to mid-Pliocene (Mitchell and Balce, 1990; Mitchell and Leach, 1991; Yumul et al., 2003). The geochemical nature of epithermal mineralisations in the Philippines and geochemical anomalies (which include As) associated with such mineralisations are described by UNDP (1987). The epithermal Au deposits are largely in the form of veins or vein breccias and stockworks, indicating strong structural controls. There is no direct evidence of genetic association between magmatism and the epithermal systems, although a causal relationship is implied by analogy with epithermal Au deposits elsewhere (e.g., Hedenquist and Henley, 1985; Singer, 2000). In most of the epithermal gold districts in the Philippines, including those in the Aroroy district, the lithostratigraphic succession comprises predominantly clastic andesitic or dacitic rocks lying unconformably on folded basement rocks. Almost all epithermal gold deposits in the Philippines are hosted by andesitic rocks and are commonly associated with minor intrusions of andesitic porphyry plutons or less commonly with dacitic porphyry plutons.

Based on further review of the structural controls of epithermal gold deposits and the geotectonic settings in the Philippines, epithermal mineralisations in the archipelago, including those in the Aroroy district, have no apparent genetic association and display lack of spatial relationship with regional fault systems, such as the sinistral strike-slip Philippine Fault system (Fig. 6-7A), which actually cuts through most of the Philippine archipelago (Aurelio et al., 1991). However, the epithermal mineralisations are commonly situated along subsidiary faults or splays of regional fault systems (Mitchell and Balce, 1990; Mitchell and Leach, 1991). This is also apparently the case for the epithermal Au deposits in the Aroroy district. Many of the faults/fractures in the Aroroy district (Fig. 5-13) are plausibly subsidiary structures of the regional sinistral strike-slip Sibuyan Sea Fault, which is a branch of the Philippine Fault system (Fig. 6-7A). The far-field stress (i.e., principal regional stress axis) that generated the Philippine Fault is generally oriented east-west (Aurelio et al., 1997). By inference from the directions of the strike-slip motions of the Sibuyan Sea Fault and the Philippine Fault, the near-field stress (i.e., principal district-scale stress axis, σ_1) in the vicinity of the Aroroy district is probably oriented towards about 150° (or 330°) (Fig. 6-7A). If this is the case, and in accordance with theoretical wrench tectonics or fault mechanics (Jaeger and Cook, 1976; Mandl, 1988), then the north-northwest and northwest trending faults/fractures in the Aroroy district (Fig. 5-13) are subsidiary conjugate faults/fractures likely associated directly with the Sibuyan Sea Fault and to a lesser extent with the Philippine Fault. In addition, according to Aurelio et al. (1991), the sinistral strike-slip motion along the Philippine Fault initiated in about late Early Pliocene. This, in turn, suggests that the Philippine Fault can be implicated in the emplacement of the Pliocene Nabongsoran Andesite porphyry intrusions (Baybayan and Matos, 1986; JICA-MMAJ, 1986), which

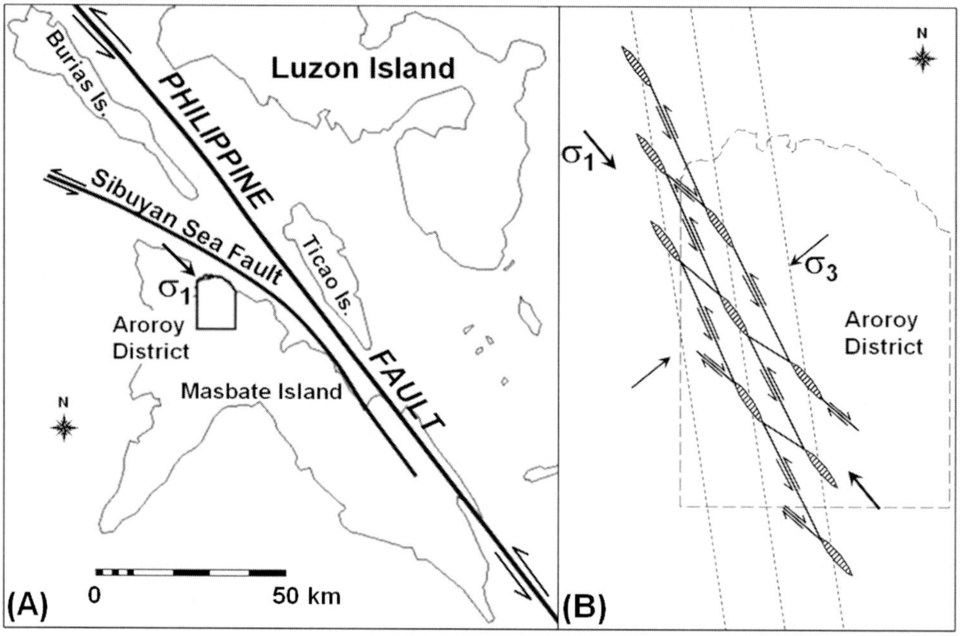

Fig. 6-7. (**A**) Sketch of regional faults near Masbate Island (Philippines) (adapted from Aurelio et al., 1991, 1997), where Aroroy district is situated (see also insets in Fig. 3-9). Based on the relative motions associated with the Philippine Fault and the Sibuyan Sea Fault, it can be inferred that orientation of the principal stress axis (σ_1) at or near Aroroy district varies about 150° (or 330°). (**B**) Schematic model of stress-controlled fault-fracture mesh comprising of interlinked shear faults/fractures (linear features) and extension fractures (hatched features) (after Sibson, 1996, 2000), as plausible spatial controls on epithermal mineralisations in the Aroroy district (in dashed outline). The outline of the Aroroy is scalable to the map in Fig. 6-7A, but it is not meant to indicate scale of the fault-fracture mesh model. Parallel north-northwest trending dotted lines are approximate axes of imaginary corridors in the Fry plots (see Fig. 6-6A).

are possibly heat source controls on epithermal mineralisations in the Aroroy district (Mitchell and Balce, 1990; Mitchell and Leach, 1991).

Based on a review of stress-controlled hydrothermal fluid circulation, Hill (1977) proposed a mechanical model for earthquake swarms involving migration of fluids in volcanic terranes through a 'honeycomb' fault-fracture mesh of interlinked shear faults/fractures and extension fractures/faults. The fault-fracture mesh model has been researched and expounded extensively by Sibson (1996, 2000, 2001) to explain fault-related fracture systems as host structures for hydrothermal mineralisation. The interpretation of a near-field stress (i.e., principal district-scale stress axis, σ_1) in the vicinity of Aroroy oriented towards about 150° (or 330°) (Fig. 6-7A) supports a fault-fracture mesh model in this district (Fig. 6-7B). Extension faults/fractures (trending about 130-150° (or 310-330°)) are developed at tips of shear faults/fractures (trending about 150-170° (or 330-350°) or about 120-140° (or 300-320°)) or, conversely, shear faults/fractures terminate at extension fractures, such that the development of lengthy through-going faults/fractures is inhibited. Instead, en echelon shear faults/fractures and en echelon extension fractures are developed (Fig. 6-7B).

The interpretation of a fault-fracture model in the Aroroy district is an explanation of the results of the Fry analysis. In the condition of near-field stress depicted in the fault-fracture mesh model in the Aroroy district (Fig. 6-7B), hydrothermal fluids (that brought about, say, epithermal mineralisations) were re-distributed towards the extension faults/fractures. This interpretation is supported by the knowledge that (a) shear stresses [i.e., $\frac{1}{2}(\sigma_1-\sigma_3)$] are highest at discontinuities or tips of en echelon shear faults/fractures (Segall and Pollard, 1980) and (b) rate of fluid flow (or strain rate) is directly proportional to shear stress (White, 1991). Thus, en echelon (or a regular pattern of) extension faults/fractures (i.e., at discontinuities of en echelon shear faults/fractures), toward which hydrothermal fluids were focused and whereabouts epithermal Au deposits were formed, are plausibly disposed along north-northwest trending corridors that are (sub-)parallel and spaced at about 2 km away from each other (see Figs. 6-6A and 6-7B) as implied from the Fry analysis.

A plausible explanation for the results of the fractal analysis of spatial distribution of the epithermal Au deposit occurrences in the study area is this. Whereas the directions of strike-slip motions of the Sibuyan Sea Fault and the Philippine Fault were the bases for interpretation of the orientation of 'district-scale' near-field stress in the Aroroy district (Fig. 6-7), the directions of inferred strike-slip motions of north-northwest and northwest trending shear faults/fractures in the study area can also be used to infer orientations of 'local-scale' near-field stresses within the district. Accordingly, local-scale fault-fracture mesh models can also be generated, thereby modeling fractal patterns of fault-fracture meshes from district scale to local scale.

Thus, in the case study area, there is strong agreement between published empirical and theoretical models of geologic controls on epithermal mineralisations and the results of analyses of spatial distributions of epithermal Au deposit occurrences. Syntheses of published empirical and theoretical models of geologic controls on mineralisation and results of analyses of the spatial distribution of mineral deposit occurrences therefore provide insights to certain geological features that are spatial evidence of prospectivity for mineral deposits of the type sought. Accordingly, based on the observed geological characteristics of the epithermal Au deposits in the case study area (see Chapter 3), the results of analyses of the spatial distribution of the epithermal Au deposits and the fault-fracture mesh model (Fig. 6-7B), north-northwest trending and northwest trending faults/fractures (Fig. 5-13) are plausible spatial evidence of prospectivity for epithermal Au deposits in the study area. In addition, based on the fault-fracture mesh model (Fig. 6-7B), intersections of north-northwest trending and northwest-trending faults/fractures are possibly whereabouts extension faults/fractures are situated and are therefore further plausible spatial evidence of prospectivity for epithermal Au deposits in the study area. These interpretations require, nonetheless, further evaluation by applications of methods for analysis of spatial associations between two sets of geo-objects, which are discussed in the following section.

SPATIAL ASSOCIATION OF MINERAL DEPOSITS AND GEOLOGIC FEATURES

Certain types of mineral deposits exhibit spatial associations with certain geological features because the latter play a role in localisation of the former. For example, igneous intrusions provide heat that causes formation of hydrothermal convection cells whilst faults/fractures provide plumbing systems for circulation of convective hydrothermal fluids (from which mineral deposits may precipitate). The spatial association dealt herewith refers to the distance or range of distances at which certain types of mineral deposits, which are usually represented or mapped as points, are preferentially located from certain geologic features, which can be represented or mapped as points, lines or polygons. This spatial association can be regarded as spatial dependence; that is, the occurrence of certain types of mineral deposits depends upon the locations of certain geological features. The smaller the distance of spatial association, the stronger the spatial dependence. Analysis of spatial associations between occurrences of mineral deposits of the type sought and certain geological features is thus instructive in conceptual modeling of mineral prospectivity.

Methods for analysis of spatial association between occurrences of certain types of mineral deposits and certain geologic features can be classified into two groups: (1) methods that lead directly to the creation and then integration of predictor maps in predictive modeling of mineral prospectivity; and (2) methods that are exploratory in nature and are useful mainly in conceptual modeling of mineral prospectivity. Methods belonging to the former group are explained and demonstrated in Chapter 8. Two methods belonging to the latter group are explained and demonstrated in this chapter: (1) distance distribution method; and (2) distance correlation method.

Distance distribution method

The theoretical model of the distance distribution method, which characterises spatial association between a set of point geo-objects and another set of (point, linear or polygonal) geo-objects, was formalised and demonstrated by Berman (1977). Further demonstrations, with certain adaptations, of this method in determining spatial associations between occurrences of certain types of mineral deposits and curvi-linear geological features can be found in Simpson et al. (1980), Bonham-Carter et al. (1985), Bonham-Carter (1985, 1994), Berman (1986), Carranza (2002) and Carranza and Hale (2002b). Variants of the distance distribution method are commonly called proximity or buffer analysis (e.g., Ponce and Glen, 2002; Bierlein et al., 2006; Park et al., 2007). The procedures described below for GIS-based application of the distance distribution method are adapted from Bonham-Carter et al. (1985) and Bonham-Carter (1985, 1994).

Consider the observed nearest (i.e., Euclidean) distances, $O(X)$, between individual points (in a set of point geo-objects of interest) and certain lines (in a set of linear geo-objects). Consider further the expected nearest (i.e., Euclidean) distances, $E(X)$, between individual random points (in a set of random point geo-objects representing a set of Poisson processes) and certain lines (in the same set of linear geo-objects). In order to test a null hypothesis that the set of point geo-objects of interest and the set of linear features are spatially independent (i.e., they lack spatial association), the graph or curve of cumulative proportion (or relative frequency or histogram) of measured nearest distances [$\hat{O}(X)$] is compared with the graph or curve of cumulative proportion of expected nearest distances [$\hat{E}(X)$] by computing the Kolmogorov-Smirnov statistic:

$$D=\hat{O}(X)-\hat{E}(X)\,. \tag{6.4}$$

If $D \cong 0$, then the set of point geo-objects of interest and the set of linear features are spatially independent. If D is positive (i.e., >0), which means that the curve $\hat{O}(X)$ is above or higher than the curve $\hat{E}(X)$, then there is positive spatial association between the set of point geo-objects and the set of linear features. If D is negative (i.e., <0), which means that the curve $\hat{O}(X)$ is under or lower than the curve $\hat{E}(X)$, then there is negative spatial association between the set of point geo-objects and the set of linear features.

A positive, rather than a negative, spatial association between a set of point geo-objects and a set of linear (or point or polygonal) features is important, because it suggests that the latter represents a set of plausible factors (or spatial evidence) of occurrence of the former. In order to determine graphically if $\hat{O}(X)$ is significantly greater than $\hat{E}(X)$, an upper confidence band for the $\hat{E}(X)$ curve can be calculated as:

$$\hat{U}(X)=\hat{E}(X)+(9.21(N+M)/4NM\}^{1/2} \tag{6.5}$$

where M denotes the number of random points used to estimate $\hat{E}(X)$, N denotes the number of point geo-objects used to estimate $\hat{O}(X)$ and 9.21 is a tabulated critical χ^2 (or chi-square) value for 2 degrees of freedom and significance level α=0.01 (other critical χ^2 values may be used for chosen α). In addition, the distance from the linear features in which positive D values are highest or optimal is of interest, because within this distance from the linear features there is significantly higher proportion of occurrence of point geo-objects than would be expected due to chance (i.e., a random or Poisson process). The distance from the linear geo-objects in which positive D values are highest or optimal can be determined through a test of statistical significance of positive spatial association by calculating the quantity:

$$\beta=4D^2NM/(N+M)\,, \tag{6.6}$$

which is distributed approximately as χ^2 with 2 degrees of freedom (Goodman, 1954; Siegel, 1956); M and N are as defined for equation (6.5). The distance from the linear geo-objects in which the estimated values of β exceed a critical χ^2 value at a certain significance level (α) represents the distance of optimal positive spatial association between the point geo-objects and the set of linear features.

The following sequence of procedures, which are illustrated schematically in Fig. 6-8, can be followed in a raster-based GIS in order to implement the distance distribution method.

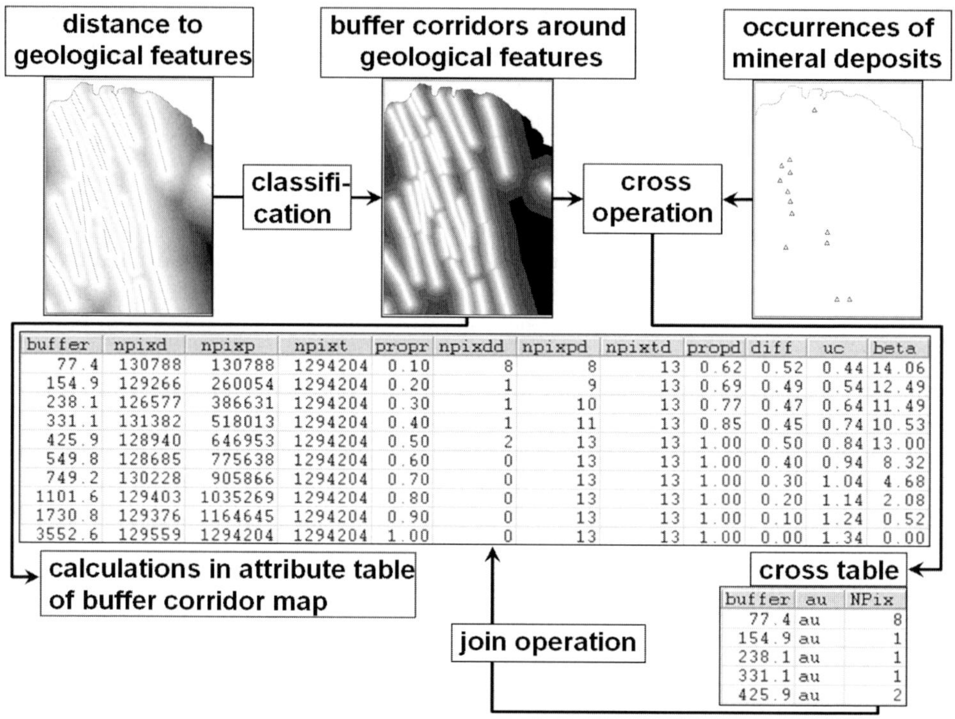

buffer	npixd	npixp	npixt	propr	npixdd	npixpd	npixtd	propd	diff	uc	beta
77.4	130788	130788	1294204	0.10	8	8	13	0.62	0.52	0.44	14.06
154.9	129266	260054	1294204	0.20	1	9	13	0.69	0.49	0.54	12.49
238.1	126577	386631	1294204	0.30	1	10	13	0.77	0.47	0.64	11.49
331.1	131382	518013	1294204	0.40	1	11	13	0.85	0.45	0.74	10.53
425.9	128940	646953	1294204	0.50	2	13	13	1.00	0.50	0.84	13.00
549.8	128685	775638	1294204	0.60	0	13	13	1.00	0.40	0.94	8.32
749.2	130228	905866	1294204	0.70	0	13	13	1.00	0.30	1.04	4.68
1101.6	129403	1035269	1294204	0.80	0	13	13	1.00	0.20	1.14	2.08
1730.8	129376	1164645	1294204	0.90	0	13	13	1.00	0.10	1.24	0.52
3552.6	129559	1294204	1294204	1.00	0	13	13	1.00	0.00	1.34	0.00

buffer	au	NPix
77.4	au	8
154.9	au	1
238.1	au	1
331.1	au	1
425.9	au	2

Fig. 6-8. Schematic diagram of raster-based GIS operations for implementation of the distance distribution method for analysis of spatial association between a set of geological features and a set of mineral deposit occurrences of the type sought. See text for explanation. For values in columns of the attribute table of the buffer corridor map: `buffer` = upper limit of distance class; `npixd` = number of pixels in buffer; `npixp` = cumulative number of pixels in buffer classes in order of increasing buffer distance; `npixt` = total number of pixels in the map; `propr` = cumulative proportion of pixels in buffer classes; `npixd` = number of deposit pixels in a buffer class; `npixpd` = cumulative number of deposits in buffer class in order of increasing distance; `npixtd` = total number of deposit pixels; `propd` = cumulative proportion of deposit pixels in buffer classes; `diff` = D (see equation (6.4)); `uc` = upper confidence values for D (see equation (6.5)); `beta` = β (see equation (6.6)). For values in columns of the cross table: `buffer` = upper limit of distance class; `au` = point geo-objects of interest (i.e., mineral deposit occurrences) contained in a buffer class; `Npix` = number of pixels of point geo-objects of interest (i.e., mineral deposit occurrences) contained in a buffer class.

1. Create a map of distances to a univariate set of geological features. These geological features may or may not necessarily be those geological features inferred from an analysis of the spatial distribution of mineral deposits to be plausible spatial evidence of prospectivity for mineral deposits of the type sought. Areas outside a study area must be masked out.

2. Discretise or classify the map of distances to geological features into a number of classes. A classification using narrow equal intervals of percentiles or cumulative proportions of distances is advisable. That is because the type (i.e., normal, log-normal, etc.) of empirical density distribution of distances to a set of geological features is usually unknown but percentiles of the distance data are robust for classification and directly represent the cumulative proportions of distances sought in the distance distribution analysis.
3. In the attribute table associated with classified map of distances to geological features, perform the following table operations (see Fig. 6-8).
   a. Determine the upper limit of each distance (`buffer`) class. In some cases, this variable is already given in the attribute table associated with a classified map of distances to geological features.
   b. Determine the number (or frequency) of pixels (`npixd`) in each distance class. In many cases, this variable is already given in the attribute table associated with a classified map of distances to geological features.
   c. Determine the cumulative number (or cumulative frequency) of distance class pixels (`npixp`) in the order of increasing distances.
   d. Determine the total number of pixels (`npixt`).
   e. Determine $\hat{E}(X)$ by dividing values of `npixp` with the value of `npixt`. As shown in Fig. 6-8, values of $\hat{E}(X)$ are stored in column `propr` (which stands for cumulative `proportion` of `'random'` points) of the attribute table. Note that, in principle, a very large number of random points must be generated to properly estimate $\hat{E}(X)$. Note also that any one of all the pixels in a map could probably represent a random point (or a Poisson process), so that (i) each pixel in a map represents a Euclidean distance $E(X)$ to its nearest geological features and (ii) using all pixels (i.e., `npixt`) leads to a reasonable estimate of the cumulative frequency distribution of expected distances $\hat{E}(X)$ from a set of linear features under examination (Bonham-Carter, 1994, pp. 163-164).
4. Perform a cross or intersect or zonal statistics operation using the classified map of distances to a set of geological features and a map of the locations (or occurrences) of mineral deposits of the type sought.
5. Join the geo-information of number (or frequency) of deposit pixels contained in certain classes of distances from the cross table output to the attribute table of the map of classified distances of a set of geological features. Note that the cross operation functions (i) to determine the Euclidean distance [i.e., $O(X)$] between each deposit pixel and then (ii) to classify the values of $O(X)$ in the same way as the values of $E(X)$ were classified.
6. In the attribute table of the map of classified distances of a set of geological features, perform further the following table operation (see Fig. 6-8).
   a. Determine the cumulative number (or cumulative frequency) of deposit pixels (`npixpd`) contained in every distance (`buffer`) class in the order of increasing distance.
   b. Determine the total number of deposit pixels (`npixtd`).

c. Determine $\hat{O}(X)$ by dividing values of `npixpd` with the value of `npixtd`. As shown in Fig. 6-8 the values of $\hat{O}(X)$ are stored in the column `propd` (which stands for cumulative proportion of deposits) of the attribute table.
d. Calculate D (`diff`) by subtracting values of `propr` from values of `propd` (i.e., according to equation (6.4)).
e. Calculate an upper confidence (`uc`) value for each value of D according to equation (6.5). The values in column `uc` shown in Fig. 6-8 were obtained using a critical χ^2 value of 9.21 (i.e., at α=0.01). Note that to calculate `uc`, M=`npixt` and N=`npixtd`.
f. Calculate β (`beta`) according to equation (6.6). Note also that M=`npixt` and N=`npixtd`.

After the calculations in the attribute table of the map of classified distances to a set of geological features, the values in columns `buffer`, `propr`, `propd`, `diff`, `uc` and `beta` (as shown in Fig. 6-8) can then be illustrated as graphs. For example, in Figs. 6-9 to 6-11, `buffer` (distance to a set of geological features) is used as variable for the x-axis in both types of graphs whilst `propr` (buffer pixels), `propd` (deposit pixels), `diff` (D) and `uc` (confidence band) are used as variables for the y-axis in one type of graph and `beta` is used for the y-axis in the other type of graph.

Fig. 6-9 shows the results of analyses of the spatial association of the epithermal Au deposits in the Aroroy district (Philippines) with north-northwest (NNW) trending faults/fractures and with northwest (NW) trending faults/fractures (see also Fig. 5-13). The epithermal Au deposit occurrences have statistically significant (at α=0.01) positive spatial association with NNW-trending faults/fractures and the positive spatial association is optimal within 0.45 km from NNW-trending faults/fractures (Figs. 6-9A and 6-9B). Within this distance from NNW-trending faults/fractures, all the epithermal Au deposit occurrences in the study area are present and, according to the curve for D, there is about 50% higher occurrence of epithermal Au deposits than would be expected due to chance (Fig. 6-9A).

The epithermal Au deposit occurrences have statistically significant (at α=0.10) positive spatial association with NW-trending faults/fractures and the positive spatial association is optimal within about 1 km from NW-trending faults/fractures (Figs. 6-9C and 6-9D). Within this distance from NW-trending faults/fractures, 85% of the epithermal Au deposit occurrences in the study area are present and, according to the curve for D, there is 35% higher occurrence of epithermal Au deposits than would be expected due to chance (Fig. 6-9C). Thus, the epithermal Au deposit occurrences have stronger spatial association with NNW-trending faults/fractures than with NW-trending faults/fractures. These results, which are consistent with the results of the Fry analyses (Fig. 6-6), further support the fault-fracture mesh model (Fig. 6-7) and the hypothesis that NNW- and NW-trending faults/fractures are plausible spatial evidence of prospectivity for epithermal Au deposits in the case study area.

Because intersections of NNW- and NW-trending faults/fractures are possibly whereabouts extension faults/fractures are situated (Fig. 6-7B), their spatial association with the epithermal Au deposit occurrences is also analysed. The epithermal Au deposit

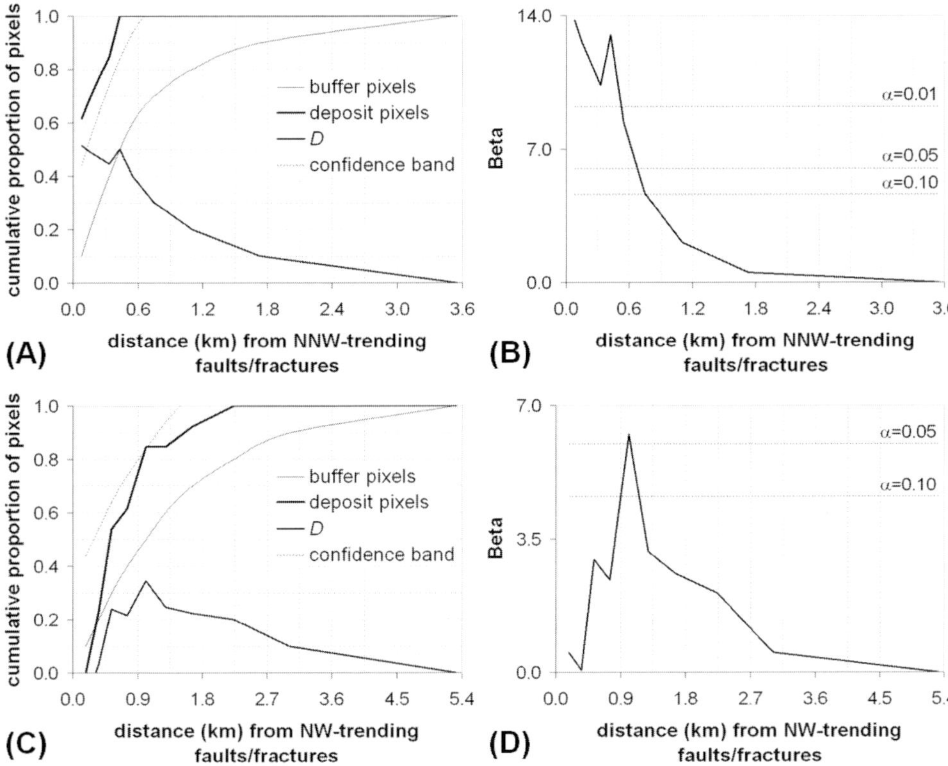

Fig. 6-9. Graphs of cumulative proportions of distance buffer and deposit pixels around faults/fractures and corresponding graphs of β-statistics of differences (*D*) between the cumulative proportion curves, Aroroy district (Philippines). Confidence bands are for α=0.05. Analysis of spatial association between epithermal Au deposit occurrences and [(**A**), (**B**)] north-northwest (NNW) trending faults/fractures and [(**C**), (**D**)] northwest (NW) trending faults/fractures.

occurrences in the case study area have statistically significant (at α=0.01) positive spatial association with intersections of NNW- and NW-trending faults/fractures and the positive spatial association is optimal within 1.1 km intersections of NNW- and NW-trending faults/fractures (Figs. 6-10A and 6-10B). Within this distance from intersections of NNW- and NW-trending faults/fractures, 85% of the epithermal Au deposit occurrences in the study area are present and, according to the curve for *D*, there is 45% higher occurrence of epithermal Au deposits than would be expected due to chance (Fig. 6-10A). These results support the fault/fracture mesh model (Fig. 6-7) and the hypotheses that intersections between NNW- and NW-trending faults/fractures are (a) possibly where extension faults/fractures are situated, toward which hydrothermal fluids are circulated and thus where epithermal Au deposits are likely formed and (b) therefore are plausible spatial evidence of prospectivity for epithermal Au deposits in the study area. In order to support further these hypotheses, it is instructive to perform an analysis

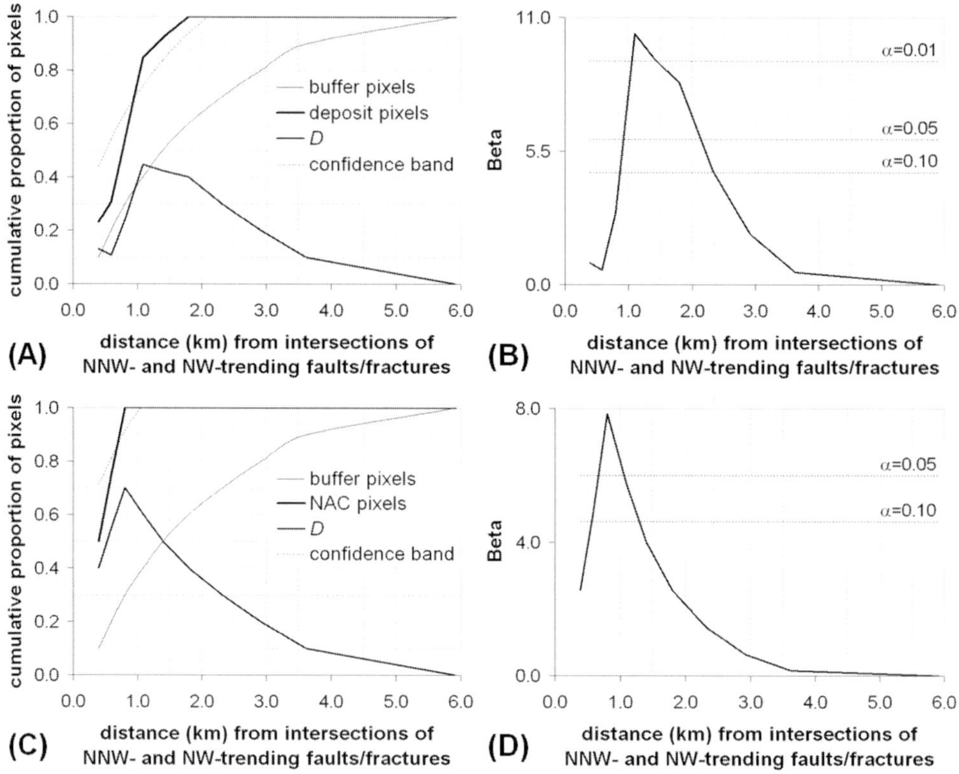

Fig. 6-10. Graphs of cumulative proportions of distance buffer pixels, deposit pixels and Nabongsoran Andesite centroid (NAC) pixels around intersections of north-northwest (NNW) and northwest (NW) trending faults/fractures and corresponding graphs of β-statistics of differences (*D*) between the cumulative proportion curves, Aroroy district (Philippines). Confidence bands are for α=0.05. Analysis of spatial association of intersections of NNW- and NW-trending faults/fractures with [(**A**), (**B**)] epithermal Au deposit occurrences and [(**C**), (**D**)] Nabongsoran Andesite centroids.

of spatial association between the epithermal Au deposit occurrences and Nabongsoran Andesite porphyry intrusions (Fig. 3-9) because the Nabongsoran Andesite porphyry intrusions are possibly heat source controls on epithermal mineralisations in the Aroroy district (Mitchell and Balce, 1990; Mitchell and Leach, 1991). Even though most of the epithermal Au deposit occurrences are within 3 km from the mapped units of Nabongsoran Andesite porphyry intrusions (Fig. 3-9), it is appealing, nonetheless, to perform an analysis of spatial association between intersections of NNW- and NW-trending faults/fractures and Nabongsoran Andesite porphyry intrusions in order to make an inference about the role of the latter as heat source controls on driving hydrothermal fluids toward the former.

The centroid of each of the four mapped units of Nabongsoran Andesite porphyry intrusions (Fig. 3-9) can be digitised as a point and the four digitised points can be used in a spatial association analysis via the distance distribution method. The results of this analysis (Figs. 6-10C and 6-10D) suggest that there is statistically significant (at α=0.05) positive spatial association between centroids of Nabongsoran Andesite porphyry intrusions and intersections of NNW- and NW-trending faults/fractures and the positive spatial association is optimal within 0.8 km of intersections of NNW- and NW-trending faults/fractures. Within this distance from intersections of NNW- and NW-trending faults/fractures, all of the centroids of Nabongsoran Andesite porphyry intrusions in the study area are present and, according to the curve for *D*, there is about 70% higher occurrence of Nabongsoran Andesite porphyry intrusions than would be expected due to chance (Fig. 6-10C). These results suggest that (a) emplacement of Nabongsoran Andesite porphyry intrusions in the case study area are controlled by intersections of NNW- and NW-trending faults/fractures and (b) Nabongsoran Andesite porphyry intrusions contributed to circulation of hydrothermal fluids toward intersections of NNW-trending and NW-trending faults/fractures. These inferences provide a link, albeit indirect, between Nabongsoran Andesite porphyry intrusions and epithermal mineralisations in the case study area as discussed below.

Porphyry intrusions in geological environments characterised by island-arc strike-slip fault systems are usually accompanied by porphyry Cu mineralisations (Titley and Beane, 1981). In these geological environments, such as in the Philippine archipelago, porphyry intrusions and/or porphyry Cu deposits have strong positive spatial associations with strike-slip fault discontinuities (Carranza and Hale, 2002c), which are often sites of intersections of strike-slip faults. In addition, there are strong spatial and temporal relationships between porphyry Cu and epithermal Au mineralisations not only in island-arc strike-slip fault systems (e.g., Cooke and Bloom, 1990; Arribas et al., 1995; Hedenquist et al., 1998) but also in continental-arc strike-slip fault systems (e.g., Muntean and Einaudi, 2001; Berger and Drew, 2002; Billa et al., 2004). In these geological environments, epithermal Au mineralisations are found superjacent to (i.e., stratigraphically although not necessarily vertically above) porphyry Cu mineralisations, because the latter are formed at relatively higher temperatures than the former. Based on these collections of pieces of knowledge, the following hypotheses can be formulated. In the Aroroy district, it is plausible that there exist porphyry Cu and epithermal Au mineralisations associated with the Nabongsoran Andesite porphyry intrusions. It is plausible that porphyry Cu mineralisations associated with the mapped units of Nabongsoran Andesite porphyry intrusions (Fig. 3-9) have already eroded, meaning that associated superjacent epithermal Au mineralisations have also already been eroded. It is plausible, however, that there exist blind Nabongsoran Andesite porphyry intrusions and blind porphyry Cu mineralisation at sites below the surface where NNW- and NW-trending faults/fractures intersect, such that only associated epithermal Au mineralisations are exposed. Therefore, it is plausible that proximity to intersections of NNW- and NW-trending faults fractures constitutes a proxy spatial evidence for heat

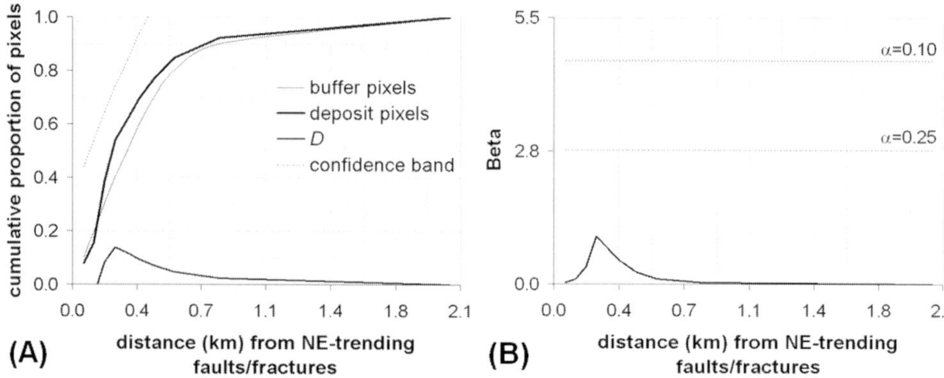

Fig. 6-11. (**A**) Graphs of cumulative proportions of distance buffer pixels and epithermal Au deposit pixels around northeast (NE) trending faults/fractures, Aroroy district (Philippines). Confidence band is for α=0.05. (**B**) Corresponding graph of β-statistic of difference (D) between the cumulative proportion curves.

source controls (aside from being plausible structural controls) on the occurrence of epithermal Au mineralisations in the Aroroy district.

In contrast to NNW- and NW-trending faults/fractures and to intersections of NNW- and NW-trending faults/fractures, northeast (NE) trending faults/fractures in the case study area do not exhibit statistically significant (at α=0.25) positive spatial association with occurrences of epithermal Au deposits (Fig. 6-11). Although it seems that, based on the D and β curves, the epithermal Au deposit occurrences in the area are preferentially located within 0.25 km of NE-trending faults/fractures, the results suggest that epithermal Au deposits in the case study area are almost randomly distributed around NE-trending faults/fractures because the 'deposit' curve closely follows the 'buffer' curve. These results are, nevertheless, consistent with the results of the Fry analysis (Fig. 6-6), which do not show NE trends in the spatial distribution of the occurrences of epithermal Au deposits. Thus, NE-trending faults/fractures do not constitute a satisfactory spatial evidence of prospectivity for epithermal Au deposits in the case study area.

The distance distribution method can also be applied to determine spatial associations between occurrences of mineral deposits of the type sought and geochemical anomalies. Whereas the application of the distance distribution method to geological features such as faults/fractures is an ascending approach (i.e., increasing distances, from minimum to maximum, are used in the analysis), the application of the distance distribution method to geochemical anomalies is a descending approach (i.e., decreasing geochemical attributes, from maximum to minimum, are used in the analysis), because prospective areas for mineral deposits of the type sought are invariably characterised by high concentrations of elements or metals associated with the deposits. The analysis is demonstrated here using the following three sets of derivative stream sediment geochemical data (SSGD) representing anomalous multi-element associations: (1) the

PC2 and PC3 scores derived in the application of exploratory data analysis or EDA (Chapter 3) and shown in Figs. 3-20B and 3-20D, respectively; (2) the integrated PC3 and PC4 scores derived in the application of fractal analysis or FA (Chapter 4) and shown in Fig. 4-21B; and (3) the integrated PC2 and PC3 scores derived in the application of catchment basin analysis or CBA (Chapter 5) and shown in Fig. 5-12. The set (1) of derivative SSGD is integrated first in the same way as the sets (2) and (3) of derivative SSGD were integrated (see Chapters 4 and 5) so that these sets of derivative SSGD can be compared properly with respect to the occurrences of epithermal Au deposits in the case study area.

Fig. 6-12 shows the results of analyses of spatial association of the epithermal Au deposits in the Aroroy district (Philippines) with the three sets of derivative SSGD. The epithermal Au deposits have positive spatial associations with intermediate to high values in each set of derivative SSGD. This is consistent with the fact that raw or derivative SSGD represent transported materials whilst the epithermal Au deposits are in situ so that highest values of the former are in many case located downstream and, thus, not spatially coincident with the former. The positive spatial association between the epithermal Au deposits and intermediate (to high) values in set (2) of derivative SSGD (Figs. 6-12C and 6-12D) has the lowest statistical significance compared to the positive spatial association between the epithermal Au deposits and intermediate (to high) values in set (1) of derivative SSGD (Figs. 6-12A and 6-12B) and in set (3) of derivative SSGD (Figs. 6-12E and 6-12F). However, considering that stream sediment anomalies represent allochthonous materials whilst the epithermal Au deposits represent autochthonous materials, using statistical significance as criterion to select which set of derivative SSGD constitutes a set of optimal spatial evidence of mineral prospectivity may not be appropriate. Alternatively, using the maximum value of β as reference, the ratio of cumulative proportion of deposit pixels to cumulative proportion of sample catchment basin (SCB) pixels, which represents conditional probability of deposit occurrence given intermediate (to high) values of derivative SSGD, is a better criterion to select which set of derivative SSGD constitutes a set of optimal spatial evidence of mineral prospectivity. Thus, based on actual data used create the graphs in Fig. 6-12, set (1) of derivative SSGD gives a ratio of 1.889 (i.e., 0.667÷0.353; data from Fig. 6-12A), set (2) of derivative SSGD gives a ratio of 1.937 (i.e., 0.583÷0.301; data from Fig. 6-12C) and set (3) of derivative SSGD gives a ratio of 1.962 (i.e., 0.667÷0.340; data from Fig. 6-12E). Therefore, among the three sets of derivative SSGD, the integrated PC2 and PC3 scores obtained from the CBA (Chapter 5) constitute the optimum spatial evidence of epithermal Au prospectivity in the case study area, followed by the integrated PC3 and PC4 scores obtained from the FA (Chapter 4) and then the integrated PC2 and PC3 scores obtained from the EDA (Chapter 3). These results indicate that CBA presented in this volume, which is actually a methodology (i.e., a collection of methods) rather than a method, is improved when EDA and FA are incorporated in the methodology.

The preceding discussions demonstrate the usefulness of the distance distribution method in determining spatial associations between occurrences of mineral deposits of the type sought and various sets of geological features. However, it is important to take

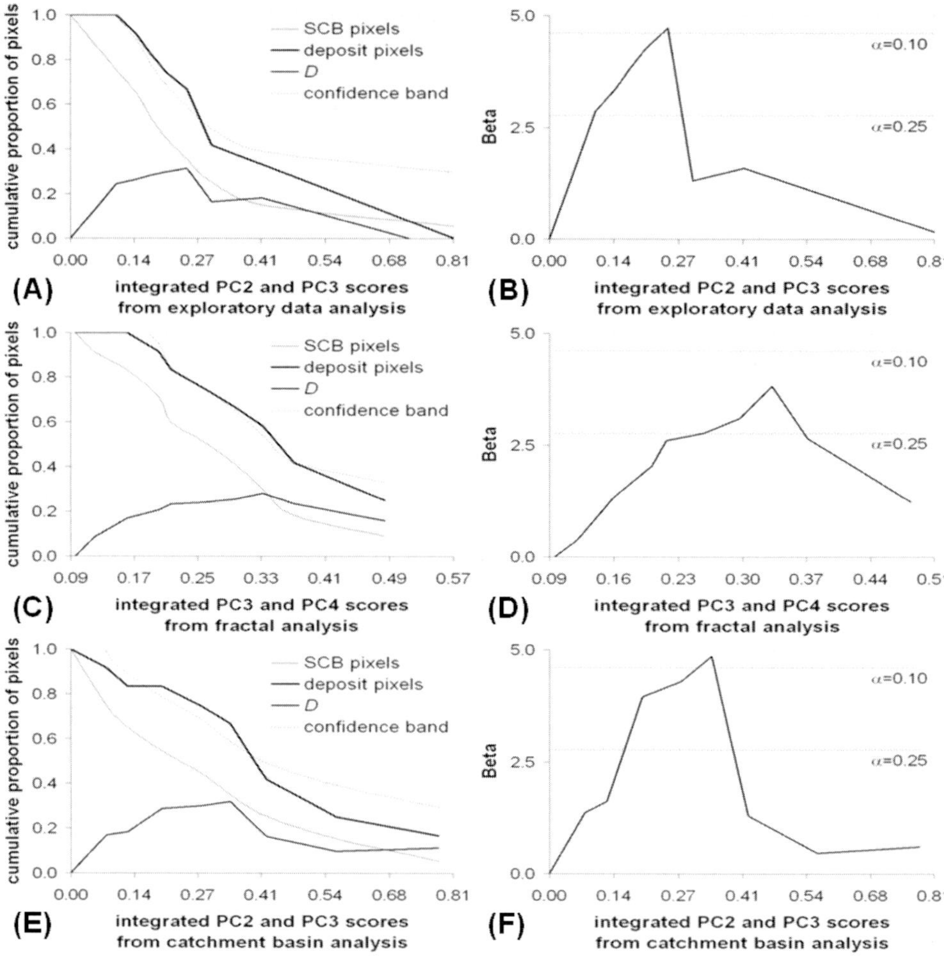

Fig. 6-12. Graphs of cumulative proportions of sample catchment basin (SCB) pixels and deposit pixels in SCBs defined by descending values of derivative stream sediment geochemical data (SSGD) and corresponding graphs of β-statistics of differences (*D*) between the cumulative proportion curves, Aroroy district (Philippines). Confidence bands are for α=0.25. Analysis of spatial association between epithermal Au deposit occurrences and [(**A**), (**B**)] integrated PC2-PC3 scores obtained in the exploratory data analysis (see Chapter 3), [(**C**), (**D**)] integrated PC3-PC4 scores obtained in the fractal analysis (see Chapter 4) and [(**E**), (**F**)] integrated PC2-PC3 scores obtained in the catchment basin analysis (see Chapter 5).

note of the following caveats in the interpretation of results of the distance distribution method. Firstly, the distance distribution method assumes that the point geo-objects of interest are Poisson-distributed (Berman, 1977). However, the results of point pattern analysis and fractal analyses presented earlier show that mineral deposit occurrences are usually not Poisson-distributed. Secondly, the distance distribution method assumes that the linear (or point) geological features under examination have both uniform and random distribution in a study area (Berman, 1977). Certainly, in many cases, this assumption is inapplicable; linear (or point) geological features may exhibit clustering in some parts of a study area and/or are sparse in other parts of a study area. Anyhow, the problem associated with this assumption about the distribution of linear (or point) geological features is avoided by using either a very large number of uniformly distributed random points (Bonham-Carter, 1985; Berman, 1986) or all pixels in a study area (Bonham-Carter, 1994). Finally, one wonders why all lines in a set of lines (e.g., all NNW-trending faults) are used in the analysis even if mineral deposits are associated with only some of these lines. The following section explains another method, in which only lines (or points) nearest to points of interests are used in the spatial association analysis.

Distance correlation method

The concept of the distance correlation method was developed and demonstrated by Carranza (2002) and Carranza and Hale (2002b) to characterise quantitatively spatial association between a set of points of interest (i.e., occurrences of mineral deposits of the type sought and a set of lines (e.g., faults/fractures) or points (e.g., centroids of porphyry stocks). This method is a non-parametric test of spatial association between a set of point geo-objects and a set of linear (or point) geo-objects because it does not involve testing statistical significance of spatial association. However, as demonstrated by Carranza (2002) and Carranza and Hale (2002b) and by the results of analyses in this volume, the method provides results that are similar to the results obtained by application of the distance distribution method.

Consider points P_{jx} (j=1, 2,…, n points) of interest, each at a certain distance X_j from a nearest line L_i (i=1, 2,…, m lines), and their corresponding nearest neighbour points P_{j0} on line L_i, and an arbitrary point AP (Figure 6-13). Hence, there are two sets of measured distances, d_{jx} and d_{j0}, from AP to P_{jx} and to P_{j0}, respectively. If all P_{jx} points lie exactly on L_i (i.e., P_{jx}=P_{j0} and X_j=0), then $d_{jx} = d_{j0}$ and the Pearson correlation coefficient $r_{d_{jx}d_{j0}}$ is unity or equal to 1, which implies a 'perfect' spatial association between P_{jx} and L_i. However, if some or every P_{jx} does not lie exactly on L_i and if X_j is variable (as in most, if not all, cases of distances between mineral deposit occurrences and curvi-linear or point geologic features), then $d_{jx} \neq d_{j0}$ and $r_{d_{jx}d_{j0}} \neq 1$. In this latter case, the spatial association can be qualified as 'imperfect' and needs to be characterised quantitatively.

Now consider points P_{jy}, lying along a line segment $\perp L_i$ (i.e., perpendicular to L_i) from P_{j0} passing through P_{jx}, at equal distances Y_j from line L_i. So, for every Y_j there are also two sets of measured distances, d_{jx} and d_{jy}, from AP to P_{jx} and to P_{jy}, respectively. Suppose further that most of the points P_{jx} lie preferentially, due to intrinsic controls, within a certain range of distances Y_j from their respective nearest L_i neighbours. One

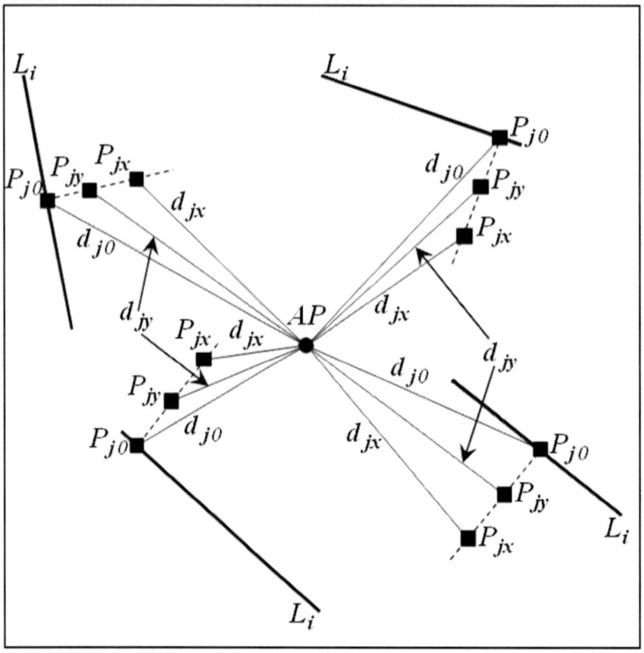

Fig. 6-13. Diagram to illustrate the distance correlation method for quantifying spatial association between a set of points (P_{jx}) of interest and their respective nearest lines L_i (or points P_{j0}). The method compares correlations ($r_{d_{jx}d_{jy}}$) between distances d_{jx} from an arbitrary point (AP) to points P_{jx} and distances d_{jy} from the same AP to points P_{jy} at regular intervals along a line starting from the nearest neighbour points P_{j0} on L_i and passing through P_{jx}. The spatial association between points P_{jx} and nearest neighbour lines L_i is characterised by the maximum $r_{d_{jx}d_{jy}}$ (see text for further explanation).

can then imagine that the Pearson correlation coefficient $r_{d_{jx}d_{jy}}$ increases as P_{jy} points are moved away from P_{j0} towards the range of distances Y_j where most of points P_{jx} lie preferentially away from their respective L_i neighbours. One can further imagine that the Pearson correlation coefficient $r_{d_{jx}d_{jy}}$ decreases as P_{jy} points are moved away from P_{j0} from the range of distances Y_j where most of points P_{jx} lie preferentially away from their respective L_i neighbours towards the other end of the line segment $\perp L_i$. Thus, in a graph with values $r_{d_{jx}d_{jy}}$ in the vertical axis and values of Y_j in the horizontal axis, a concave-downward curve is obtained. The distance correlation method endeavours, therefore, to find a distance Y_j along line segments $\perp L_i$ at which the value of the Pearson correlation coefficient $r_{d_{jx}d_{jy}}$ is highest (i.e., the apex of the downward-concave curve). This distance is hereafter referred to as $Y_j\text{max}_r$. At distance $Y_j\text{max}_r$, the differences between d_{jy}

and d_{jx} are minimal (i.e., P_{jx} and P_{jy} are close enough), which implies that most of the points P_{jx} lie preferentially at distances about $Y_j\max_r$ from their respective nearest L_i neighbours. Hence, the distance $Y_j\max_r$ is considered the distance at which there is optimum spatial association between a set of points and a set of lines (or points).

Within a certain study area, a very large but finite number of possible *AP*s can be used for characterising spatial association between P_{jx} and L_i via the distance correlation method. Because this method is sensitive to the position of an *AP* with respect to line segments $\perp L_i$, as shown below, it is recommended to use not just one but many *AP*s in order to properly characterise the optimum spatial association between P_{jx} and L_i. For simplicity, Carranza (2002) recommends using nine *AP*s within and along the boundaries of a study area: its centre (C); the mid-point of its north (N) boundary; its northeast (NE) corner; the mid-point of its east (E) boundary; its southeast (SE) corner; the mid-point of its south (S) boundary; its southwest (SW) corner; the mid-point of its west (W) boundary; and its northwest (NW) corner. Based on at least one of these nine *AP*s, it is shown below (by application to the case study area) that the longest $Y_j\max_r$ is the distance of optimum spatial association between P_{jx} and L_i. In the case of a positive spatial association, more P_{jx} points would lie at distances less than or equal to $Y_j\max_r$. Hence, it can be expected that the mean $r_{d_{jx}d_{jy}}$ at distances less than or equal to $Y_j\max_r$ (hereafter denoted as proximal-$r_{d_{jx}d_{jy}}$) is greater than the mean $r_{d_{jx}d_{jy}}$ at distances greater than $Y_j\max_r$ (hereafter denoted as distal-$r_{d_{jx}d_{jy}}$). In the case of a negative spatial association, more P_{jx} points would lie at distances less than or equal to $Y_j\max_r$. Hence, it can be expected that the proximal-$r_{d_{jx}d_{jy}}$ is less than the distal-$r_{d_{jx}d_{jy}}$.

The following sequence of procedures can be followed in a raster-based GIS in order to quantify spatial association between P_{jx} and L_i via the distance correlation method.

1. On each L_i nearest to a point P_{jx}, determine P_{j0} visually. (Note that in case the geological feature is a 'point', then the 'point' becomes P_{j0}).
2. Digitise line segments $\perp L_i$ (broken lines in Fig. 6-13) starting from P_{j0} and passing through P_{jx}. Note that each $\perp L_i$ is perpendicular to its respective L_i. The length of each $\perp L_i$ should be at least equal to the length of the longest X_j plus twice the standard deviation of all X_j. This is mainly based on experience or empiricism but not on any formalism. That is, points P_{jx} may lie preferentially at a certain narrow range (i.e., small standard deviation) of distances X_j from L_i and such distance range may happen to be very close to the maximum X_j. Thus, to find $Y_j\max_r$ properly, the length of each $\perp L_i$ should be at most twice the length of the longest X_j in order to compare the proximal-$r_{d_{jx}d_{jy}}$ with the distal-$r_{d_{jx}d_{jy}}$.
3. Create points P_{jy} along the line segments at regular distance intervals Y_j from P_{j0}. Hence, P_{jy} and P_{j0} are the same when Y_j =0. This can be achieved by converting the line segments to points at a specified distance interval.
4. Digitise nine *AP*s as recommended above. Create a map of distances from each *AP*.

5. For each point P_{jx} (i.e., points of interest), determine d_{jx} from an AP. Likewise, for each point P_{jy} (i.e., points at regular intervals along $\perp L_i$), determine d_{jy} from an AP. These can be achieved via an overlay operation between point map P_{jx} (or P_{jy}) and a map of distances from an AP.
6. Calculate $r_{d_{jx}d_{jy}}$ for every pair of d_{jx} and d_{jy}. Create plots of $r_{d_{jx}d_{jy}}$ versus Y_j.
7. To test if values of $r_{d_{jx}d_{jy}}$ are significantly different from zero, a t-value can be calculated as (Davis, 2002):

$$t=\frac{r_{d_{jx}d_{jy}}\sqrt{n-2}}{\sqrt{1-(r_{d_{jx}d_{jy}})^2}}, \tag{6-4}$$

   which has n-2 degrees of freedom, where n is equal to the number of P_{jx} and thus equal to the number of P_{jy}. A calculated t-value is then compared with a critical t-value at a certain significance level from published statistical tables.
8. Repeat steps 5-8 each time for a different AP.
9. Calculate the mean proximal-$r_{d_{jx}d_{jy}}$ and the mean distal-$r_{d_{jx}d_{jy}}$. This procedure aims to determine whether the spatial association between P_{jx} and L_i is positive (i.e., mean proximal-$r_{d_{jx}d_{jy}}$ > mean distal-$r_{d_{jx}d_{jy}}$) or negative (i.e., mean proximal-$r_{d_{jx}d_{jy}}$ < mean distal-$r_{d_{jx}d_{jy}}$).

The application of the distance correlation method to quantify the spatial associations of the 13 epithermal Au deposit occurrences with the different sets of linear and point structural features in the case study area yielded satisfactory results (discussed below) from all of the nine APs (Tables 6-III to 6-VI and Fig. 6-14).

Table 6-III and Fig. 6-14A indicate that there is positive spatial association between the epithermal Au deposit occurrences and NNW-trending faults/fractures in the study area and the positive spatial association is optimal within 0.2 km of NNW-trending faults/fractures. Inspection of the individual distances between the epithermal Au deposits and NNW-trending faults/fractures shows that eight (or 62%) and nine (or 69%) of the 13 epithermal Au deposits are within 0.1 km and 0.2 km of NNW-trending faults/fractures, respectively. These results are similar to the results of the distance distribution analysis (see Figs. 6-9A and 6-9B).

Table 6-IV and Fig. 6-14B indicate that there is positive spatial association between the epithermal Au deposit occurrences and NW-trending faults/fractures in the study area and the positive spatial association is optimal within 0.8 km of NW-trending faults/fractures. Inspection of the individual distances between the epithermal Au deposits and NW-trending faults/fractures shows that seven (or 54%) and nine (or 69%) of the 13 epithermal Au deposits are within 0.6 km and 0.8 km of NW-trending faults/fractures, respectively. These results are similar to the results of the distance distribution analysis (see Figs. 6-9C and 6-9D).

TABLE 6-III

Results of distance correlation calculations to quantify spatial association between occurrences of epithermal Au deposits and NNW-trending faults/fractures, Aroroy district (Philippines). Table entries in bold pertain to optimal spatial association

AP	Y_j $\max_r$ (m)	$r_{d_{jx}d_{jy}}$	*t*-value*	Proximal-$r_{d_{jx}d_{jy}}$	Distal-$r_{d_{jx}d_{jy}}$
C	**200**	**0.9981**	**53.123**	**0.9962**	**0.9771**
N	**200**	**0.9994**	**98.819**	**0.9994**	**0.9963**
NE	100	0.9984	58.081	0.9978	0.9864
E	**200**	**0.9914**	**25.193**	**0.9857**	**0.9249**
SE	**200**	**0.9996**	**112.630**	**0.9992**	**0.9949**
S	**200**	**0.9999**	**308.139**	**0.9999**	**0.9989**
SW	**200**	**0.9989**	**69.957**	**0.9985**	**0.9905**
W	**200**	**0.9988**	**68.604**	**0.9983**	**0.9960**
NW	100	0.9999	233.427	0.9998	0.9988

*Calculated values of $r_{d_{jx}d_{jy}}$ are significantly different from zero; the critical *t*-value at α=0.001 and ν=13-2=11 degrees of freedom is 4.025 (Davis, 2002).

Table 6-V and Fig. 6-14C indicate that there is positive spatial association between the epithermal Au deposits occurrences and intersections of NNW and NW-trending faults/fractures in the study area and the positive spatial association is optimal within 0.9 km of intersections of NNW- and NW-trending faults/fractures. Inspection of individual distances between the epithermal Au deposits and intersections of NNW- and NW-

TABLE 6-IV

Results of distance correlation calculations to quantify spatial association between occurrences of epithermal Au deposits and NW-trending faults/fractures, Aroroy district (Philippines). Table entries in bold pertain to optimal spatial association.

AP	Y_j $\max_r$ (m)	$r_{d_{jx}d_{jy}}$	*t*-value*	Proximal-$r_{d_{jx}d_{jy}}$	Distal-$r_{d_{jx}d_{jy}}$
C	**800**	**0.9739**	**14.219**	**0.9526**	**0.8497**
N	600	0.9893	22.519	0.9811	0.9230
NE	600	0.9872	20.513	0.9787	0.8917
E	**800**	**0.9752**	**14.624**	**0.9501**	**0.8932**
SE	600	0.9969	41.905	0.9945	0.9815
S	700	0.9956	35.309	0.9925	0.9565
SW	600	0.9854	19.215	0.9727	0.8961
W	700	0.9878	20.841	0.9765	0.9372
NW	600	0.9975	46.962	0.9954	0.9598

*Calculated values of $r_{d_{jx}d_{jy}}$ are significantly different from zero; the critical *t*-value at α=0.001 and ν=13-2=11 degrees of freedom is 4.025 (Davis, 2002).

TABLE 6-V

Results of distance correlation calculations to quantify spatial association between occurrences of epithermal Au deposits and intersections of NNW- and NW-trending faults/fractures, Aroroy district (Philippines). Table entries in bold pertain to optimal spatial association.

AP	$Y_j\ \max_r$ (m)	$r_{d_{jx}d_{jy}}$	*t*-value*	Proximal-$r_{d_{jx}d_{jy}}$	Distal-$r_{d_{jx}d_{jy}}$
C	800	0.9765	15.075	0.9443	0.8399
N	800	0.9928	27.528	0.9837	0.9446
NE	600	0.9916	25.491	0.9835	0.9197
E	600	0.9776	15.404	0.9437	0.7355
SE	800	0.9958	35.928	0.9913	0.9705
S	**900**	**0.9969**	**42.306**	**0.9923**	**0.9749**
SW	800	0.9908	24.291	0.9765	0.9212
W	800	0.9869	20.250	0.9741	0.9191
NW	800	0.9980	52.751	0.9952	0.9758

*Calculated values of $r_{d_{jx}d_{jy}}$ are significantly different from zero; the critical *t*-value at α=0.001 and ν=13-2=11 degrees of freedom is 4.025 (Davis, 2002).

trending faults/fractures shows that seven (or 54%) and ten (or 77%) of the 13 epithermal Au deposits are within 0.8 km and 0.9 km of intersections between NNW- and NW-trending faults/fractures, respectively. These results are similar to the results of the distance distribution analysis (see Figs. 6-10A and 6-10B).

Table 6-VI and Fig. 6-14D indicate that there is positive spatial association between

TABLE 6-VI

Results of distance correlation calculations to quantify spatial association between occurrences of epithermal Au deposits and NE-trending faults/fractures, Aroroy district (Philippines). Table entries in bold pertain to optimal spatial association.

AP	$Y_j\ \max_r$ (m)	$r_{d_{jx}d_{jy}}$	*t*-value*	Proximal-$r_{d_{jx}d_{jy}}$	Distal-$r_{d_{jx}d_{jy}}$
C	**300**	**0.9926**	**27.094**	**0.9902**	**0.9513**
N	**300**	**0.9996**	**118.041**	**0.9993**	**0.9959**
NE	**300**	**0.9995**	**102.544**	**0.9993**	**0.9953**
E	**300**	**0.9857**	**19.433**	**0.9822**	**0.9138**
SE	**300**	**0.9986**	**62.821**	**0.9977**	**0.9893**
S	**300**	**0.9990**	**74.858**	**0.9983**	**0.9926**
SW	**300**	**0.9993**	**92.141**	**0.9989**	**0.9968**
W	**300**	**0.9989**	**71.849**	**0.9973**	**0.9901**
NW	**300**	**0.9991**	**76.211**	**0.9983**	**0.9917**

*Calculated values of $r_{d_{jx}d_{jy}}$ are significantly different from zero; the critical *t*-value at α=0.001 and ν=13-2=11 degrees of freedom is 4.025 (Davis, 2002).

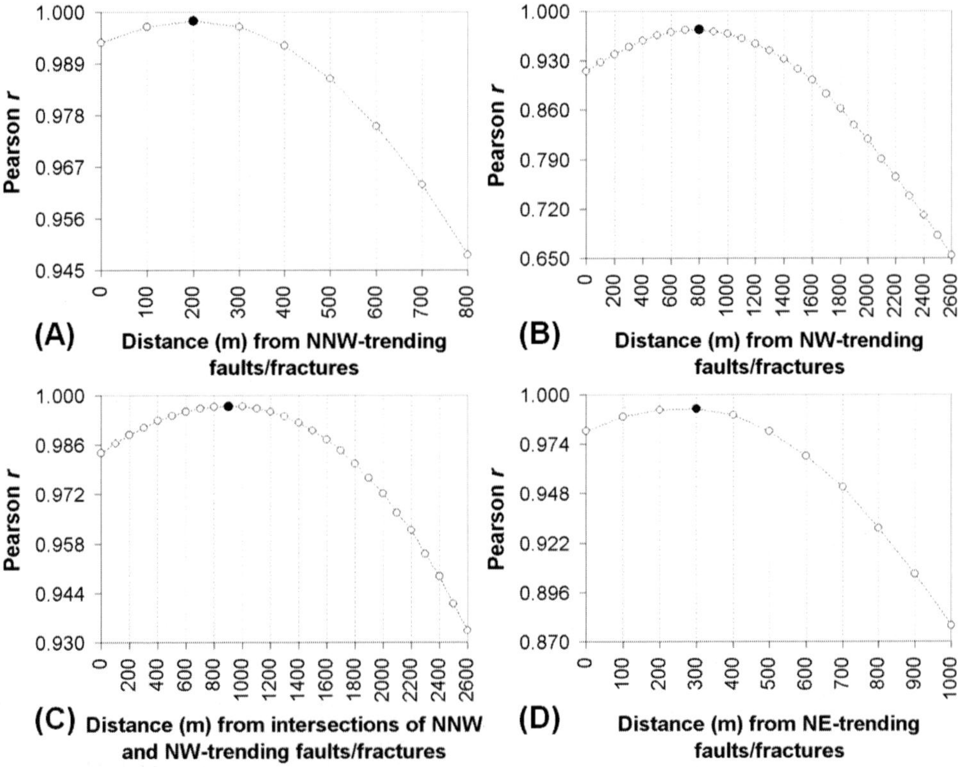

Fig. 6-14. Plots of $r_{d_{jx}d_{jy}}$ (Pearson *r*) versus Y_j (distances from linear geological features), with respect to an arbitrary point (*AP*), depicting spatial associations between epithermal Au deposits occurrences and faults/fractures in the Aroroy district (Philippines). Analysis for (**A**) NNW-trending faults/fractures using centre *AP* (see Table 3-III), (**B**) NW-trending faults/fractures using centre *AP* (see Table 3-IV), (**C**) intersections between NNW- and NW-trending faults/fractures using south *AP* (see Table 3-V) and (**D**) NE-trending faults/fractures using centre *AP* (see Table 3-VI). A dot along the plots represents the distance $Y_j \max_r$ at which $r_{d_{jx}d_{jy}}$ is highest and thus the distance of optimal spatial association between epithermal Au deposit occurrences and a set of faults/fractures.

the epithermal Au deposits occurrences and NE-trending faults/fractures in the study area and the positive spatial association is optimal within 0.3 km of NE-trending faults/fractures. Inspection of the individual distances between the epithermal Au deposits and NE-trending faults/fractures shows that eight (or 62%) of the 13 epithermal Au deposits are within 0.3 km of NE-trending faults/fractures. These results are similar to the results of the distance distribution analysis (see Fig. 6-11).

Although application of the distance correlation method produces similar results to the application of the distance distribution method, the former does not indicate the statistical significance of spatial association between point geo-objects of interest and linear (or point) features. In addition, the results shown in Tables 6-III to 6-VI indicate that the distance correlation method is sensitive to the position of an *AP* with respect to line segments $\perp L_i$, so that not just one but at least nine *AP*s are recommended in order to properly characterise the optimum spatial association between P_{jx} and L_i. This is demonstrated further with the analysis of the spatial association between the centroids of mapped units of Nabongsoran Andesite porphyry and intersections of NNW- and NW-trending faults/fractures in the case study area.

There are very few (i.e., four) mapped units of Nabongsoran Andesite and they are situated only in the southwest quadrant of the study area (see Fig. 3-9). The distances between each centroid and its nearest neighbour intersection of NNW- and NW-trending faults/fractures vary from 230 to 700 m, with a mean of 460 m and standard deviation of 205 m. The length of each line segment $\perp L_i$, which passes through each of the centroids starting from their respective nearest neighbour intersection of NNW- and NW-trending faults/fractures, is set at 1200 m. The results of the analyses based on each of the nine recommended *AP*s are illustrated in Fig. 6-15. The north, northeast and southwest *AP*s resulted in 'flat' graphs of $r_{d_{jx}d_{jy}}$ versus Y_j (Figs. 6-15B, 6-15C and 6-15G, respectively), although peak values of $r_{d_{jx}d_{jy}}$ at 800 m can be observed if the resolutions of the y-axes of these graphs are enhanced. The southeast and west *AP*s resulted in downward-concave graphs of $r_{d_{jx}d_{jy}}$ versus Y_j with peak values of $r_{d_{jx}d_{jy}}$ at 1000 m and 900 m, respectively (Figs. 6-15E and 6-15H, respectively). These results are unsatisfactory because the distances $Y_j\max_r$ obtained are well beyond the maximum X_j (i.e., 700 m). Nevertheless, the results obtained based on the centre, east, south and northwest *AP*s (Figs. 6-15A, 6-15D, 6-15F and 6-15I, respectively) are satisfactory, because they clearly show downward-concave graphs of $r_{d_{jx}d_{jy}}$ versus Y_j with peak values of $r_{d_{jx}d_{jy}}$ mostly at 700 m.

The factors for the unsatisfactory results obtained based on the north, northeast, southwest, southwest and west *AP*s and the satisfactory results obtained based on the centre, east, south and northwest *AP*s are given in Table 6-VII. Satisfactory results are obtained when an *AP* forms intermediate angles (i.e., mostly between 30° and 60°) with $\perp L_i$ and when the distances (d_{jy}) between an *AP* and every point P_{jy} along $\perp L_i$ have relatively high coefficients of variation (i.e., ratio of standard deviation to mean). Note that both of these two factors apply to the north *AP*, which forms angles with every $\perp L_i$ similar to that formed by the centre AP although distances (d_{jy}) from the former to points P_{jy} along line segments $\perp L_i$ have relatively low coefficients of variation compared to distances from the former. Note further that *AP*s forming, on average, low angles (<20°) with $\perp L_i$ result in 'flat' graphs of $r_{d_{jx}d_{jy}}$ versus Y_j, whereas *AP*s forming, on average, high angles (>63°) with $\perp L_i$ result in over-estimation of $Y_j\max_r$.

Based on the results shown in Fig. 6-15 and summarised in Table 6-VIII for the *AP*s providing the satisfactory results, there is positive spatial association between the four

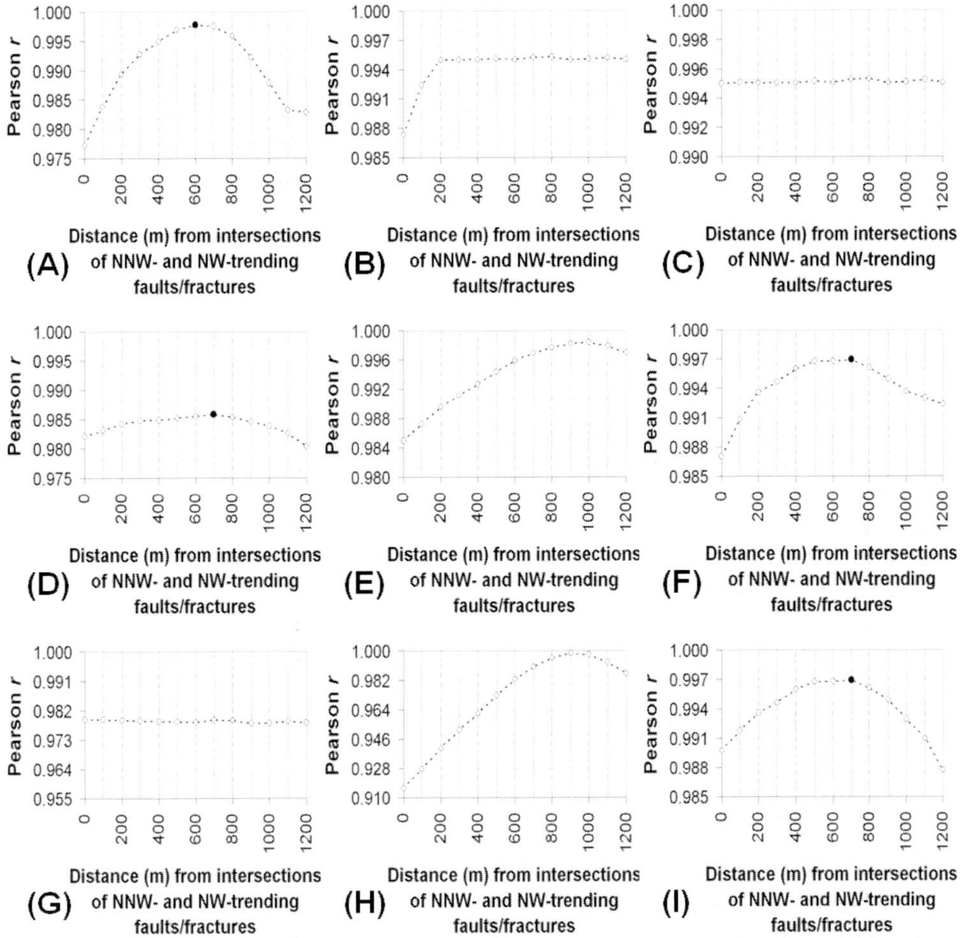

Fig. 6-15. Plots of $r_{d_{jx}d_{jy}}$ (Pearson r) versus Y_j (distances from linear geological features) with respect to an arbitrary point (*AP*), depicting spatial associations between centroids of Nabongsoran Andesite porphyry and intersections of NNW- and NW-trending faults/fractures in the Aroroy district (Philippines), based on (**A**) centre *AP*, (**B**) north *AP*, (**C**) northeast *AP*, (**D**) east *AP*, (**E**) southeast *AP*, (**F**) south *AP*, (**G**) southwest *AP*, (**H**) west *AP* and (**I**) northwest *AP*. A dot along some of the plots represents the distance $Y_j \max_r$ at which $r_{d_{jx}d_{jy}}$ is highest and, thus, the distance of optimal spatial association between centroids of Nabongsoran Andesite porphyry and intersections of NNW- and NW-trending faults/fractures.

centroids of mapped units of Nabongsoran Andesite porphyry and their respective nearest neighbour intersection of NNW- and NW-trending faults/fractures. The positive spatial association is optimal within 0.7 km of intersections NNW- and NW-trending

TABLE 6-VII

Factors related to position of an *AP* (arbitrary point) with respect to every $\perp L_i$ that render satisfactory results (table entries in bold or highest final ranks) and unsatisfactory results in the application of the distance correlation method. Results shown are based on analysis of the spatial association between centroids of mapped units of Nabongsoran Andesite porphyry and intersections of NNW- and NW-trending faults/fractures in Aroroy district (Philippines).

AP	Factor 1: mean of angles formed by an *AP* with $\perp L_i$		Factor 2: mean coefficients of variation* of d_{jy} along $\perp L_i$		Combination of factors 1 and 2	
	Value	Rank (1)	Value	Rank (2)	(1)×(2)	Final rank
S	**61°**	**7**	**0.049**	**7**	**98**	**9**
E	**31°**	**5**	**0.038**	**6**	**60**	**8**
C	**26°**	**3**	**0.077**	**9**	**54**	**7**
NW	**47°**	**6**	**0.029**	**4**	**48**	**6**
N	29°	4	0.034	5	40	5
W	63°	8	0.015	2	32	4
SE	72°	9	0.008	1	18	3
SW	11°	1	0.076	8	16	2
NE	18°	2	0.021	3	12	1

*Ratio of standard deviation to mean.

faults/fractures (i.e., proximal-$r_{d_{jx}d_{jy}}$ > distal-$r_{d_{jx}d_{jy}}$), within which all the centroids of mapped units of Nabongsoran Andesite porphyry are present. These results are similar to the results of the distance distribution analysis (see Figs. 6-10C and 6-10D).

Synthesis and discussion of results

Table 6-IX shows that, in the applications of the distance distribution method and the distance correlation method in the case study area, the former method results in larger distances of optimum positive spatial associations between the point geo-objects of interest and individual sets of geological features. This is usually not the case as shown by previous comparisons between the two methods (Carranza, 2002; Carranza and Hale, 2002b). The differences between the results of the two methods are attributable to the discretisation of the distance data (see step 2) in the application of the distance distribution method, whereas there is no discretisation of the distance data in application of the distance correlation method.

The similar results of applications of the distance distribution method and the distance correlation method presented in this volume further demonstrates the usefulness of the former method in characterising spatial associations of a set mineral deposit occurrences and individual sets of linear or point geological features. An important advantage of the distance correlation method over the distance distribution method is that the former provides meaningful results even though it makes use of only linear (or point) geological features that are closest to the known occurrences mineral deposits of interest. This is an advantage because, in many cases, localities of mineral deposit

TABLE 6-VIII

Results of distance correlation calculations to quantify spatial association between centroids of Nabongsoran Andesite porphyry and intersections of NNW- and NW-trending faults/fractures, Aroroy district (Philippines). Table entries in bold pertain to optimal spatial association.

AP	Y_j $\max_r$ (m)	$r_{d_{jx}d_{jy}}$	*t*-value*	Proximal-$r_{d_{jx}d_{jy}}$	Distal-$r_{d_{jx}d_{jy}}$
C	600	0.9978	21.320	0.9904	0.9899
E	**700**	**0.9858**	**8.306**	**0.9845**	**0.9833**
S	**700**	**0.9969**	**17.996**	**0.9941**	**0.9940**
NW	**700**	**0.9969**	**13.453**	**0.9945**	**0.9925**

* Calculated values of $r_{d_{jx}d_{jy}}$ are significantly different from zero; the critical *t*-value at α=0.01 and ν=4-2=2 degrees of freedom is 6.965 (Davis, 2002).

occurrences are relatively well-mapped than other portions of a study area. Thus, the distance correlation method is a useful exploratory analytical tool in determining which relevant geological features should be given preferential attention in further field geological mapping or geological remote sensing of poorly explored portions of a study area. The main disadvantage of the distance correlation method is that, being a non-parametric method, it does not provide a statistical test of significance of spatial association. However, it allows for empirical testing of spatial association by estimation of the proportion of mineral deposit occurrences within the derived distance of optimal

TABLE 6-IX

Summary of results of analyses of spatial associations between epithermal Au deposit occurrences and individual sets of structural features in Aroroy district (Philippines). Table entries in bold relate to the results of analyses of the spatial distribution of the epithermal Au deposit occurrences in the case study area (see above).

Geological features*	Results of distance distribution analysis		Results of distance correlation analysis		Average of results	
	*D*** (m)	Percentage of Au (or NAC) within *D*	*D*** (m)	Percentage of Au (or NAC) within *D*	*D*** (m)	Percentage of Au (or NAC) within *D*
Au vs. NNW	**450**	**100**	**200**	**69**	**325**	**84.5**
Au vs. NW	**1000**	**85**	**800**	**69**	**900**	**77.0**
Au vs. FI	**1100**	**85**	**900**	**77**	**1000**	**81.0**
Au vs. NE	250	55	300	62	275	58.5
NAC vs. FI	800	100	700	100	750	100

*Au = epithermal Au deposit occurrences; NAC = centroids of mapped units of Nabongsoran Andesite porphyry; NNW = NNW-trending faults/fractures; NW = NW-trending faults/fractures; FI = intersections of NNW- and NW-trending faults/fractures; NE = NE-trending faults/fractures.
***D* = distance of optimum positive spatial association.

spatial association with nearest neighbour linear or point geological features. A limitation of the distance correlation method is that it is not applicable to the analysis of spatial association between mineral deposit occurrences and geochemical anomalies.

The observed differences in the results of the distance distribution method and the distance correlation method illustrate, however, that spatial associations of mineral deposit occurrences and certain geological features can only be quantitatively explored but cannot be definitely confirmed. Nevertheless, the results of analyses of spatial associations between mineral deposit occurrences and certain geological features can provide ideas for new investigations that may lead to further understanding of geologic controls on occurrences of mineral deposits of the type sought in a particular area. For example, the similar results of the applications of the distance distribution method and distance correlation method that there is positive albeit very weak spatial association between the epithermal Au deposit occurrences and NE-trending faults/fractures in the case study area (Table 6-IX) suggest that the set of all NE-trending faults/fractures (see Fig. 5-13) probably represents various processes or structural regimes. However, because the distance correlation method makes use of only linear (or point) geological features that are closest to the known occurrences of mineral deposits of interest, it is plausible that the epithermal Au deposits in the study area are associated with only one of the possible various structural regimes represented by the NE-trending faults/fractures. More importantly, the results of analyses of spatial associations between mineral deposit occurrences and certain geological features in conjunction with the results of analyses of spatial distribution of mineral deposit occurrences are useful in defining a conceptual model of mineral prospectivity.

From the results of analyses of spatial associations between the epithermal Au deposit occurrences and individual sets of geological features shown in Table 6-IX, it is possible to rank the examined geological features according to their relative importance to the occurrence of mineral deposits of the type sought. Note that, as stated earlier, the smaller the distance of positive spatial association, the stronger the spatial dependence. This proposition can be supported by estimation of likelihood of mineral deposit occurrence as the ratio of the proportion (or percentage) of deposits in zones within the distance of optimum positive spatial association with a set of certain geological features to the proportion (or percentage) of zones in a study area within that distance from the same set of geological features. This analysis is supported directly by the distance distribution method [i.e., dividing $\hat{O}(X)$ by $\hat{E}(X)$] but can also be derived based on the results of the distance correlation method. Thus, the following recognition criteria of prospectivity for epithermal Au deposits, arranged according to decreasing importance, are postulated for the case study area:

- proximity to NNW-trending faults/fractures (representing structural controls);
- proximity to intersections of NNW- and NW-trending faults/fractures (representing structural controls and proxy for heat source controls); and
- proximity to NW-trending faults/fractures (representing structural controls).

By integrating these interpretations with those derived from the analyses of spatial distribution of the epithermal Au deposit occurrence, a conceptual model of structural

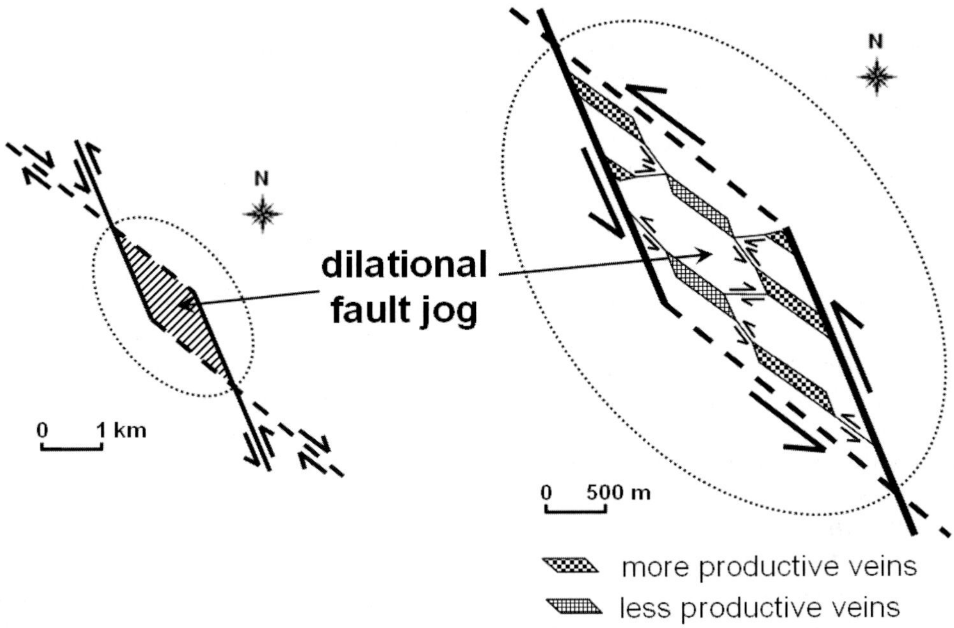

Fig. 6-16. Conceptual model of structural control on epithermal mineralisation in the Aroroy district (Philippines). Dilational jogs or zones of extension arise from en echelon strike-slip fault/fracture segments or from intersections of strike-slip faults/fractures (see Fig. 6-7). In case of the former in the study area, a set of conjugate shears (in this case the dashed NW-trending faults) is possibly relatively passive. Within a dilational jog, a mesh of interlinked shear faults/fractures and extension faults/fractures is developed and the extension faults/fractures are favourable sites for deposition of epithermal veins. See text for further explanation.

control on epithermal mineralisation in the case study area can be drawn (Fig. 6-16) and postulated as follows.

The stronger positive spatial association of the epithermal Au deposit occurrences with the NNW-trending faults/fractures than with the NW-trending faults/fractures suggests that the former were the more active structural controls on the formation of the epithermal Au deposits. Dilational jogs or zones of extension faults/fractures, arising from en echelon NNW-trending strike-slip faults/fractures or from intersections of the NNW- and NW-trending strike-slip faults/fractures (see Fig. 6-7), provided conditions (i.e., structural permeability and hydrothermal fluid flow) that favoured epithermal mineralisation (Fig. 6-16). On regional- to district-scales, hydrothermal fluids were circulated or focused towards dilational jogs or intersections of strike-slip faults/fractures (e.g., Laing, 2004) during seismic activity (cf. Sibson, 1987, 1996, 2000, 2001; Bellot, 2008). Then, on a local-scale, hydrothermal fluids were focused towards extension faults/fractures in a mesh of interlinked shear faults/fractures and extension faults/fractures within dilational jogs (cf. Blenkinsop and Kadzviti, 2006; Nelson, 2006). These interpretations of presence of regional- to local-scale dilational jogs and regional-to local-scale intersections of strike-slip fault/fractures represent the results of the fractal analysis. The hydrothermal fluids were further dispersed, however, outwards from the center of a dilational jog, but they were trapped by the major NNW-trending strike-slip faults/fractures bordering a dilational jog. Thus, it can be hypothesised that less productive or barren epithermal veins formed at the central parts of a dilational jog (or intersection of NNW- and NW-trending faults/fractures) whereas more productive epithermal veins formed close to major NNW-trending strike-slip faults/fractures bordering a dilational jog. These interpretations represent the quantified spatial associations between epithermal Au deposit occurrences and the NNW-trending faults/fractures, NW-trending faults/fractures and intersections of these two sets of faults/fractures and the field observations that Au-bearing quartz veins in the study area are associated mostly with NNW-trending faults/fractures (Mitchell and Leach, 1991) .

In addition to the above-mentioned prospectivity recognition criteria, which constitute a conceptual model of geologic controls in the case study area, the following non-geologic control is an important prospectivity recognition criterion to consider:

- presence of multi-element stream sediment geochemical anomalies (representing surficial expressions or evidence).

A conceptual model of geologic controls and surficial expressions of mineral deposits of the type sought, which can be referred to as a deposit exploration model, provides the theoretical framework for mineral prospectivity mapping.

CONCLUSIONS

Because the geological processes involved in mineralisation are too complex to be modeled in a GIS in order to predict prospective areas for further exploration, a conceptual model of geologic controls on mineralisation forms the basis of GIS-based modeling of mineral prospectivity. A conceptual model of geologic controls on mineralisation is usually a synthesis of exploration experience, qualitative analysis (i.e., review of existing knowledge about mineral deposit formation) and quantitative analyses of spatial distributions of mineral deposit occurrences and their spatial associations with certain geological features. The quantitative (GIS-based) analyses are based on the general geological characteristics of mineral deposits of interest described in mineral deposit models and on the specific geological characteristics of known occurrences of mineral deposits of the type sought in a particular area. In the literature about mineral deposit geology, several recent studies demonstrate applications of similar but different GIS-based techniques to support conceptual modeling of geologic controls on mineralisation (e.g., Groves et al., 2000; Coolbaugh et al., 2002; Porwal et al., 2006c; Bierlein et al., 2008; Hronsky and Groves, 2008).

The GIS-based methods explained and demonstrated in this volume, as well as those in the literature about mineral deposit geology, illustrate the utility of exploratory spatial data analyses in studying patterns of mineral deposit occurrences and the plausible factors or controls on such spatial patterns. These analyses constitute an *inductive* process because they lead to conceptualisation of geologic controls on mineralisation based on a number of occurrences of mineral deposits of the type sought. The derived components of a conceptual model of geologic controls and surficial expressions of mineral deposits of the type sought (i.e., prospectivity recognition criteria) may not represent the whole story of mineral deposit occurrence. However, they describe spatial relationships among certain geological features that may have been involved in mineralisation and thus provide insights in part as to how and mainly as to where mineral deposits of the type sought might occur.

The representation and integration of mineral prospectivity recognition criteria as spatial evidence maps in a GIS constitute a *deductive* process of developing a spatial model (i.e., a map) of mineral prospectivity. This can be either *knowledge-driven* (i.e., based on qualitative analysis) or *data-driven* (i.e., based on quantitative analysis). The types of techniques for modeling mineral prospectivity are explained and demonstrated in the succeeding chapters.

Geochemical Anomaly and Mineral Prospectivity Mapping in GIS
by E.J.M. Carranza
Handbook of Exploration and Environmental Geochemistry, Vol. 11 (M. Hale, Editor)

Chapter 7

KNOWLEDGE-DRIVEN MODELING OF MINERAL PROSPECTIVITY

INTRODUCTION

Knowledge-driven mineral prospectivity mapping is appropriate in frontier or less-explored (or so-called 'greenfields') geologically permissive areas where no or very few mineral deposits of the type sought are known to occur. Knowledge of empirical spatial associations between the mineral deposits and indicative geological features in moderately- to well-explored areas is the basis of knowledge-driven mineral prospectivity mapping in frontier geologically permissive areas with similar, if not the same, geological settings as the former. This means that a conceptual model of mineral prospectivity developed in moderately- to well-explored areas is applied to mineral prospectivity mapping in frontier geologically permissive areas. This conceptual model of mineral prospectivity is considered in the creation of evidential maps (i.e., estimation of evidential map weights and evidential class scores) and the integration of these evidential maps according to the proposition that "*this location is prospective for mineral deposits of the type sought*". Thus, the term 'knowledge-driven' refers to the qualitative assessment or weighting of evidence with respect to a proposition. The estimates of weights for every evidential map and estimates of scores for every class in an evidential map reflect one's 'expert' judgment of the spatial association between mineral deposits of the type sought and every set of indicative geologic features. Accordingly, knowledge-driven mineral prospectivity mapping is also known as expert-driven mineral prospectivity mapping.

The 'expert' knowledge one applies in knowledge-driven mineral prospectivity mapping may have been obtained via substantial field experiences in mineral exploration or via substantial experiences in the application of spatial analytical techniques to study spatial distributions of mineral deposits of the type sought and their spatial associations with certain geological features (Chapter 6). Alternatively, one may elicit knowledge from other experts, who have profound expertise in exploration of mineral deposits of the type sought. The process of eliciting expert knowledge for GIS-based mineral prospectivity mapping is not well established and is not further treated in this volume. In this regard, readers are referred to Schuenemeyer (2002) for elicitation of expert knowledge needed in assessment of fossil fuel resources or to Hodge et al. (2001) for elicitation of knowledge for engineering applications.

Knowledge-driven mineral prospectivity in frontier geologically permissive areas may employ either binary or multi-class evidential maps depending on the (a) degree of applicability of knowledge from well-explored areas to frontier geologically permissive areas and/or (b) degree of accuracy of exploration data available in frontier geologically permissive areas. If the degree of one or both of these two factors is considered high, then it is appropriate to model mineral prospectivity by using multi-class evidential maps; otherwise, it is appropriate to model mineral prospectivity by using binary evidential maps.

This chapter explains the concepts of different modeling techniques for knowledge-driven mapping of mineral prospectivity, which employ either binary or multi-class evidential maps. Each of the modeling techniques is then demonstrated in mapping of prospectivity for epithermal Au deposits in the case study Aroroy district (Philippines) (Fig. 3-9). It is assumed that (a) there are very few known occurrences of epithermal Au deposits in the case study area and (b) the following prospectivity recognition criteria represent knowledge of epithermal Au prospectivity developed in other areas having very similar geological settings as the case study area.

- Proximity to NNW-trending faults/fractures.
- Proximity to NW-trending faults/fractures.
- Proximity intersections of NNW- and NW-trending faults/fractures.
- Presence of multi-element stream sediment geochemical anomalies.

The common spatial data sets used in the applications of the individual modeling techniques are: (a) distance to NNW-trending faults/fractures; (b) distance to NW-trending faults/fractures; (c) distance to intersections of NNW- and NW-trending faults/fractures; and (d) integrated PC2 and PC3 scores obtained from the catchment basin analysis of stream sediment geochemical data (see Fig. 5-12). For the first three prospectivity recognition criteria and the first three data sets, the threshold distances to geologic structures that are used for the case demonstrations below are (a) 0.35 (rounded-off from 0.325) km of NNW-trending faults/fractures (see Table 6-IX), (b) 0.9 km of NW-trending faults/fractures (see Table 6-IX) and 1 km of intersections of NNW- and NW-trending faults/fractures (see Table 6-IX). For the last prospectivity recognition criterion and the last data set, the threshold value of multi-element geochemical anomaly that is used for the case demonstrations below is 0.34 (see Figs. 6-12E and 6-12F).

The threshold values of spatial data specified by the prospectivity recognition criteria form the bases for assignment of evidential class scores in individual evidential maps, particularly binary evidential maps, in knowledge-driven mineral prospectivity mapping. This means that the conceptual model of mineralisation controls is effectively a *prescriptive* model. The implementation of this prescriptive model in knowledge-driven mineral prospectivity mapping, especially in new exploration areas, results in a *predictive* model. The performance of a knowledge-driven predictive map of mineral prospectivity can then be evaluated against the (very) few known occurrences of mineral deposits of the type sought in a study area (Agterberg and Bonham-Carter, 2005; Chung and Fabbri, 2005). Because the known occurrences of mineral deposits of the type sought are not directly used (i.e., they are presumed undiscovered) in the creation of evidential maps (i.e., estimation of evidential class scores and evidential map weights), as in data-driven modeling (see next chapter), the performance evaluation or *cross-*

validation of a knowledge-driven mineral prospectivity map yields estimates of its *prediction-rate*.

Although this chapter discusses the performances of knowledge-driven mineral prospectivity maps derived via applications of the individual modeling techniques explained, it does not mean that the examples of evidential class scores, evidential maps weights and output maps presented portray the 'best' possible prediction models of mineral prospectivity in the case study area. The general ways of deriving an optimal prediction model of mineral prospectivity (i.e., predictive model calibration) are discussed in Chapter 1. One must note, however, that calibration of knowledge-driven predictive modeling of mineral prospectivity is possible only when cross-validation deposits are available. Considering that this is the case, some additional guidelines for calibration of GIS-based knowledge-driven predictive modeling of mineral prospectivity are given here.

GENERAL PURPOSE APPLICATIONS OF GIS

The types of GIS operations principally used in knowledge-driven mineral prospectivity mapping include retrieval, (re-)classification and map overlay (see Chapter 2). The first two operations are concerned with spatial evidence representation (i.e., evidential map creation) whilst the last operation is concerned with spatial evidence integration. In the case when certain prospectivity recognition criteria are represented by input spatial data of continuous fields (e.g., distances to faults/fractures), the classification operation results in an evidential map of either binary or multi-class discrete geo-objects (e.g., classes of proximity) (Fig. 7-1). When certain prospectivity recognition criteria are represented by input spatial data of discrete fields (e.g., derivative data obtained from geochemical data analysis; see Chapters 3 to 5), the re-classification operation also results in an evidential map of either binary or multi-class discrete geo-objects (e.g., ranges of derivative geochemical data) (Fig. 7-1). The scores for the evidential classes are then assigned in the attribute tables associated with individual evidential maps. The assignment of evidential class scores and evidential map weights and the integration of evidential maps vary depending on which modeling technique is applied (see further below).

In order to obtain a mineral prospectivity map, evidential maps are combined via certain computational functions considered by the modeler as appropriately representing the interactions or inter-relationships among the various geologic controls and surficial manifestations of mineral occurrence portrayed by the individual evidential maps, thus:

$$prospectivity\,map = f\,(evidential\ maps)\ .$$

There are different forms of the computational function *f*. In knowledge-driven modeling of mineral prospectivity, *f* can be either logical functions (e.g., AND and/or OR operators, etc.) or arithmetic functions. The applications of these functions, which are

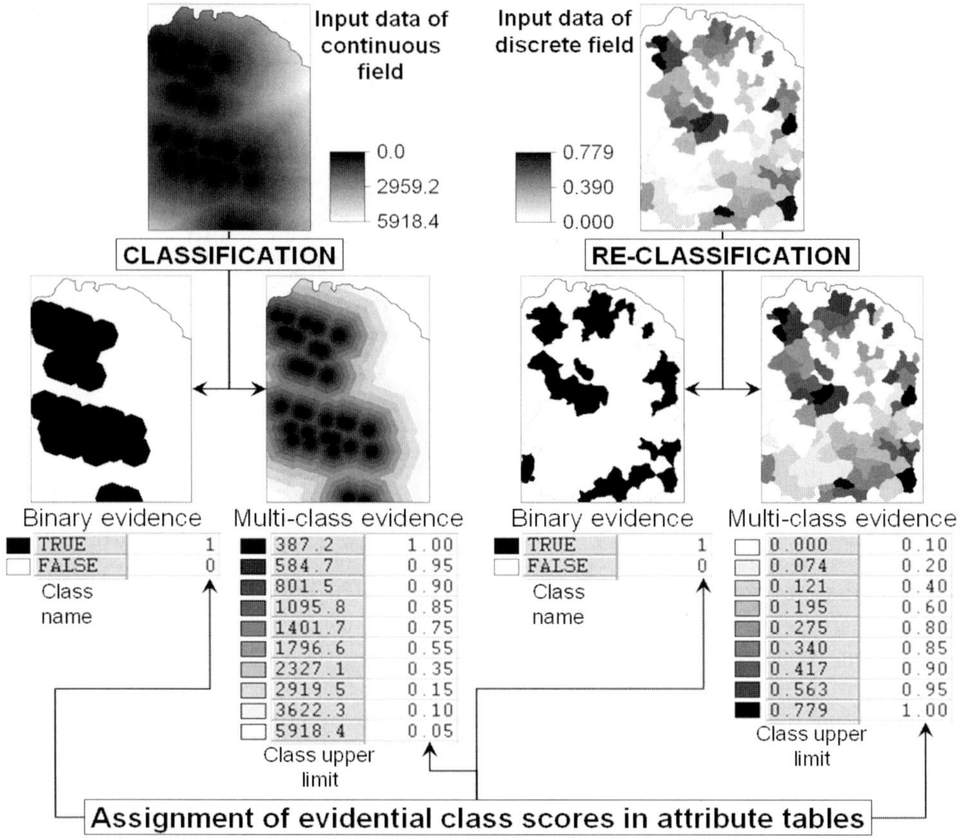

Fig. 7-1. Schematic GIS-based procedures for creating binary or multi-class evidential maps from input spatial data of continuous or discrete fields. Evidential class scores are assigned and stored in attribute tables associated with evidential maps.

sometimes called *aggregation functions*, in order to combine evidential maps result in a mineral prospectivity model.

The process of evaluating or cross-validating a mineral prospectivity map involves a number of steps (Fig 7-2). Firstly, a table histogram of descending classes of prospectivity values (`proscl`) is obtained in order to determine the number of unit cells or pixels per class (`npixcl`), cumulative number of class pixels (`npixclc`), total number of pixels (`npixclt`) and proportion of prospective areas (`proparea`). Values in the column `proparea` are derived by dividing values in the column `npixclc` with corresponding values in the column `npixclt`. Descending or decreasing prospectivity values are used in the table histogram because the objective is to study the predictive performances of increasing proportions of prospective areas (i.e., from high to low prospectivity values). In the creation of a table histogram of descending prospectivity

Prospectivity map

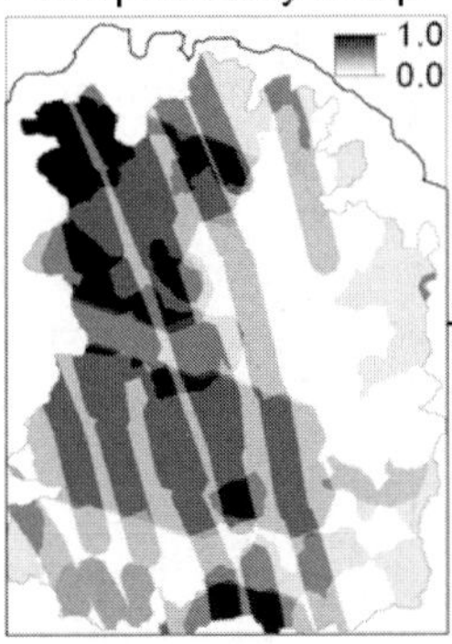

Cross-validation deposits

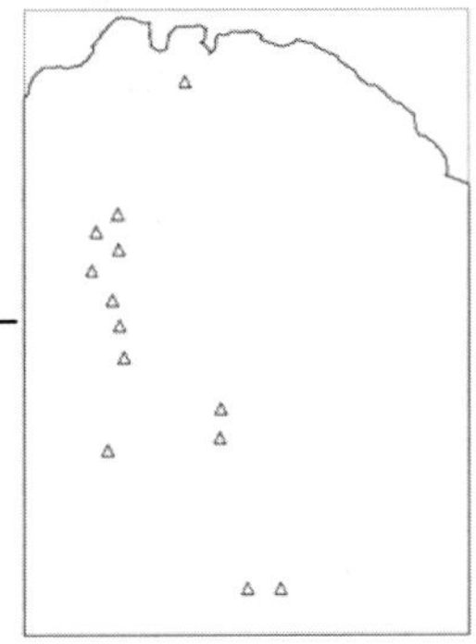

Cross operation

proscl	au	NPix
0.251 – 0.375	au	1
0.501 – 0.625	au	3
0.626 – 0.750	au	4
0.876 – 1.000	au	4

Output cross table

Table join operation

Table histogram of prospectivity values

proscl	npixcl	npixclc	npixclt	proparea	ndep	ndepc	ndept	propdep
0.876 – 1.000	100092	100092	1007704	0.099	4	4	12	0.333
0.751 – 0.875	5794	105886	1007704	0.105	0	4	12	0.333
0.626 – 0.750	182900	288786	1007704	0.287	4	8	12	0.667
0.501 – 0.625	109623	398409	1007704	0.395	3	11	12	0.917
0.376 – 0.500	33243	431652	1007704	0.428	0	11	12	0.917
0.251 – 0.375	200927	632579	1007704	0.628	1	12	12	1.000
0.126 – 0.250	125340	757919	1007704	0.752	0	12	12	1.000
0.001 – 0.125	48453	806372	1007704	0.800	0	12	12	1.000
0.000	201332	1007704	1007704	1.000	0	12	12	1.000

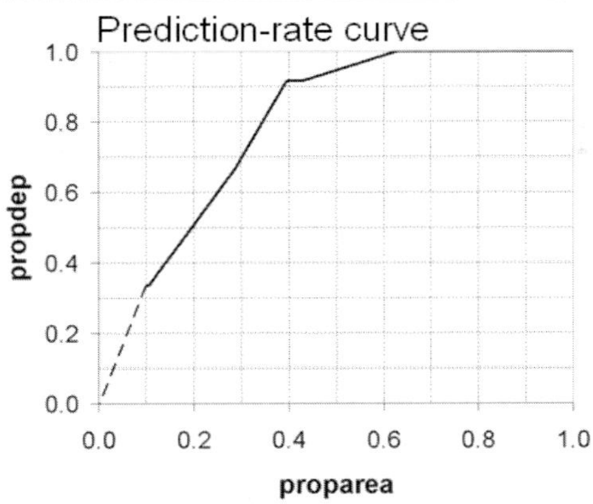

Fig. 7-2. Schematic GIS-based procedures for cross-validating the performance of a mineral prospectivity map. The objective is to create a prediction-rate curve, which allows mapping of optimal prospective areas. See text for further explanations, especially the variables in the tables.

values, it is instructive to start as much as possible with several narrow equal-area or equal-percentile classes and then to combine them, as necessary, into wider equal-area or equal-percentile classes in order to achieve a monotonically descending histogram. Secondly, the mineral prospectivity map is crossed with or intersected by the map of cross-validation deposits. The output cross table contains information about number (`Npix`) of cross-validation known deposits (`au`) contained in a class of prospectivity values (`proscl`). Values in the `Npix` column of the cross table are joined to a column (`ndep`) in the table histogram of prospectivity values for subsequent calculation of the cumulative number of deposits (`ndepc`), total number of deposits (`ndept`) and the proportion of deposits per prospectivity class (`propdep`). Values in the column `propdep` are then derived by dividing values in the column `ndepc` with corresponding values in the column `ndept`. Finally, a prediction-rate graph of `propdep` values versus `proparea` values is created.

The prediction-rate curve allows estimation of likelihood of mineral deposit discovery according to the prospectivity map. Any point along the prediction-rate curve represents a prediction of prospective zones with a corresponding number of delineated deposits and number of unit cells or pixels, so the ratio of the former to the latter is related to the degree of likelihood of mineral deposit occurrence (or discovery) in the delineated prospective zones. This means that, the higher the value of `propdep÷proparea` of predicted prospective zones (Fig. 7-2), the better is the prediction. It is, therefore, ideal to obtain a mineral prospectivity map with a steep prediction-rate curve. However, the performance of a mineral prospectivity map is influenced by (a) the quality of the input spatial data and (b) the way by which evidential maps are created (i.e., the number of evidential classes per evidential map) and integrated and, thus, by the modeling technique applied to create a mineral prospectivity map. We now turn to the concepts of individual modeling techniques that are applicable to knowledge-driven mapping of mineral prospectivity.

MODELING WITH BINARY EVIDENTIAL MAPS

In this type of modeling, evidential maps representing prospectivity recognition criteria contain only two classes of evidential scores – maximum evidential score and minimum evidential score (Figs. 7-1 and 7-3). Maximum evidential score is assigned to spatial data representing presence of indicative geological features and having optimum positive spatial association with mineral deposits of the type sought. Minimum evidential score is assigned to spatial data representing absence of indicative geological features and lacking positive spatial association with mineral deposits of the type sought. There are no intermediate evidential scores in modeling with binary evidential maps. This knowledge-based representation is usually inconsistent with real situations. For example, whilst certain mineral deposits may actually be associated with certain faults, the locations of some mineral deposit occurrences indicated in maps are usually, if not always, the surface projections of their positions in the subsurface 3D-space, whereas the locations of faults indicated in maps are more-or-less their 'true' surface locations (Fig. 7-3). Thus, for locations within the range of distances to such faults where positive spatial association with mineral deposits is optimal, the evidential scores should not be uniformly equal to the maximum evidential score. Likewise, for locations beyond the distance to faults with threshold optimum positive spatial association to the mineral

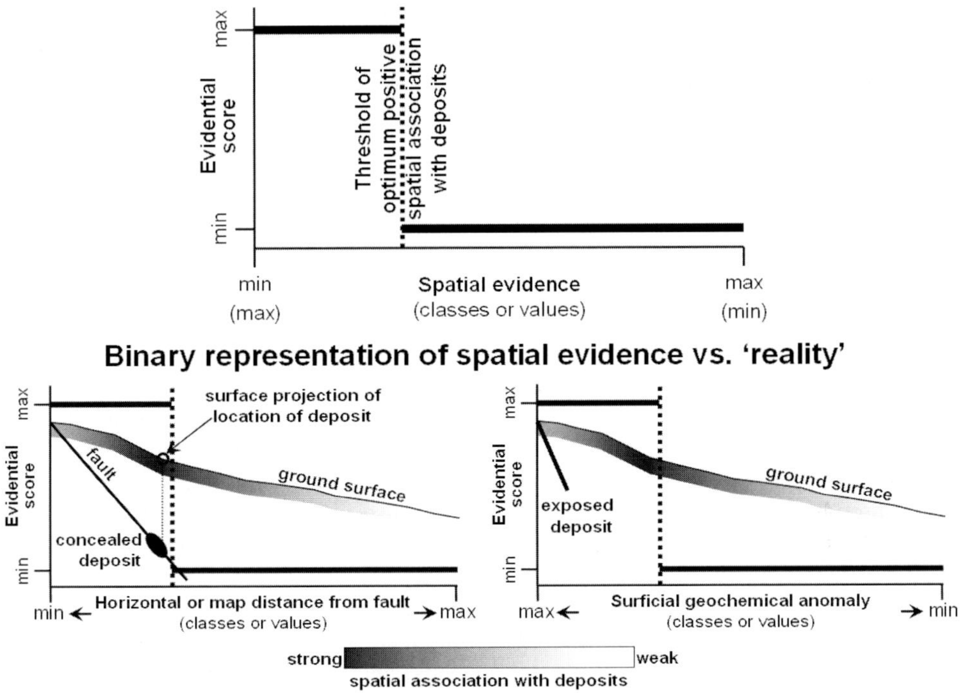

Fig. 7-3. Knowledge-based binary representation of spatial evidence of mineral prospectivity. Knowledge of spatial association between mineral deposits of the type sought and spatial data of indicative geological features is applied to assign binary evidential scores (upper part of the figure). If values or classes of values of spatial data have optimum positive spatial association with mineral deposits of the type sought, they are given a maximum evidential score of mineral prospectivity; otherwise, they are given a minimum evidential score of mineral prospectivity. These scores are discontinuous, meaning there are no intermediate evidential scores of mineral prospectivity. Binary representation of spatial evidence is inconsistent with real situations of spatial associations between mineral deposits and indicative geological features. For visual comparison, the graph in the upper part of the figure is overlaid on schematic cross-sections of ground conditions (lower part of the figure), but the y-axis of the graph does represent vertical scale of the cross-sections. See text for further explanation.

deposits, the evidential scores should not be uniformly equal to the minimum evidential score. The same line of reasoning can be accorded to the binary representation of evidence for presence of surficial geochemical anomalies, which may be significant albeit allochthonous (i.e., located not directly over the mineralised source) (Fig. 7-3). Note also that the graph of binary evidential scores versus data of spatial evidence is inconsistent with the shapes of the *D* curves (Figs. 6-9 to 6-12) in the analyses of spatial associations between epithermal Au deposit occurrences and individual sets of spatial evidential data in the case study area. Nevertheless, binary representation of evidence of mineral prospectivity is suitable in cases when the level of knowledge applied is lacking or incomplete and/or when the accuracy or resolution of available spatial data is poor. We now turn to the individual techniques for knowledge-based binary representation and integration of spatial evidence that can be used in order to derive a mineral prospectivity map.

Boolean logic modeling

Ample explanations of Boolean logic applications to geological studies can be found in Varnes (1974) and/or Robinove (1989), whilst examples of Boolean logic applications to mineral prospectivity mapping can be found in Bonham-Carter (1994), Thiart and De Wit (2000) and Harris et al. (2001b).

In the application of Boolean logic to mineral prospectivity mapping, attributes or classes of attributes of spatial data that meet the condition of a prospectivity recognition criterion are labelled TRUE (or given a class score of 1); otherwise, they are labelled FALSE (or given a class score of 0). Thus, a Boolean evidential map contains only class scores of 0 and 1. Every Boolean evidential map has equal weight in providing support to the proposition under examination; that is because in the concept of Boolean logic there is no such thing as, say, "2×truth". Thus, the class scores of 0 and 1 in a Boolean evidential map are only symbolic and non-numeric.

Boolean evidential maps are combined logically according to a network of steps, which reflect inferences about the inter-relationships of processes that control the occurrence of a geo-object (e.g., mineral deposits) and spatial features that indicate the presence of that geo-object. The logical steps of combining Boolean evidential maps are illustrated in a so-called inference network. Every step, whereby at least two evidential maps are combined, represents a hypothesis of inter-relationship between two sets of processes that control the occurrence of a geo-object (e.g., mineral deposits) and/or spatial features that indicate the presence of the geo-object. A Boolean inference network makes use of set operators such as AND and OR. The AND (or intersection) operator is used if it is considered that at least two sets of spatial evidence must be present together in order to provide support to the proposition under examination. The OR (or union) operator is used if it is considered that either one of at least two sets of spatial evidence is sufficient to provide support to the proposition under examination. Boolean logic modeling is not exclusive to using only the AND and OR operators, although the other Boolean operators (e.g., NOT, XOR, etc.) are not commonly applied in GIS-based knowledge-driven mineral prospectivity mapping. The output of combining evidential maps via Boolean logic modeling is a map with two classes, one class represent locations where all or most of the prospectivity recognition criteria are satisfied, whilst the other class represents locations where none of the prospectivity recognition criteria is satisfied.

For the case study area, Boolean evidential maps are prepared from the individual spatial data sets according to the prospectivity recognition criteria given above. Fig. 7-4 shows the inference network for combining the evidential maps based on the given spatial data sets. The Boolean evidential maps of proximity to NNW-trending faults/fractures and proximity to NW-trending faults/fractures are first combined by

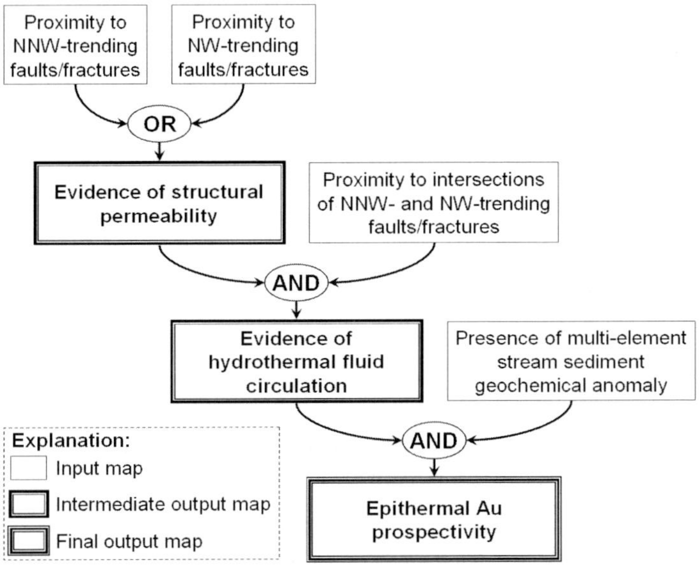

Fig. 7-4. An inference network for combining input Boolean evidential maps for modeling of epithermal Au prospectivity in the Aroroy district (Philippines).

using the OR operator in order to represent a hypothesis of structural permeability. The OR operator is used because it is plausible that at certain locations either set of faults/fractures can be a predominant control on structural permeability required for the plumbing system in epithermal mineralisation. The intermediate Boolean evidential map of structural permeability is then combined with the Boolean evidential map of proximity to intersections of NNW- and NW-trending faults by using the AND operator. The latter map is considered a proxy evidence for heat source control on hydrothermal fluid circulation (see Chapter 6). Thus, the AND operator is used because heat source and structural permeability controls are both essential controls on hydrothermal fluid circulation. Finally, the evidential map representing hydrothermal fluid circulation is combined with the Boolean evidential map of multi-element stream sediment geochemical anomalies by using the AND operator. The AND operator is used because the presence of indications of hydrothermal fluid circulation (e.g., hydrothermal altered rocks) does not necessarily mean the presence of mineralisation and the presence of stream sediment anomalies does not necessarily indicate the presence of mineral deposit occurrences, so that the presence of both types of evidence would be more indicative of the presence of a mineral deposit occurrence in the vicinity.

The final Boolean output map of epithermal Au prospectivity (Fig. 7-5A) shows strong influence of the multi-element geochemical anomaly evidence, which is a consequence of using the AND operator in the final step of the inference network (Fig. 7-4). The pattern of prospective areas shown in Fig. 7-5A is similar to the pattern of highly significant anomalies shown in Fig. 5-14; the latter map is derived as a product

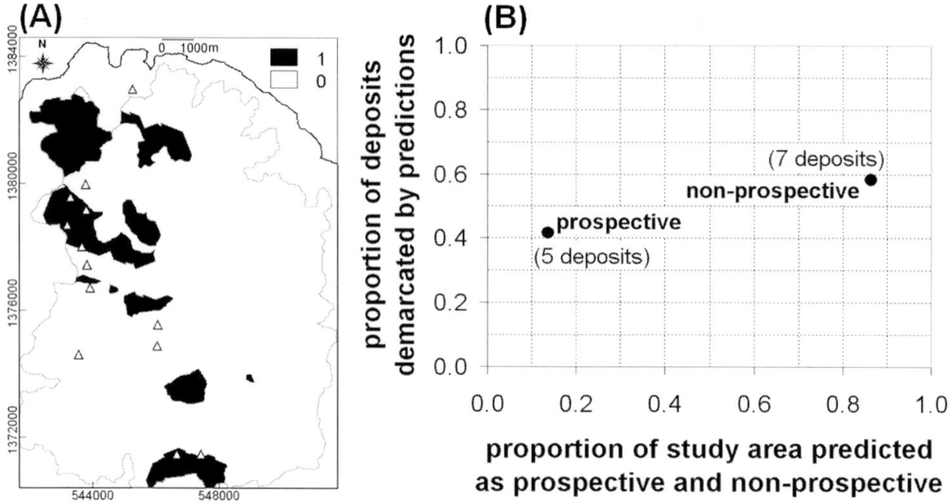

Fig. 7-5. (**A**) An epithermal Au prospectivity map obtained by application of an inference network (Fig. 7-4) for Boolean logic modeling, Aroroy district (Philippines). 1 = prospective zones; 0 = non-prospective zones. Triangles are locations of known epithermal Au deposits; whilst polygon outlined in grey is area of stream sediment sample catchment basins (see Fig. 4-11). (**B**) Plots of proportion of deposits demarcated by the predictions versus proportion of study area predicted as prospective and non-prospective. The numbers of cross-validation deposits delineated in prospective and non-prospective zones are indicated in parentheses.

(i.e., by multiplication) of the same multi-element geochemical anomaly evidence and fault/fracture density. Thus, the Boolean AND operator has a multiplicative net effect. In contrast, the Boolean OR operator has an additive net effect, which is unsuitable is many cases, such as in combining the Boolean evidential map of hydrothermal fluid circulation and the Boolean evidential map of multi-element stream sediment geochemical anomalies. Doing so results in a large prospective area with a low prediction-rate. Note, however, that the application of the Boolean AND operator returns an output value only for locations with available data in both input evidential maps. Thus, for the case study area, locations with missing stream sediment geochemical data do not take on predicted prospectivity values by application of Boolean logic modeling (Fig. 7-5A).

Cross-validation of a Boolean mineral prospectivity map results in a plot with only two points, one representing completely prospective areas and the other representing completely non-prospective areas, which should not be connected to form a prediction-rate curve (Fig. 7-5B). In the Boolean epithermal Au prospectivity map of the case study area, prospective zones contain about 42% of the cross-validation deposits and occupy about 14% of the study area, whereas non-prospective zones contain about 58% of the cross-validation deposits and occupy about 86% of the study area. Estimates of ratios of the proportion of cross-validation deposits delineated to the corresponding proportion of the prospective or non-prospective areas suggest that there is at least four times higher likelihood of epithermal Au deposit occurrence in the predicted prospective areas than in the predicted non-prospective areas. The significance of this performance of the Boolean epithermal Au prospectivity map of the case study area can be appreciated by comparing it with the performances of other mineral prospectivity maps derived via the other modeling techniques.

Binary index overlay modeling

In binary index overlay modeling, attributes or classes of attributes of spatial data that satisfy a prospectivity recognition criterion are assigned a class score of 1; otherwise, they are assigned a class score of 0. Therefore, a binary index map is similar to a Boolean map, except that the values in the former are both symbolic and numeric. So, a binary index map is amenable to arithmetic operations. Consequently, each binary evidential map B_i (i=1,2,…,n) can be given (i.e., multiplied with) a numerical weight W_i based on 'expert' judgment of the relative importance of a set of indicative geological features represented by an evidential map with respect to the proposition under examination.

The weighted binary evidential maps are combined using the following equation, which calculates an average score, S, for each location (cf. Bonham-Carter, 1994):

$$S=\frac{\sum_{i}^{n} W_i B_i}{\sum_{i}^{n} W_i} \tag{7.1}$$

where W_i is weight of each B_i (i=1,2,…,n) binary evidential map. In the output map S, each location or pixel takes on values ranging from 0 (i.e., completely non-prospective) to 1 (i.e., completely prospective). So, although the input maps only have two classes, the output map can have intermediate prospectivity values, which is more intuitive than the output in Boolean logic modeling. Examples of mapping mineral prospectivity via binary index overlay modeling can be found in Bonham-Carter (1994), Carranza et al. (1999), Thiart and De Wit (2000) and Carranza (2002).

Assignment of meaningful weights to individual evidential maps is a highly subjective exercise and it may involve a trial-and-error procedure, even in the case when 'real expert' knowledge is available particularly from different experts. The difficulty lies in deciding objectively and simultaneously how much more important or how much less important is one evidential map compared to every other evidential map. This difficulty may be overcome by making pairwise comparisons among the evidential maps in the context of a decision making process known as the analytical hierarchy process (AHP). The concept of the AHP was developed by Saaty (1977, 1980, 1994) for pairwise analysis of priorities in multi-criteria decision making. It aims to derive a hierarchy of criteria based on their pairwise relative importance with respect to the objective of a decision making process (e.g., evaluation of the mineral prospectivity proposition). Most GIS-based applications of the AHP concern land-use allocations (e.g.,

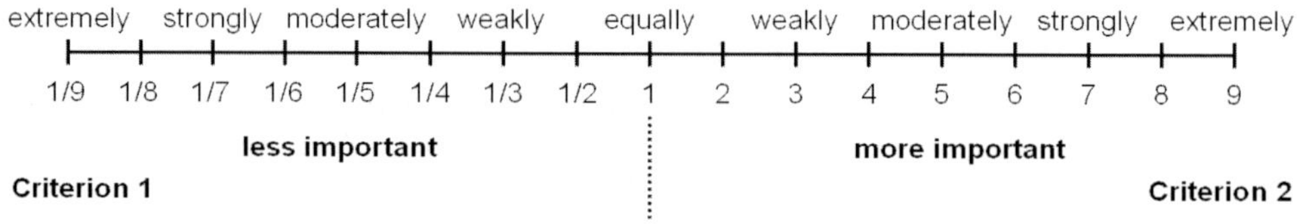

Fig. 7-6. Continuous rating scale for pairwise comparison of relative importance of one criterion versus another criterion with respect to a proposition (adapted from Saaty, 1977).

Eastman et al., 1995). De Araújo and Macedo (2002), Moreira et al. (2003) and Hosseinali and Alesheikh (2008) provide case applications of the AHP to derive criteria or evidential map weights for mineral prospectivity mapping. The application of the AHP is useful not only for binary index overlay modeling but also for multi-class index overlay modeling (see further below) and fuzzy logic modeling.

The method of deriving criteria weights via the AHP involves pairwise comparisons of criteria according to their relative importance with respect to a proposition. The method adopts a 9-point continuous pairwise rating scale for judging whether Criterion X is less important or more important than Criterion Y (Fig. 7-6). The relative importance rating is read from the either the left or right extremity of the scale, depending (a) on which extremity each of two criteria is positioned and (b) on which criterion is compared to the other. In addition, the importance rating of one criterion is always the inverse of the importance rating of the other criterion. For example, if Criterion 1 and Criterion 2 are positioned at the left extremity and right extremity, respectively, and if Criterion 1 is 'weakly' less important compared to Criterion 2, then the relative importance rating of Criterion 1 compared to Criterion 2 is 1/3. It follows that the relative importance rating of Criterion 2 compared to Criterion 1 is 3. Reversing the positions of Criterion 1 and Criterion 2 does not change their relative importance ratings. Thus, if Criterion 1 and Criterion 2 are now positioned at the right extremity and left extremity, respectively, and if Criterion 2 is 'weakly' more important compared Criterion 1, then the relative importance rating of Criterion 2 compared to Criterion 1 is 3. It follows that the relative importance rating of Criterion 1 compared to Criterion 2 is 1/3. The pairwise relative importance ratings of all possible pairs of criteria are then entered into a pairwise comparison matrix.

For the epithermal Au prospectivity recognition criteria in our case study area, the pairwise comparisons of relative importance of each criterion (Table 7-I) are based on the results of the spatial analyses in Chapter 6. Proximity to NNW-trending faults/fractures is considered to be between moderately and strongly more important than proximity to intersections of NNW- and NW-trending faults/fractures; thus, a rating of 6 is given to the former. This is because, in the case study area, the NNW-trending faults/fractures seem to have more influence in the formation of dilational jogs (Fig. 6-16), which are known to be favourable sites for mineralisation. Proximity to NNW-trending faults/fractures is considered moderately more important than proximity to NW-trending faults/fractures; thus a rating of 5 is given to the former. This is because, in the

TABLE 7-I

Example of a matrix of pairwise ratings (see Fig. 7-6) of relative importance of recognition criteria for epithermal Au prospectivity in Aroroy district (Philippines). Values in bold are used for demonstration in Table 7-II, whilst values in bold italics are used for demonstrations in Tables 7-II and 7-III.

Criteria[1]	NNW	FI	NW	ANOMALY
NNW	***1***	***5***	***6***	***1/2***
FI	***1/5***	1	5	1/2
NW	***1/6***	1/5	1	1/2
ANOMALY	***2***	2	2	1
Sum[2]	**3.37**	**8.2**	**14**	**2.5**

[1]Criteria: NNW = proximity to NNW-trending faults/fractures; NW = proximity to NW-trending faults/fractures; FI = proximity to intersections of NNW- and NW-trending faults/fractures: ANOMALY = integrated PC2 and PC3 scores obtained from the catchment basin analysis of stream sediment geochemical data (see Chapter 3). [2]Sum of ratings down columns.

case study area, the known epithermal Au deposit occurrences are more strongly spatially associated with NNW-trending faults/fractures than with NW-trending faults/fractures. Proximity to intersections of NNW- and NW-trending faults/fractures is considered moderately more important that proximity to NW-trending faults/fractures; thus a rating of 5 is given to the former. This is because dilational jogs in the case study area, which generally coincide with intersections of NNW- and NW-trending faults/fractures, seem to be more associated with NNW-trending faults/fractures rather than with NW-trending faults/fractures. The catchment basin anomalies of stream sediment geochemical data are considered to be between moderately more important than and equally important as proximity to individual sets of structures; thus, a rating of 2 is given to the former.

When a matrix of pairwise importance ratings for all possible pairs of criteria is obtained, the next step is to estimate the eigenvectors of the matrix (cf. Boroushaki and Malczewski, 2008). Good approximations of the eigenvectors of the pairwise comparison matrix can be achieved by normalising the pairwise ratings down each column and then by calculating criterion weight as the average of the normalised pairwise ratings across each row (Tables 7-I and 7-II). For example, in column NNW in Table 7-I, the sum of the pairwise ratings is 3.37. By dividing each pairwise rating in that column by 3.37, we obtain the normalised pairwise ratings for the same column (Table 7-II). This procedure is repeated for all the columns in the matrix. Then, the fractional weight of each criterion is obtained by averaging the normalised pairwise ratings across a row (Table 7-II). The sum of the fractional criteria weights is approximately equal to 1 (Table 7-II), reflecting approximately 100% of the explained variance of the values in the matrix.

The fractional criteria weights obtained can then be used in equation (7.1). Alternatively, instead of using the fractional criteria weights, they can be converted into

TABLE 7-II

Example of calculation of weights of recognition criteria for epithermal Au prospectivity in Aroroy district (Philippines). Values in bold and bold italics are taken from Table 7-I. Underlined values are used for demonstration in Table 7-III.

Criteria[1]	NNW	FI	NW	ANOMALY	Fractional weight[2] (W_f)	Integer weight[3] (W_i)
NNW	***1*** **÷ 3.37** = 0.30	**5 ÷ 8.2** = 0.61	**6 ÷ 14** = 0.43	***1/2*** **÷ 2.5** = 0.2	0.39	4
FI	***1/5*** **÷ 3.37** = 0.06	0.36	0.12	0.2	0.19	2
NW	***1/6*** **÷ 3.37** = 0.05	0.07	0.02	0.2	0.09	1
ANOMALY	**2 ÷ 3.37** = 0.59	0.14	0.24	0.4	0.34	4

[1]See footnotes to Table 7-I. [2]Example: fractional weight$_{NNW}$ = (0.30+0.61+0.43+0.2) ÷ 4 = 0.39. [3]$W_i = W_f \div [\min(W_f)]$.

integers or whole numbers by dividing each of the fractional criteria weights by the smallest fractional criterion weight (Table 7-II). The integer criteria weights are more intuitive than the fractional criteria weights. Before using either the fractional or integer criteria weights obtained via the AHP, it is important to determine if the pairwise rating matrix and thus the derived weights are consistent, which also reflects the consistency of the 'expert' judgment applied in assigning the pairwise relative importance ratings.

A matrix is consistent if every value across each row is a multiple of every other value in the other rows. This is not the case of the matrix in Table 7-I, meaning that there is some degree of inconsistency among the pairwise ratings in the matrix. In addition, pairwise ratings are consistent if they are transitive. This means, for example from Table 7-I, that because the weight for NNW is 5× the weight for FI (or, NNW=5×FI) and the weight for NNW is 6× the weight of NW (or, NNW=6×NW), then the weight for FI should be 6/5× but not 5× the weight for NW (or, FI=6/5×NW≠5×NW). However, one may argue that transitive pairwise ratings are not intuitively representative of knowledge or judgment of inter-play of geological processes involved in a complex phenomenon such as mineralisation. Nevertheless, when applying the AHP, it is imperative to quantify and determine whether inconsistencies in a pairwise comparison matrix are within acceptable limits.

A $n \times n$ matrix (n = number of factors or criteria), such as a pairwise comparison matrix, is consistent if it has one eigenvalue with a value equal to n; otherwise it has at most n eigenvalues with values varying around n (Saaty, 1977). The inconsistency of a matrix is then related to how much the mean of eigenvalues (λ) of such matrix deviates from n. According to Saaty (1977), the eigenvalues of the pairwise comparison matrix may be estimated from the pairwise importance ratings (Table 7-I) and the estimates of the eigenvectors or criteria weights (Table 7-II). Approximations of the eigenvalues can be referred to as the consistency vectors (CV) of the individual criteria (Table 7-III). The

TABLE 7-III

Estimation of consistency ratio (CR) for judgment of consistency of pairwise ratings of recognition criteria for epithermal Au prospectivity in Aroroy district (Philippines). Values in bold italics are taken from Table 7-I. Underlined values are taken from Table 7-III.

Criteria[1]	NNW	FI	NW	ANOMALY	Sum[2]	Consistency vector (CV)[3]
NNW	0.39 × ***1*** = 0.39	0.19 × ***5*** = 0.95	0.09 × ***6*** = 0.54	0.34 × ***1/2*** = 0.07	1.95	5.00
FI	0.39 × ***1/5*** = 0.08	0.19	0.45	0.07	0.79	4.16
NW	0.39 × ***1/6*** = 0.07	0.07	0.09	0.07	0.27	3.00
ANOMALY	0.39 × ***2*** = 0.78	0.38	0.18	0.14	1.48	4.35

λ = mean of CV = 16.51/4 = 4.13

Consistency Index (CI) = $(\lambda - n_{criteria})/(n_{criteria} - 1)$ = (4.13-4)/(4-1) = 0.043

Consistency Ratio (CR) = CI/RI[4] = 0.043/0.9 = 0.048

[1]See footnotes to Table 7-I. [2]Example: Sum_{NNW} = 0.39 + 0.95 + 0.54 + 0.07 = 1.95. [3]Example: $CV_{NNW} = Sum_{NNW} \div Weight_{NNW}$ = 1.95 ÷ 0.39 = 5.00. [4]RI = Random inconsistency index. See Table 7-IV for RI values corresponding to $n_{criteria}$.

mean of the CVs is an estimate of λ. The deviation of λ from n (i.e., maximum eigenvalue) in a situation of a consistent $n \times n$ matrix is used to define a consistency index (CI), which in turn is used to define a consistency ratio (CR) (Table 7-III). The CR is obtained by comparing CI to a random inconsistency index (RI) of a $n \times n$ matrix of randomly generated pairwise importance ratings (Table 7-IV). If the value of CR is greater than 0.1 (i.e., CI > 10% of RI), then the inconsistencies in a pairwise comparison matrix are unacceptable so that a re-evaluation of the pairwise relative importance ratings of the criteria is in order. The results shown in Table 7-III indicate that inconsistencies in the pairwise comparison matrix (Table 7-I) for the epithermal Au prospectivity recognition criteria in the case study area are within acceptable limits, meaning that the values of W_i (Table 7-II) can be relied upon in deriving a meaningful epithermal Au prospectivity map via application of the binary index overlay modeling.

The epithermal Au prospectivity map of the case study area derived via binary index overlay modeling and shown in Fig. 7-6A was obtained by assigning the following weights (values in bold) to the individual evidential maps:

- **4**: presence of multi-element stream sediment geochemical anomalies.
- **4**: proximity to (within 0.35 km of) NNW-trending faults/fractures.
- **2**: proximity to (within 1 km of) intersections of NNW- and NW-trending faults/fractures.
- **1**: proximity to (within 0.9 km of) NW-trending faults/fractures.

TABLE 7-IV

Some random inconsistency indices (RI) generated by Saaty (1977) for a large number of $n \times n$ matrices of randomly generated pairwise comparison ratings

n	1	2	3	4	5	6	7	8	9	10
RI	0	0	0.58	0.9	1.12	1.24	1.32	1.41	1.45	1.49

The heavier weights on proximity to NNW-trending faults/fractures and stream sediment multi-element anomalies are evident in the output map of the binary index overlay modeling (Fig. 7-7A), although the pattern of the predicted most prospective areas (in black) is strongly similar to the pattern of the prospective areas predicted by application of Boolean logic modeling (Fig. 7-5A). This indicates that the pairwise ratings given to the stream sediment anomalies with respect to the individual structural criteria are consistent with the importance given to the former in the way the evidential maps are combined via Boolean logic modeling (see inference network in Fig. 7-4).

Like the application of Boolean logic modeling, the application of binary index overlay modeling returns an output value only for locations with available data in all input evidential maps. Thus, for the case study area, locations with missing stream sediment geochemical data do not take on prospectivity values by application of binary index overlay modeling (Fig. 7-7A). So, one of the 13 known epithermal Au deposit occurrence is not considered in the cross-validation of the prospectivity map.

The prospectivity map derived via binary index overlay modeling is better than the prospectivity map derived via Boolean logic modeling because the former delineates all cross-validation deposits in about 75% of the study area (Fig. 7-7B) whilst the latter delineates all cross-validation deposits in at least 85% of the case study area (Fig. 7-5B). However, both prospectivity maps are similar in terms of prediction-rate (roughly 40%) of a prospective area equal in size (about 14% of the case study area) to that predicted by application of the Boolean logic modeling.

Calibration of predictive modeling with binary evidential maps

Because of the very limited range of evidential class scores that can be assigned to classes in a Boolean or binary map, probably the best method calibration is to perform a number of changes in the threshold values specified by the conceptual model of mineral prospectivity. This results in changing the areas of evidential classes in a Boolean or binary map. In Boolean logic modeling, another method for calibration is to modify the inference network, whilst in binary index overlay modeling another method for calibration is to modify the evidential map weights. Because any set of changes would, in turn, correspond to a change in prediction-rate of prospective areas, the objective of any strategy for predictive model calibration is to find a set of modifications in the modeling that corresponds with prospective areas having the highest prediction-rate. However, in any strategy for calibration of predictive modeling using binary evidential

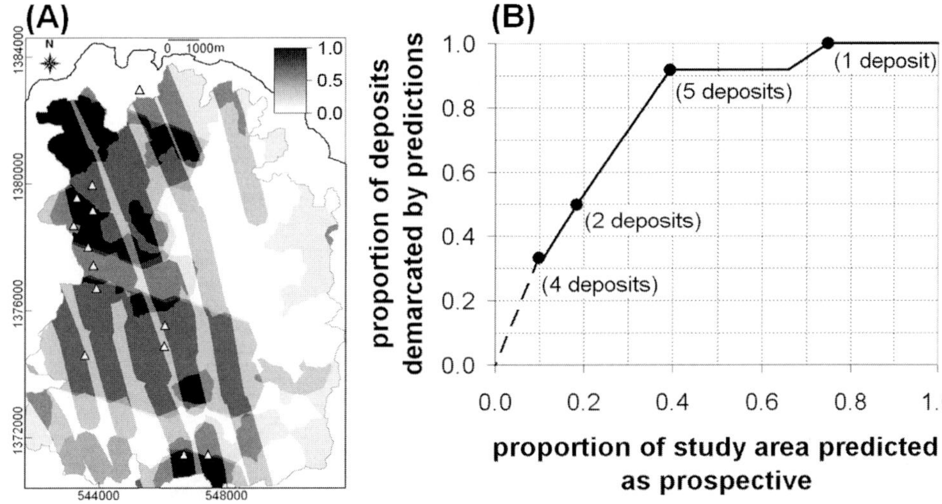

Fig. 7-7. (**A**) An epithermal Au prospectivity map obtained via binary index overlay modeling, Aroroy district (Philippines) (see text for explanations about the input evidential maps used). Triangles are locations of known epithermal Au deposits; whilst polygon outlined in grey is area of stream sediment sample catchment basins (see Fig. 4-11). (**B**) Prediction-rate curve of proportion of deposits demarcated by the predictions versus proportion of study area predicted as prospective. The dots along the prediction-rate curve represent classes of prospectivity values that correspond spatially with a number of cross-validation deposits (indicated in parentheses).

maps, one must consider the significance of any modifications made in the modeling in terms of the geologic controls on mineral occurrence and/or spatial features that indicate the presence of mineral deposit occurrence.

It is clear from the above examples that binary index overlay modeling is more advantageous than Boolean logic modeling, especially in terms of producing a realistic multi-class output instead of a synthetic binary output. We now turn to predictive modeling techniques, whereby evidential maps can take on more than two classes of evidence of mineral prospectivity. These 'multi-class' modeling techniques provide more flexibility in assignment of evidential class scores than the 'two-class' techniques described so far.

MODELING WITH MULTI-CLASS EVIDENTIAL MAPS

In this type of modeling, evidential maps representing prospectivity recognition criteria contain more than two classes (Figs. 7-1 and 7-8). Individual classes or ranges of values of evidence in an evidential map are hypothesised to have different degrees of importance relative to the proposition under consideration and therefore are given different scores depending on the concept of the spatial data modeling technique that is applied. Highest evidential scores are assigned to classes of spatial data portraying presence of indicative geological features and varying about the threshold spatial data

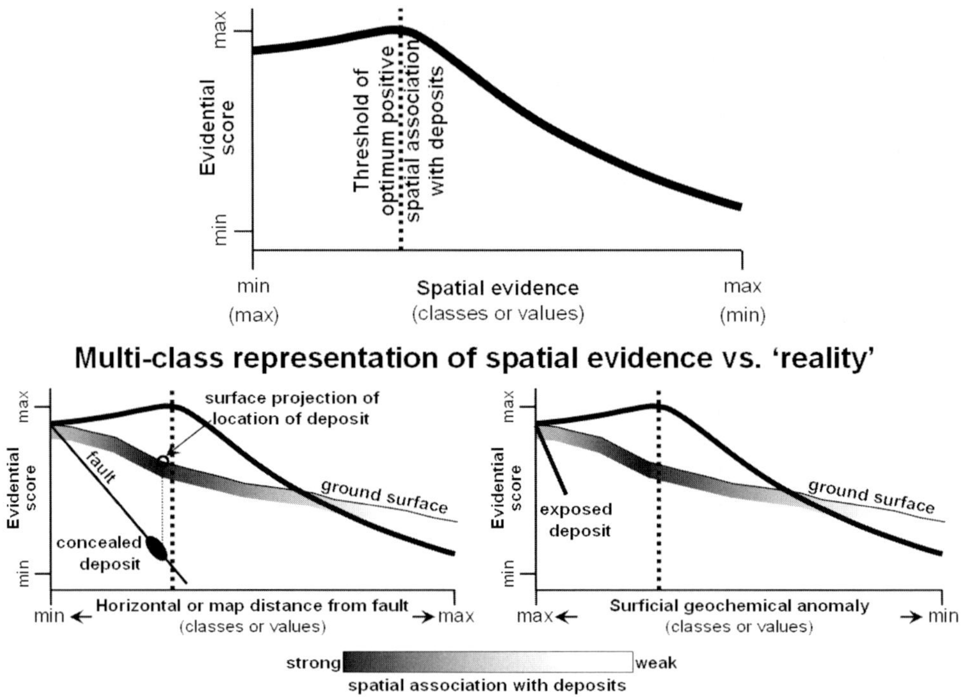

Fig. 7-8. Knowledge-based multi-class representation of spatial evidence of mineral prospectivity. Knowledge of spatial association between mineral deposits of the type sought and spatial data of indicative geological features is applied to assign evidential scores (upper part of the figure). If classes or values of spatial data vary about the threshold spatial data of optimum positive spatial association with mineral deposits of the type sought, they are given close to maximum evidential scores of mineral prospectivity; otherwise, they are given scores decreasing to the minimum evidential score of mineral prospectivity. These scores are continuous (i.e., they vary from minimum to maximum). Multi-class representation of spatial evidence is more-or-less consistent with real situations of spatial associations between mineral deposits and indicative geological features. For visual comparison, the graph in the upper part of the figure is overlaid on schematic cross-sections of ground conditions (lower part of the figure), but the y-axis of the graph does represent vertical scale of the cross-sections. See text for further explanation.

value representing optimum positive spatial association with mineral deposits of the type sought. Reduced and lowest evidential scores are assigned to spatial data representing increasing degrees of absence of indicative geological features and increased lack of positive spatial association with mineral deposits of the type sought. So, there is a continuous range of minimum-maximum evidential scores in modeling with multi-class evidential maps. This knowledge-based representation is more-or-less consistent with real situations. For example, whilst certain mineral deposits may actually be associated with certain faults, the locations of some mineral deposits indicated in maps are usually, if not always, the surface projections of their positions in the subsurface 3D-space, whereas the locations of faults indicated in maps are more-or-less their 'true' surface locations. Thus, for locations within the range of distances to certain faults where the positive spatial association with mineral deposits is optimal, the evidential scores assigned are highest but these scores decrease slowly from maximum at the threshold distance to a lower score at the minimum distance (Fig. 7-8). For locations beyond the threshold distance to certain faults, where the positive spatial association with mineral deposits is non-optimal, the evidential scores assigned decrease rapidly from maximum at the threshold distance to the minimum evidential score at the maximum distance. The same line of reasoning can be accorded to multi-class representation of evidence for the presence of surficial geochemical anomalies, which may be significant albeit allochthonous (i.e., located not directly over the mineralised source) (Fig. 7-8). Note, therefore, that the graph of multi-class evidential scores versus data of spatial evidence is more-or-less consistent with the shapes of the *D* curves (Figs. 6-9 to 6-12) in the analyses of spatial associations between epithermal Au deposit occurrences and individual sets of spatial evidential data in the case study area. Thus, multi-class representation of evidence of mineral prospectivity is suitable in cases where the level of knowledge applied is seemingly complete and/or when the accuracy or resolution of available spatial data is satisfactory. We now turn to the individual techniques for knowledge-based multi-class representation and integration of spatial evidence that can be used in order to derive a mineral prospectivity map.

Multi-class index overlay modeling

This is an extension of binary index overlay modeling. Each of the jth classes of the ith evidential map is assigned a score S_{ij} according to their relevance to the proposition under examination. The class scores assigned can be positive integers or positive real values. There is no restriction on the range of class scores, except that the range of class scores in every evidential map must be compatible (i.e., the same minimum and maximum values). This means that it is impractical to control the relative importance of an evidential map in terms of the proposition under consideration by making the range of its class scores different from the range of class scores in another evidential map. The relative importance of an evidential map compared to each of the other evidential maps is controlled by assignment of weights W_i, which are usually positive integers. Weighted evidential maps are then combined using the following equation, which calculates an average weighted score ($\overline{S}$) for each location (Bonham-Carter, 1994):

$$\overline{S}=\frac{\sum_{i}^{n} S_{ij}W_i}{\sum_{i}^{n} W_i} \tag{7.2}$$

The output value $\overline{S}$ for each location is the sum of the products of S_{ij} and W_i in each evidential map divided by the sum of W_i for each evidential map. Recent examples of

TABLE 7-V

Examples of scores assigned to evidential classes in individual evidential maps portraying the recognition criteria for epithermal Au prospectivity, Aroroy district (Philippines). Range of values in bold include the threshold value of spatial data of optimum positive spatial associations with epithermal Au deposits in the case study area.

Proximity to NNW[1]		Proximity to FI[2]		Proximity to NW[3]		ANOMALY[4]	
Range (km)	Score	Range (km)	Score	Range (km)	Score	Range	Score
0.00 – 0.08	8.0	0.00 – 0.39	8.0	0.00 – 0.18	8.0	0.00 – 0.06	1.0
0.08 – 0.15	8.5	0.39 – 0.58	9.0	0.18 – 0.36	8.5	0.06 – 0.10	2.0
0.15 – 0.23	9.0	0.58 – 0.80	9.5	0.36 – 0.54	9.0	0.10 – 0.16	4.0
0.23 – 0.32	9.5	**0.80 – 1.09**	10.0	0.54 – 0.75	9.5	0.16 – 0.25	6.0
0.32 – 0.41	10.0	1.09 – 1.40	9.0	**0.75 – 1.01**	10.0	0.25 – 0.29	8.0
0.41 – 0.52	9.0	1.40 – 1.80	7.0	1.01 – 1.29	8.0	**0.29 – 0.37**	10.0
0.52 – 0.71	7.0	1.80 – 2.32	5.0	1.29 – 1.65	6.0	0.37 – 0.49	9.5
0.71 – 1.06	5.0	2.32 – 2.92	3.0	1.65 – 2.24	4.0	0.49 – 0.78	9.0
1.06 – 1.73	3.0	2.92 – 3.62	2.0	2.24 – 3.02	2.0		
1.73 – 3.55	1.0	3.62 – 5.92	1.0	3.02 -5.32	1.0		

[1]NNW-trending faults/fractures. [2]Intersections of NNW- and NW-trending faults/fractures. [3]NW-trending faults/fractures. [4]Integrated PC2 and PC3 scores obtained from the catchment basin analysis of stream sediment geochemical data (Chapter 3).

multi-class index overlay modeling applied to mineral prospectivity mapping can be found in Harris et al. (2001b), Chico-Olmo et al. (2002), De Araújo and Macedo (2002) and Billa et al. (2004).

Examples of scores assigned to classes in individual evidential maps representing the recognition criteria for epithermal Au prospectivity in the case study area are listed in Table 7-V. The evidential maps of proximity to structures each contain 10 classes obtained by classifying the range of map distances to a set of structures into 10-percentile intervals. The scores assigned to proximal classes within the range of distance of optimum positive spatial association with known epithermal Au deposits are substantially higher than the scores assigned to distal classes. The evidential map of multi-element stream sediment geochemical anomalies contains only eight classes because the lowest values represent 30 percentile of the data and the remaining values are classified into 10-percentile intervals. The scores assigned to the upper 30 percentile classes, which have optimum positive spatial association with known epithermal Au deposits, are substantially higher than the scores assigned to the lower 70 percentile classes. The individual evidential maps are assigned the same weights (W_i) as those used in the binary index overlay modeling (see Table 7-II).

The prospectivity map derived via multi-class index overlay modeling (Fig. 7-9A) shows the influence of the catchment basin stream sediment multi-element anomalies and NNW-trending linear patterns indicating the influence of proximity to the NNW-trending faults. These indicate consistency of the slightly higher pairwise importance rating given to the stream sediment anomalies compared to the proximity to NNW-

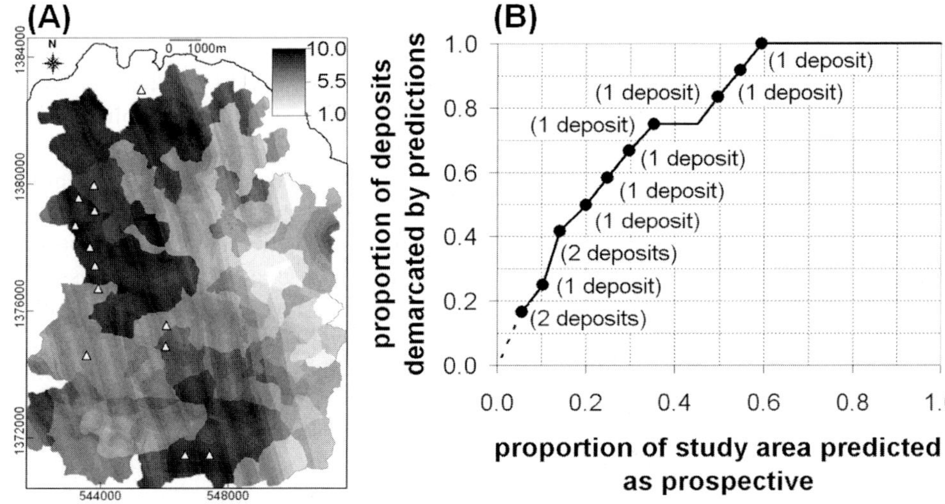

Fig. 7-9. (**A**) An epithermal Au prospectivity map obtained via multi-class index overlay modeling, Aroroy district (Philippines) (see text for explanations about the input evidential maps used). Triangles are locations of known epithermal Au deposits; whilst polygon outlined in grey is area of stream sediment sample catchment basins (see Fig. 4-11). (**B**) Prediction-rate curve of proportion of deposits demarcated by the predictions versus proportion of study area predicted as prospective. The dots along the prediction-rate curve represent classes of prospectivity values that correspond spatially with a number of cross-validation deposits (indicated in parentheses).

trending faults/fractures and the higher pairwise importance rating given to the proximity to NNW-trending faults/fractures compared to the proximity to the other structures (see Table 7-I). Like the application of Boolean logic modeling and binary index overlay modeling, the application of the multi-class index overlay modeling returns an output value only for locations with available data in all input evidential maps. Thus, one of the 13 known epithermal Au deposit occurrence is not considered in the cross-validation of the prospectivity map.

The prospectivity map derived via multi-class index overlay modeling is better than the prospectivity map derived via binary index overlay modeling because the former delineates all cross-validation deposits in 60% of the study area (Fig. 7-9B) whilst the latter delineates all cross-validation deposits in about 75% of the case study area (Fig. 7-7B). However, the prospectivity map derived via multi-class index overlay modeling is similar to the prospectivity maps derived via Boolean logic modeling (Fig. 7-5B) and binary index overlay modeling (Fig. 7-7B) in terms of prediction-rate (roughly 40%) of a prospective area of about 14% of the case study area.

The prospectivity maps derived via application of binary and multi-class index overlay modeling are different mainly in terms of predicted prospective areas occupying 15-40% of the case study area. In this case, the prospectivity map derived via binary index overlay modeling is slightly better than the prospectivity map derived via multi-class index modeling. However, this does not indicate the advantage of former technique over the latter technique because the objective of prospectivity mapping is to delineate small prospective zones with high prediction-rates. On the contrary, the slightly poorer performance of the prospectivity map derived via application of multi-class index overlay modeling compared to that of prospectivity map derived via application of binary index overlay modeling, with respect to prospective areas occupying 15-40%, indicates the caveat associated with binarisation of spatial evidence because this is prone to both Type I (i.e., false positive) and Type II (i.e., false negative) errors. That is, in the binarisation process, the classification (or 'equalisation') of evidential values to either 1 (rather than less than 1 but greater than 0) or to 0 (rather than greater than 0 but less than 1) means that some evidential values are 'forced' to become favourable evidence even if they are not (thus, leading to false positive error) and that some evidential values are 'forced' to become non-favourable even if they are not (thus, leading to false negative error).

Thus, in addition to its flexibility of assigning evidential class scores, multi-class index overlay modeling is advantageous compared to binary index overlay modeling in terms of suggesting uncertain predictions. It is therefore instructive to apply both of these two techniques together instead of applying only either one of them. A disadvantage of both of these techniques is the linear additive nature in combining evidence, which does not intuitively represent the inter-play of geological processes involved in mineralisation. We now turn to fuzzy logic modeling, which, like the Boolean logic modeling, allows integration of evidence in an intuitive and logical way and, like the multi-class index overlay modeling, allows flexibility in assigning evidential class scores.

Fuzzy logic modeling

Fuzzy logic modeling is based on the fuzzy set theory (Zadeh, 1965). Demicco and Klir (2004) discuss the rationale and illustrate the applications of fuzzy logic modeling to geological studies; unfortunately, they do not provide examples of fuzzy logic applications to mineral prospectivity mapping. Recent examples of applications of fuzzy logic modeling to mineral prospectivity mapping are found in D'Ercole et al. (2000), Knox-Robinson (2000), Porwal and Sides (2000), Venkataraman et al. (2000), Carranza and Hale (2001a), Carranza (2002), Porwal et al. (2003b), Tangestani and Moore (2003), Ranjbar and Honarmand (2004), Eddy et al. (2006), Harris and Sanborn-Barrie (2006), Rogge et al. (2006) and Nykänen et al. (2008a, 2008b). Typically, application of fuzzy logic modeling to knowledge-driven mineral prospectivity mapping involves three main feed-forward stages (Fig. 7-10): (1) fuzzification of evidential data; (2) logical integration of fuzzy evidential maps with the aid of an inference network and appropriate fuzzy set operations; and (3) defuzzification of fuzzy mineral prospectivity output in order to aid its interpretation. Each of these stages in fuzzy logic modeling of mineral prospectivity is reviewed below with demonstrations of their applications to epithermal Au prospectivity mapping in the case study area.

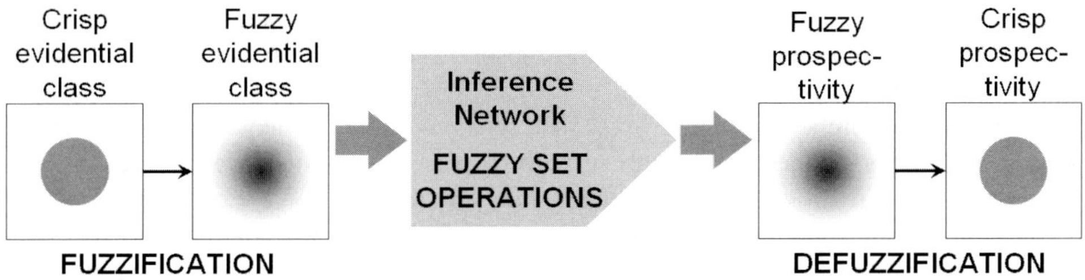

Fig. 7-10. Main stages in fuzzy logic modeling.

Fuzzification is the processes of converting individual sets of spatial evidence into fuzzy sets. A fuzzy set is defined as a collection of objects whose grades of membership in that set range from complete (=1) to incomplete (=0). This contrasts with the classical set theory, whereby the grade of membership of an object in a set of objects is either complete (=1) or incomplete (=0), which is applied in the binary representation of evidence demonstrated above. Thus, in Fig. 7-10, the abstract idea behind the illustration of fuzzification is to determine the varying degrees of greyness of every pixel in a binary (or Boolean) image of a grey object.

Fuzzy sets are represented by means of membership grades. If X is a set of object attributes denoted generically by x, then a fuzzy set A in X is a set of ordered pairs of object attributes and their grades of membership in A $(x, \mu_A(x))$:

$$A = \{(x, \mu_A(x)) | x \in X\} \tag{7.3}$$

where $\mu_A(x)$ is a membership grade function of x in A. A membership grade function, $\mu_A(x)$, is a classification of the fuzzy membership of x, in the unit interval [0,1], from a universe of discourse X to fuzzy set A; thus

$$\{\mu_A(x) | x \in X\} \rightarrow [0,1] .$$

In mineral prospectivity mapping, an example of a universe of discourse X is distances to geological structures. An example of a set of fuzzy evidence from X is a range of distances to intersections of NNW- and NW-trending faults/fractures (denoted as FI) in the case study area. Hence, a fuzzy set of 'favourable distance to FI' with respect to the proposition of mineral prospectivity, d, translates into a series of distances (x), each of which is given fuzzy membership grade, thus:

$$d = \{(x, \mu_d(x)) | x \in X\} \tag{7.4}$$

where $\mu_d(x)$ is a mathematical function defining the grade of membership of distance x in the fuzzy set 'favourable distance to FI'.

For a particular object or value of spatial evidence, the more completely it belongs to the fuzzy set of favourable evidence, the closer its membership grade is to 1. Thus, individual objects or values of spatial evidence portrayed in maps can be evaluated in terms of their membership in a fuzzy set of favourable evidence based on expert judgment. Grades of membership are usually represented by a mathematical function that may be linear or continuous; indeed, many fuzzy sets have extremely nonlinear membership grade functions (Zimmerman, 1991). The evaluation of fuzzy membership grades always relates to a certain proposition. In mineral prospectivity mapping, the grade of membership of a class or value of evidence (i.e., geological attributes) in a fuzzy set of favourable evidence is evaluated according to the proposition "*this location is prospective for mineral deposits of the type sought*".

Fuzzification is thus carried out by application of a membership function $\mu_A(x)$ to a set or map of values of classes of values of spatial evidence. Robinson (2003) has reviewed several types of fuzzy membership functions that are applicable to geographical analysis with the aid of a GIS. In knowledge-driven mineral prospectivity mapping, the choice or definition of a fuzzy membership function in order to fuzzify a spatial evidence of mineral prospectivity must be based on sound perception or judgment of spatial association between geological features represented by the evidence and occurrence of mineral deposits of the type sought. For example, based on the results of analysis of spatial association between FI and epithermal Au deposit occurrences in the case study area (see Table 6-IX), the following membership function may be defined for the fuzzy set 'favourable distance to FI':

$$\mu_d(x)=\begin{cases} 1 & \textit{for } x<1 \\ (4-x)/(4-1) & \textit{for } 1\le x\le 4 \\ 0 & \textit{for } x>4 \end{cases} \tag{7.5}$$

where x is distance (km) to FI. The graph and generic form of this function are illustrated in Fig. 7-11A. The parameters of the function (i.e., 1 and 4, which are α and γ, respectively, in Fig. 7-11A) are based on (a) range to distances to FI with optimum positive spatial association with the epithermal Au deposit occurrences, which is 1 km (see Table 6-IX) and (b) the minimum of the range of distances to FI (e.g., 4 km; see Fig. 6-10B) considered to be completely unfavourable for the occurrence of mineral deposits of the type sought. The function parameters are chosen arbitrarily based on subjective judgment or knowledge of spatial association between mineral deposits of interest and the types of geological features under consideration. The fuzzy membership function in equation (7.5) or Fig. 7-11A is linear and, thus, inconsistent with the conceptual knowledge-based representation of spatial evidence illustrated in Fig. 7-8. Alternatively, the following membership function may be defined for the set 'favourable distance to FI':

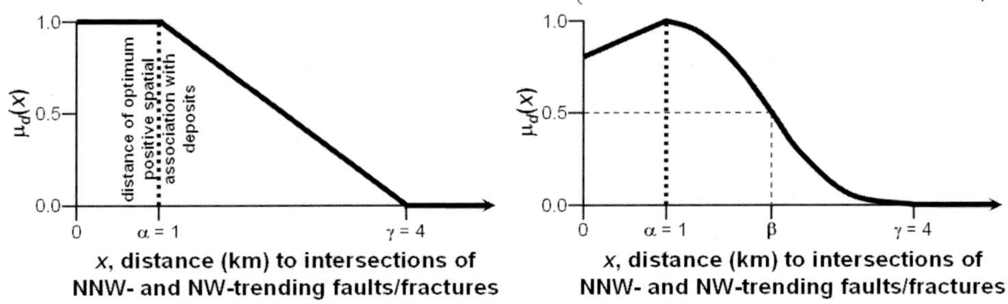

$$\mu_d(x)=\begin{cases}1 & \text{for } x<\alpha\\ (\gamma-x)/(\gamma-\alpha) & \text{for } \alpha\le x\le\gamma\\ 0 & \text{for } x>\gamma\end{cases}$$

$$\mu_d(x)=\begin{cases}[0.2(x-\min)(\alpha-\min)]+0.8 & \text{for } x<\alpha\\ 1-[(x-\alpha)/(\gamma-\alpha)]^2 & \text{for } \alpha\le x\le\beta\\ [(\gamma-x)/(\gamma-\alpha)]^2 & \text{for } \beta\le x\le\gamma\\ 0 & \text{for } x>\gamma\end{cases}$$

Fig. 7-11. Two examples of fuzzy membership functions for knowledge-based representation of proximity to intersections of NNW- and NW-trending faults/fractures as spatial evidence of mineral prospectivity in the case study area. (**A**) A linear fuzzy membership function defined by parameters α and γ, which are two different distances describing the spatial association between mineral deposits and structural features (see text for further explanation). (**B**) A fuzzy membership function consisting of a linear function and a nonlinear function defined, respectively, by the first condition and the last three conditions of the equation above the graph. The parameters α and γ used in (**A**) are also used in (**B**). The linear function represents decreasing fuzzy scores from 1.0 for $x = \alpha$ to 0.8 for $x = 0$. The nonlinear S^- function represents decreasing fuzzy scores from 1.0 for $x = \alpha$ to 0.0 for $x \geq \gamma$. The S^- function requires another parameter, β, at which the function is forced through the cross-over point [i.e., $\mu_d(x) = 0.5$] (see text for further explanation). The slope of the nonlinear function changes with different values of β.

$$\mu_d(x)=\begin{cases}[0.2(x-\min)/(\alpha-\min)]+0.8 & \text{for } x<1\\ 1-[(x-1)/(4-1)]^2 & \text{for } 1\le x\le\beta\\ [(x-4)/(4-1)]^2 & \text{for } \beta\le x\le 4\\ 0 & \text{for } x>4\end{cases}. \tag{7.6}$$

The graph and generic form of this function are illustrated in Fig. 7-11B. The function in equation (7.6) consists of a linear part (i.e., the first condition) and a continuous nonlinear part (i.e., the last three conditions). The latter is called a left-shoulder *S* function (denoted as S^- in Fig. 7-11B). The parameters of the fuzzy membership function in equation (7.6) are the same as those of the function in equation (7.5). However, the S^- function in equation (7.6) requires another parameter, β, which is a value of x that forces the function to equal the cross-over point (i.e., fuzzy membership equal to 0.5; Fig. 7-11B). Specification of a suitable value of x to represent β requires expert judgment. A value close to but greater than the maximum distance to FI, for example, within which all known deposits are present would be a suitable choice for β. From Fig. 6-10A, this distance could be about 2 km. This means, for example, that one considers locations beyond 2 km of FI to be considerably less prospective than locations within 2 km of FI. The fuzzy membership function in equation (7.6) or Fig. 7-11B is, thus, apparently consistent with the conceptual knowledge-based representation of spatial evidence illustrated in Fig. 7-8.

Another example of a universe of discourse *Y* in mineral prospectivity mapping is a variety of geochemical anomalies. A set of fuzzy evidence from this *Y* is multi-element stream sediment anomalies defined by, say, catchment basin analysis (hereafter denoted as ANOMALY). Based on the results of analysis of spatial association between ANOMALY and epithermal Au deposit occurrences in the case study area (see Figs. 6-12E and 6-12F), the following membership function may be defined for the fuzzy set 'favourable ANOMALY' (*g*):

$$\mu_g(y)=\begin{cases} 0 & \textit{for } y<0.14 \\ (y-0.14)/(0.34-0.14) & \textit{for } 0.14\le y\le 0.34 \\ 1 & \textit{for } y>0.34 \end{cases} \tag{7.7}$$

where *y* represents values of ANOMALY scores. The graph and generic form of this function are illustrated in Fig. 7-12A. The parameters of the function (i.e., 0.14 and 0.34, which are α and γ, respectively, in Fig. 7-12A) represent (a) the maximum of the range of ANOMALY scores (e.g., 0.14; see Figs. 6-12E and 6-12F) considered to be completely non-significant and (b) the minimum of the range to ANOMALY scores with optimum positive spatial association with the epithermal Au deposit occurrences (i.e., 0.34; see Figs. 6-12E and 6-12F). The fuzzy membership function in equation (7.7) or Fig. 7-12A is linear and, thus, inconsistent with the conceptual knowledge-based representation of spatial evidence illustrated in Fig. 7-8. Alternatively, the following membership function may be defined for the set 'favourable ANOMALY' (*g*):

$$\mu_g(y)=\begin{cases} 0 & \textit{for } y<0.14 \\ [(x-0.14)/(0.34-0.14)]^2 & \textit{for } 0.14\le y\le \beta \\ 1-[(0.34-x)/(0.34-0.14)]^2 & \textit{for } \beta\le y\le 0.34 \\ [0.1(\max-y)(\max-0.34)]+0.9 & \textit{for } y>0.34 \end{cases}. \tag{7.8}$$

The graph and generic form of this function are illustrated in Fig. 7-12B. The function in equation (7.8) consists of a linear part (i.e., the last condition) and a continuous nonlinear part (i.e., the first three conditions). The former is called a right-shoulder *S* function (denoted as S^+ in Fig. 7-12B). The parameters of the fuzzy membership function in equation (7.8) are the same as those of the function in equation (7.7). However, the S^+ function in equation (7.8) requires another parameter, β, which is a value of *y* that forces the function to equal the cross-over point (i.e., fuzzy membership equal to 0.5; Fig. 7-12B). Choosing a suitable value of *y* to represent β requires expert judgment. The median of the range of values between α and γ could, for example, be chosen for β. So,

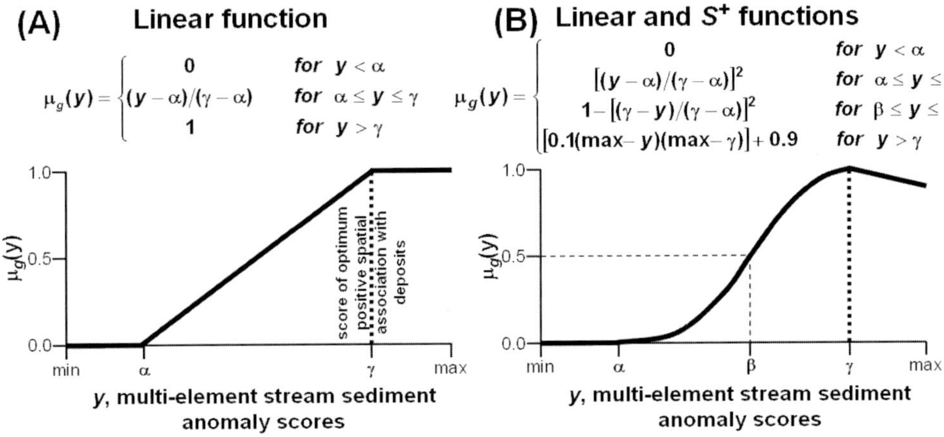

Fig. 7-12. Two examples of fuzzy membership functions for knowledge-based representation of multi-element stream sediment anomaly scores as spatial evidence of mineral prospectivity in the case study area. (**A**) A linear fuzzy membership function defined by parameters α and γ, which are two geochemical anomaly scores describing the spatial association between mineral deposits and geochemical anomalies (see text for further explanation). (**B**) A fuzzy membership function consisting of a linear function and a nonlinear functions defined, respectively, by the last condition and the first three conditions of the equation above the graph. The parameters α and γ used in (**A**) are also used in (**B**). The linear function represents decreasing fuzzy scores from 1.0 for $y = \gamma$ to, say, 0.9 for maximum y. The nonlinear S^+ function represents decreasing fuzzy scores from 1.0 for $y = \gamma$ to 0.0 for $y \le \alpha$. The S^+ function requires another parameter, β, at which the function is forced through the cross-over point [i.e., $\mu_g(y) = 0.5$] (see text for further explanation). The slope of the nonlinear function changes with different values of β.

for the present case β could be 0.24, which would mean that one considers locations with ANOMALY scores below 0.24 to be considerably less prospective than locations with anomaly scores above 0.24. The fuzzy membership function in equation (7.8) or Fig. 7-12B is, thus, apparently consistent with the conceptual knowledge-based representation of spatial evidence illustrated in Fig. 7-8.

A fuzzy membership function defined for spatial data of a continuous field (e.g., distance to certain structures, geochemical anomalies, etc.) to be used as evidence in support of the proposition of mineral prospectivity may be applied directly to such spatial data in map form. Here, for comparison with the results of the multi-class index overlay modeling, the model of fuzzy membership function depicted in equation (7.6) and illustrated in Fig. 7-11B is applied to derive fuzzy membership scores for the same classes of proximity to faults/fractures used in the multi-class index overlay modeling (see Table 7-V). The averages of distances in the classes of proximity to individual sets of structures are used in the calculation of fuzzy membership scores by application of the fuzzy membership function. Likewise, the model of fuzzy membership function depicted in equation (7.8) and illustrated in Fig. 7-12B is applied to derive fuzzy membership scores for the same classes of ANOMALY used in the multi-class index overlay

TABLE 7-VI

Examples of fuzzy membership scores assigned to evidential classes in individual evidential maps portraying the recognition criteria for epithermal Au prospectivity, Aroroy district (Philippines), derived by using (a) class means as evidential data, (b) appropriate fuzzy membership functions (Figs. 7-11B, Fig. 7-12B) and (c) function parameters (see footnotes) based on results of spatial association analyses (Chapter 6). Ranges of values in bold include the threshold value of spatial data of optimum positive spatial associations with epithermal Au deposits in the case study area.

Proximity to NNW[1]			Proximity to FI[2]		
Range (km)	Mean (km)	Fuzzy score	Range (km)	Mean (km)	Fuzzy score
0.00 – 0.08	0.05	0.80	0.00 – 0.39	0.20	0.80
0.08 – 0.15	0.11	0.84	0.39 – 0.58	0.49	0.81
0.15 – 0.23	0.19	0.89	0.58 – 0.80	0.69	0.83
0.23 – 0.32	0.27	0.95	**0.80 – 1.09**	0.95	0.99
0.32 – 0.41	0.36	1.00	1.09 – 1.40	1.25	0.82
0.41 – 0.52	0.46	0.99	1.40 – 1.80	1.60	0.58
0.52 – 0.71	0.61	0.59	1.80 – 2.32	2.06	0.33
0.71 – 1.06	0.88	0.29	2.32 – 2.92	2.62	0.12
1.06 – 1.73	1.39	0.01	2.92 – 3.62	3.27	0.01
1.73 – 3.55	2.64	0.00	3.62 – 5.92	4.77	0.00
Proximity to NW[3]			ANOMALY[4]		
Range (km)	Mean (km)	Fuzzy score	Range	Mean	Fuzzy score
0.00 – 0.18	0.10	0.80	0.00 – 0.06	0.03	0.00
0.18 – 0.36	0.27	0.84	0.06 – 0.10	0.08	0.00
0.36 – 0.54	0.45	0.89	0.10 – 0.16	0.13	0.00
0.54 – 0.75	0.64	0.94	0.16 – 0.25	0.21	0.12
0.75 – 1.01	0.88	1.00	0.25 – 0.29	0.27	0.88
1.01 – 1.29	1.15	0.99	**0.29 – 0.37**	0.35	1.00
1.29 – 1.65	1.47	0.93	0.37 – 0.49	0.43	0.96
1.65 – 2.24	1.95	0.75	0.49 – 0.78	0.58	0.90
2.24 – 3.02	2.63	0.03			
3.02 – 5.32	4.17	0.00			

[1]NNW-trending faults/fractures. Function parameters: α=0.35; β=0.8; γ=1.5. [2]Intersections of NNW- and NW-trending faults/fractures. Function parameters: α=1; β=1.9; γ=3.5. [3]NW-trending faults/fractures. Function parameters: α=0.9; β=2.3; γ=3. [4]Integrated PC2 and PC3 scores obtained from the catchment basin analysis of stream sediment geochemical data (see Chapter 3). Function parameters: α=0.14; β=0.26; γ=0.34.

modeling (see Table 7-V). The parameters (α, β, γ) chosen for a fuzzy membership function are based, as explained above, on the results of spatial association analyses in Chapter 6. Table 7-VI shows the fuzzy membership scores of classes of proximity to individual sets of faults/fractures and of classes of ANOMALY based on models of fuzzy membership functions illustrated in Figs. 7-11B and 7-12B, respectively.

The resulting fuzzy scores (Table 7-VI) are, when multiplied by 10, more-or-less similar to the multi-class index scores given in Table 7-V, although differences are evident. For example, the classes of distances farthest from the geological structures now have evidential scores equal to zero, suggesting their complete non-favourability for epithermal Au deposit occurrence. Likewise, the three lowest classes of ANOMALY are completely non-indicative of presence of epithermal Au deposit occurrences. It is uncommon practise in mineral prospectivity mapping to use minimum and maximum fuzzy scores of 0 and 1, respectively, because by using them suggests that one has complete knowledge about the spatial association of any set of spatial evidence with the mineral deposits of interest (cf. Bárdossy and Fodor, 2005). Here, the minimum and maximum fuzzy scores of 0 and 1, respectively, are used only for the purpose of demonstrating their effects on the output compared to the output of multi-class index overlay modeling described earlier and to the outputs of the other modeling techniques that follow further below.

The preceding examples of fuzzy membership functions are applicable to spatial data of continuous fields to be used as evidence in support of the proposition of mineral prospectivity. For spatial data of discrete geo-objects to be used as evidence in support of the proposition of mineral prospectivity (e.g., a lithologic map to be used as evidence of 'favourable host rocks'), discontinuous fuzzy membership functions are defined based on sound judgment of their pairwise relative importance or relevance to the proposition under examination. In this regard, the application of the AHP (Saaty, 1977) may be useful as has been demonstrated in, for example, operations research (i.e., an inter-disciplinary branch of applied mathematics for decision-making) (Triantaphyllou, 1990; Pendharkar, 2003), although the application of the AHP to assign fuzzy scores to classes of evidence (rather than to assign weights to evidential maps) for mineral prospectivity mapping has not yet been demonstrated. Proving this proposition is, however, beyond the scope of this volume. The criteria for judgment of favourability of various lithologic units as host rocks may include, for example, a-priori knowledge of host rock lithologies of mineral deposits of the type sought, chemical reactivity, age with respect to that of mineralisation of interest, etc. Knowledge of quantitative spatial associations between various mapped lithologic units and mineral deposits of interest in well-explored areas may also be considered as a criterion for judging which lithologic units are favourable host rocks for the same type of mineral deposits in frontier areas. Prudence must be exercised, nonetheless, in doing so because the degree of spatial association between known host lithologies and mineral deposits varies from one area to another depending on the present level of erosion and, therefore, on the areas of mapped lithologies and number of mineral deposit occurrences. This caveat also applies to the knowledge representation of host rock evidence via the preceding techniques as well as to interpretations of results of applications of data-driven techniques for mineral prospectivity mapping (Carranza et al., 2008a).

Although assignment of fuzzy membership grades or definition of fuzzy membership functions is a highly subjective exercise, the choice of fuzzy membership scores or the definition of fuzzy membership functions must reflect realistic spatial associations between mineral deposits of interest and spatial evidence as illustrated, for example, in Figs. 7-11 and 7-12. Because the fuzzy membership scores propagate through a model and ultimately determine the output, fuzzification is the most critical stage in fuzzy logic modeling. We now turn to the stage of logical integration of fuzzy evidence with the aid of an inference network and appropriate fuzzy set operations (Fig. 7-8).

As in classical or Boolean set theory, set-theoretic operations can be applied or performed on fuzzy sets or fuzzified evidential maps. There are several types of fuzzy operators for combining fuzzy sets (Zadeh, 1965, 1973, 1983; Thole et al., 1979; Zimmerman, 1991). Individual fuzzy operators, all of which have meanings analogous to operators for combining classical or crisp sets, portray relationships between fuzzy sets (e.g., equality, containment, union, intersection, etc.). There are five fuzzy operators, which are in fact arithmetic operators, that are useful for combining fuzzy sets representing spatial evidence of mineral prospectivity, namely, the fuzzy AND, fuzzy OR, fuzzy algebraic product, fuzzy algebraic sum and fuzzy gamma (γ) (An et al. (1991; Bonham-Carter, 1994). These fuzzy operators are useful in mineral prospectivity mapping in the sense that each of them or a combination of any of them can portray relationships between sets of spatial evidence emulating the conceptualised model of inter-play of geological processes involved in mineralisation. Thus, the choice of fuzzy operators to be used in combining fuzzy sets of spatial evidence of mineral prospectivity must be consistent with the defined conceptual model of mineral prospectivity (see Fig. 1-3).

The fuzzy AND (hereafter denoted as FA) operator, which is equivalent to the Boolean AND operator in classical theory, is defined as

$$\mu_{FA}=MIN(\mu_1,\mu_2,...,\mu_n) \tag{7.9}$$

where μ_{FA} is the output fuzzy score and μ_1, μ_2,..., μ_n are, respectively, the input fuzzy evidential scores at a location in evidence map 1, evidence map 2,..., evidence map *n*. The *MIN* is an arithmetic function that selects the smallest value among a number of input values. The output of the FA operator is, therefore, controlled by the lowest fuzzy score at every location (Fig. 7-13). So, for example, if in one evidence map the fuzzy score at a location is 0 even though in at least one of the other evidence maps the fuzzy score is 1, the output fuzzy score for that location is still zero (Fig. 7-13A). Clearly, the FA operator is appropriate in combining complementary sets of evidence, meaning that the pieces of evidence to be combined via this operator are deemed all necessary to support the proposition of mineral prospectivity at every location.

The fuzzy OR (hereafter denoted as FO) operator, which is equivalent to the Boolean OR operator in classical theory, is defined as

$$\mu_{FO}=MAX(\mu_1,\mu_2,...,\mu_n) \tag{7.10}$$

where μ_{FO} is the output fuzzy score and μ_1, μ_2,..., μ_n are, respectively, the input fuzzy evidential scores at a location in evidence map 1, evidence map 2,..., evidence map *n*. The *MAX* is an arithmetic function that selects the largest value among a number of input values. The output of the FO operator is, therefore, controlled by the highest fuzzy scores at every location (Fig. 7-13). So, for example, if in one evidence map the fuzzy

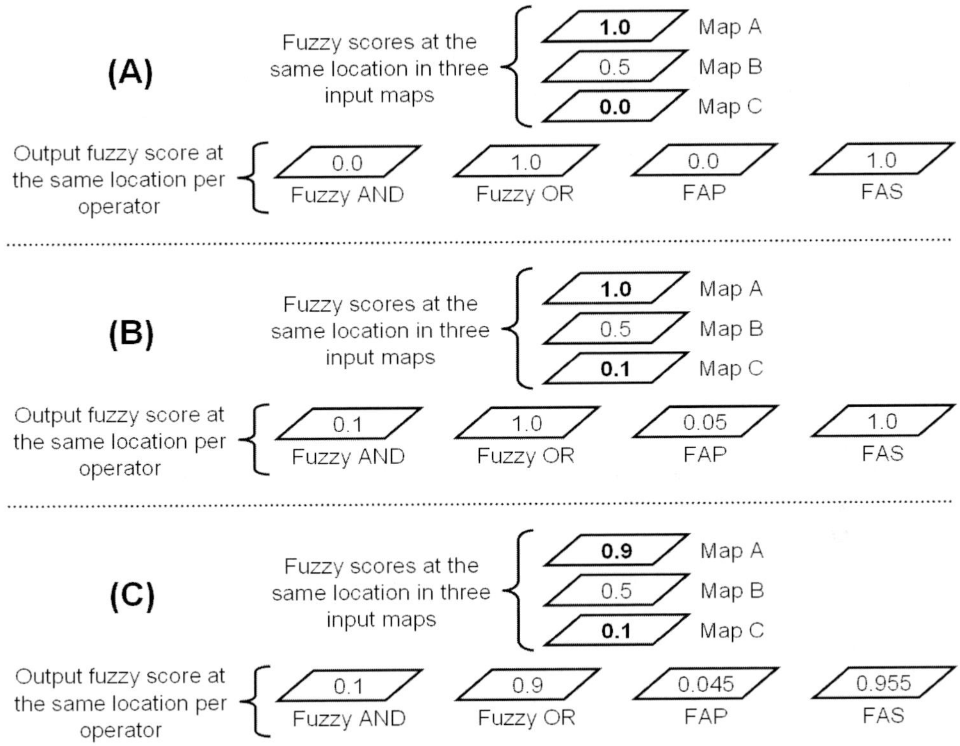

Fig. 7-13. Examples of output of the fuzzy AND, fuzzy OR, fuzzy algebraic product (FAP) and fuzzy algebraic sum (FAS) based on input fuzzy scores (**A**) inclusive of 1 and 0, (**B**) inclusive of 1 but exclusive of 0 and (**C**) exclusive of 0 and 1.

score at a location is 1 even though in at least one of the other evidence maps the fuzzy score is 0, the output fuzzy score for that location is still 1 (Fig. 7-13A). Clearly, the FO operator is appropriate in combining supplementary sets of evidence, meaning that at least one of any of the pieces of evidence to be combined via this operator is deemed necessary to support the proposition of mineral prospectivity at every location.

The fuzzy algebraic product (hereafter denoted as FAP) is defined as

$$\mu_{FAP} = \prod_{i=1}^{n} \mu_i \tag{7.11}$$

where μ_{FAP} is the output fuzzy score and μ_i represents the fuzzy evidential scores at a location in i (=1, 2,..., n) evidence maps. The output of the FAP is less than or equal to the lowest fuzzy score at every location (Fig. 7-13). So, for example, if in one evidence map the fuzzy score at a location is 0 even though in at least one of the other evidence maps the fuzzy score is 1, the output fuzzy score for that location is still zero (Fig. 7-

13A). Clearly, like the FA operator, the FAP is appropriate in combining complementary sets of evidence, meaning that all input fuzzy scores at a location must contribute to the output to support the proposition of mineral prospectivity, except in the case when at least one of the input fuzzy scores is 0 (Fig. 7-13A). In contrast to the FA operator, the FAP has a 'decreasive' effect (Figs. 7-13B and 7-13C), meaning that the presence of very low but non-zero fuzzy scores tend to deflate or under-estimate the overall support for the proposition.

The fuzzy algebraic sum (hereafter denoted as FAS) is defined as

$$\mu_{FAS} = 1 - \prod_{i=1}^{n}(1-\mu_i) \tag{7.12}$$

where μ_{FAS} is the output fuzzy score and μ_i represents the input fuzzy evidential scores at a location in i (=1, 2,..., n) evidence maps. The FAS is, by definition, not actually an algebraic sum, whereas the FAP is consistent with its definition. The output of the FAS is greater than or equal to the highest fuzzy score at every location (Fig. 7-13). So, for example, if in one evidence map the fuzzy score at a location is 1 even though in at least one of the other evidence maps the fuzzy score is 0, the output fuzzy score for that location is still 1 (Fig. 7-13A). Clearly, like the FO operator, the FAS is appropriate in combining supplementary sets of evidence, meaning that all input fuzzy scores at a location must contribute to the output to support the proposition of mineral prospectivity, except in the case when at least one of the input fuzzy scores is 1 (Figs. 7-13A and 7-13B). In contrast to the FO, the FAS has an 'increasive' effect (Fig. 7-13C), meaning that the presence of very high fuzzy scores (but not equal to 1) tend to inflate or over-estimate the overall support for the proposition.

In order to (a) regulate the 'decreasive' effect of the FAP and the 'increasive' effect of the FAS, meaning to constrain the range of the output values to the range of the input values, or (b) make use of the 'decreasive' effect of the FAP or the 'increasive' effect of the FAS, as is needed in order to derive a desirable output that is more-or-less consistent with the conceptual model of mineral prospectivity, the fuzzy γ (hereafter denoted as FG) operator can be applied. The FG operator is defined as (Zimmerman and Zysno, 1980)

$$\mu_{FG} = \left(\prod_{i=1}^{n}\mu_i\right)^{1-\gamma}\left(1-\prod_{i=1}^{n}(1-\mu_i)\right)^{\gamma} \tag{7.13}$$

where μ_{FG} is the output fuzzy score and μ_i represents the fuzzy evidential scores at a location in i (=1, 2,..., n) evidence maps. The value of γ varies in the range [0,1]. If γ = 0, then FG = FAP. If γ = 1, then FG = FAS. Finding a value of γ that contains the output fuzzy scores in the range of the input fuzzy scores, as illustrated in Fig. 7-14, entails some trials. The example shown in Fig. 7-14 is based on only two input fuzzy scores. In practise, however, the input fuzzy scores could come from two or more fuzzy evidence

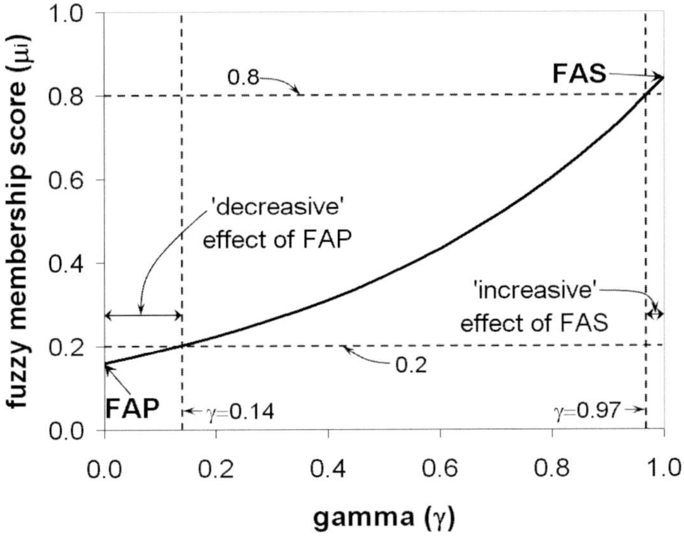

Fig. 7-14. Variation of output fuzzy membership scores (μ_i), obtained from two input fuzzy scores μ_1 and μ_2, as a function of γ in the fuzzy gamma (FG) operator. In this case, μ_1=0.8 and μ_2=0.2. If γ=0, then $\mu_{i(FG)}=\mu_{i(FAP)}$. If γ=1, then $\mu_{i(FG)}=\mu_{i(FAS)}$. When $0.97<\gamma<1$, $\mu_{i(FG)}>\mu_1$ due to the 'increasive effect of FAS (fuzzy algebraic sum). When $0<\gamma<0.14$, $\mu_{i(FG)}<\mu_2$ due to the 'decreasive effect of FAP (fuzzy algebraic product). When $0.14<\gamma<0.97$, the value of $\mu_{i(FG)}$ lies in the range of the input fuzzy scores. Thus, the value of γ that constrains the output values of μ_i in the range of the input fuzzy scores depends on the values of the input fuzzy scores. (cf. Bonham-Carter, 1994, pp. 298.)

maps and the ranges of input fuzzy scores at one location (or pixel) to another can be highly variable. So, the graph shown in Fig. 7-14 only serves to illustrate the 'decreasive' effect of the FAP and the 'increasive' effect of the FAS but it is not a device that can be used to determine a suitable value of γ. The final choice of an optimal value of γ depends on some experiments and judgment of the 'best' output of mineral prospectivity model. For example, Carranza and Hale (2001a) obtained optimal mineral prospectivity models from values of γ that vary between 0.73 and 0.79, which imply that delineated prospective areas are defined by spatial evidence that are more supplementary rather than complementary to one another.

Any one of the above-explained fuzzy operators may be applied to logically combine evidential fuzzy sets (or maps) according to an inference network, which reflects inferences about the inter-relationships of processes that control the occurrence of a geo-object (e.g., mineral deposit) and spatial features that indicate the presence of that geo-object. As in Boolean logical modeling, every step in a fuzzy inference network, in which at least two evidential maps are combined, represents a hypothesis of an inter-play of at least two sets of processes that control the occurrence of a geo-object (e.g., mineral deposit) and spatial features that indicate the presence of that geo-object. The inference network and the fuzzy operators thus form a series of logical rules that sequentially

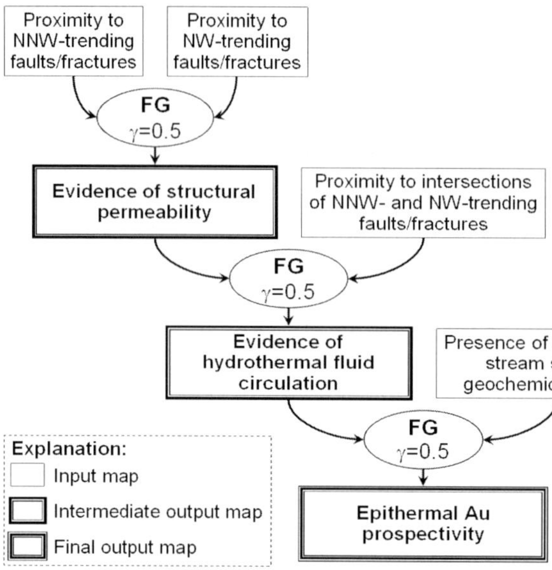

Fig. 7-15. An inference network for combining input fuzzy evidential maps for modeling of epithermal Au prospectivity in the Aroroy district (Philippines). FG = fuzzy gamma (γ) operator. A value of γ equal to 0.5 means that the output values of FG lies between the output values of fuzzy algebraic product and the output values of fuzzy algebraic sum.

combine evidential fuzzy maps. The inference network also functions to filter out the effect of ambiguous evidence. For example, all NNW-trending faults/fractures in the case study area are used to create a fuzzy evidence of favourable distance to these geological structures. However, it is certainly implausible that every NNW-trending fault/fracture is associated with mineralisation. Therefore, by logically combining a fuzzy evidence of proximity to NNW-trending faults/fractures with another fuzzy evidence, only the contributions of both or either of the two evidential fuzzy sets are transmitted to the output depending on the hypothesis. There are no general guidelines for designing a fuzzy inference network, except that as much as possible it should emulate knowledge of how the mineral deposits of the type sought were formed and what spatial features or combinations of spatial features indicate where mineral deposits of the type sought may occur. Thus, a fuzzy inference network must adequately represent the conceptual model of mineral prospectivity.

Fig. 7-15 shows an example of an inference network that can be applied to combine fuzzy evidential maps for modeling epithermal Au prospectivity in the case study area. This inference network is quite similar to, but in detail different from, the Boolean inference network in Fig. 7-4. The use of a value of γ equal to 0.5 implies that the intermediate or final output maps portray contributions of either complementary or supplementary pieces of spatial evidence, meaning that the output values lie (a) between output values of FAP and output values of FAS or (b) between output values of FA and

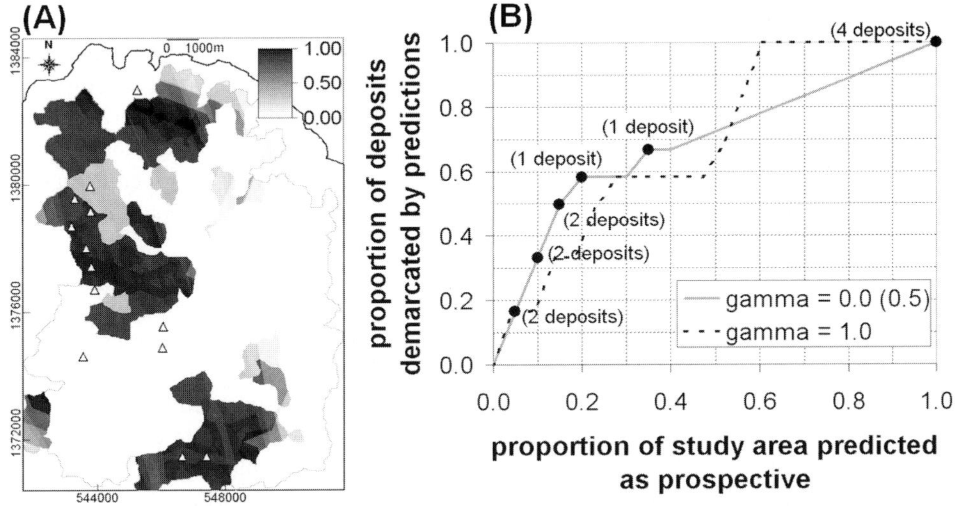

Fig. 7-16. (**A**) An epithermal Au prospectivity map of Aroroy district (Philippines) obtained via fuzzy logic modeling based on evidential maps with fuzzy evidential scores shown in Table 7-VI and on the inference network shown in Fig. 7-15. Triangles are locations of known epithermal Au deposits; whilst polygon outlined in grey is area of stream sediment sample catchment basins (see Fig. 4-11). (**B**) Prediction-rate curves of proportion of deposits demarcated by the predictions versus proportion of study area predicted as prospective. The prediction-rate curve of the map obtained by using γ=0.5 is compared with prediction-rate curves of map obtained by using γ=0 and γ=1 in the final step of the inference network in Fig. 7-15. The prediction-rate curves of the maps obtained by using γ=0.5 and γ=0 identical, meaning that their prediction-rates are equal. The dots, which pertain to the prediction-rate curve of the map derived by using γ=0.5, represent classes of prospectivity values that correspond spatially with a number of cross-validation deposits (indicated in parentheses).

output values of FO. An inference network such as shown in Fig. 7-15 reflects prudence of the modeler in combining sets of spatial evidence possibly due either to the lack of 'expert' knowledge about the inter-play of geological processes represented by individual sets of spatial evidence or to the average quality of spatial data sets used to portray the individual sets of spatial evidence.

The output of combining fuzzy sets is also a fuzzy set. For example, the final output of applying the inference network shown in Fig. 7-15 is shown in Fig. 7-16A. It represents a fuzzy set of a continuous field of mineral prospectivity values although there are sharp transitions between low and high values of fuzzy prospectivity values. In the fuzzy model of epithermal Au prospectivity shown in Fig. 7-16A, there are apparently many locations with fuzzy prospectivity values equal to zero and there are relatively less locations with high and very high fuzzy prospectivity values. The former are due to classes of evidence with fuzzy membership scores of zero, especially the classes of fuzzy ANOMALY evidence (Table 7-VI), whereas the latter are due to intersecting classes of evidence with high and very high fuzzy membership scores. The generic analysis of information (i.e., mineral prospectivity) embedded in a fuzzy set of values is via defuzzification (Fig. 7-10), so that discrete spatial entities or geo-objects representing, for example, prospective and non-prospective areas, are recognised or mapped. Hellendoorn and Thomas (1993) describe a number of criteria for defuzzification. However, in GIS-based mineral prospectivity mapping, the optimal method of defuzzifying a fuzzy model of mineral prospectivity is to construct its prediction-rate curve (see Fig. 7-2) against some cross-validation occurrences of mineral deposits of the type sought in a study area.

The prediction-rate curve of the fuzzy model of epithermal Au prospectivity in Fig. 7-16A indicates that, if 20% of the case study area is considered prospective, then it performs equally as well as the multi-class index overlay model of epithermal Au prospectivity shown in Fig. 7-9. The predictive model in Fig. 7-16A, which is obtained by using γ=0.5, performs equally as well as a predictive model obtained by using γ=0 in the final step of the inference network (Fig. 7-15). Their prediction-rate curves (Fig. 7-16B) are identical and both of them have better prediction-rates than a predictive model obtained by using γ=1 in the final step of the inference network (Fig. 7-16B). These results imply that the contributions of complementary pieces of spatial evidence provide better predictions than the contributions of supplementary pieces of spatial evidence. These results are therefore realistic because epithermal Au mineralisation requires complementary effects of both structural controls (represented by proximity to NNW- and NW-trending faults/fractures) and heat source controls (represented by proximity to intersections of NNW- and NW-trending faults/fractures; see Chapter 6). In addition, the presence of stream sediment geochemical anomalies is important in indicating locations of anomalous sources. However, the predictive model obtained by using γ=1 in the final step of the inference network is better than the predictive models obtained by using γ=0 and γ=0.5 in the final step of the inference network in the sense that the former predicts all cross-validation deposits if 60% of the case study area is considered prospective whereas the former predict all cross-validation deposits if 100% of the case study area is considered prospective (Fig. 7-16B).

The poor performance of the predictive models obtained by using γ=0 and γ=0.5 in the final step of the inference network (Fig. 7-15), in terms of correct delineation of all the cross-validation deposits, is due to classes of evidence with fuzzy membership scores of zero, especially the classes of fuzzy ANOMALY evidence (Table 7-VI). In Fig. 7-16A the locations of four cross-validation deposits have output fuzzy prospectivity values of zero. In order to demonstrate the deleterious effect using a fuzzy membership score of zero, those classes of fuzzy evidence with fuzzy membership scores of zero in Table 7-VI are re-assigned the lowest non-zero fuzzy membership scores in the individual fuzzy sets as shown in Table 7-VII.

The new predictive map (Fig. 7-17A) shows low (rather than zero) fuzzy prospectivity values at the locations of the four cross-validation deposits not delineated correctly by the predictive map in Fig. 7-16A. The new results show improvements mainly for the predictive maps obtained by using γ=0.5 and γ=0 in the final step of the inference network, which now delineate correctly all cross-validation deposits if 65%

TABLE 7-VII

Examples of fuzzy membership scores assigned to evidential classes in individual evidential maps portraying the recognition criteria for epithermal Au prospectivity, Aroroy district (Philippines). Table entries are the same as in Table 7-VI except for the values in bold italics, which are revisions of initial fuzzy scores of zero. Ranges of values in bold include the threshold value of spatial data of optimum positive spatial associations with epithermal Au deposits in the study area.

	Proximity to NNW[1]			Proximity to FI[2]	
Range (km)	Mean (km)	Fuzzy score	Range (km)	Mean (km)	Fuzzy score
0.00 – 0.08	0.05	0.80	0.00 – 0.39	0.20	0.80
0.08 – 0.15	0.11	0.84	0.39 – 0.58	0.49	0.81
0.15 – 0.23	0.19	0.89	0.58 – 0.80	0.69	0.83
0.23 – 0.32	0.27	0.95	**0.80 – 1.09**	0.95	0.99
0.32 – 0.41	0.36	1.00	1.09 – 1.40	1.25	0.82
0.41 – 0.52	0.46	0.99	1.40 – 1.80	1.60	0.58
0.52 – 0.71	0.61	0.59	1.80 – 2.32	2.06	0.33
0.71 – 1.06	0.88	0.29	2.32 – 2.92	2.62	0.12
1.06 – 1.73	1.39	0.01	2.92 – 3.62	3.27	0.01
1.73 – 3.55	2.64	***0.005***	3.62 – 5.92	4.77	***0.005***
	Proximity to NW[3]			**ANOMALY[4]**	
Range (km)	Mean (km)	Fuzzy score	Range	Mean	Fuzzy score
0.00 – 0.18	0.10	0.80	0.00 – 0.06	0.03	***0.01***
0.18 – 0.36	0.27	0.84	0.06 – 0.10	0.08	***0.03***
0.36 – 0.54	0.45	0.89	0.10 – 0.16	0.13	***0.06***
0.54 – 0.75	0.64	0.94	0.16 – 0.25	0.21	0.12
0.75 – 1.01	0.88	1.00	0.25 – 0.29	0.27	0.88
1.01 – 1.29	1.15	0.99	**0.29 – 0.37**	0.35	1.00
1.29 – 1.65	1.47	0.93	0.37 – 0.49	0.43	0.96
1.65 – 2.24	1.95	0.75	0.49 – 0.78	0.58	0.90
2.24 – 3.02	2.63	0.03			
3.02 – 5.32	4.17	***0.01***			

[1]NNW-trending faults/fractures. Function parameters: α=0.35; β=0.8; γ=1.5. [2]Intersections of NNW- and NW-trending faults/fractures. Function parameters: α=1; β=1.9; γ=3.5. [3]NW-trending faults/fractures. Function parameters: α=0.9; β=2.3; γ=3. [4]Integrated PC2 and PC3 scores obtained from the catchment basin analysis of stream sediment geochemical data (see Chapter 3). Function parameters: α=0.14; β=0.26; γ=0.34.

and 70%, respectively, of the case study area is considered prospective (Fig. 7-17B). The new results also show that the new predictive map obtained by using γ=0.5 is slightly better than the new the predictive map obtained by using γ=0 (Fig. 7-16B), indicating that supplementary but subtle pieces of spatial evidence (i.e., those with revised low fuzzy scores, especially in the fuzzy ANOMALY evidence (Table 7-VI) provide minor contributions to the improvement of the prediction. Nevertheless, both of the fuzzy mineral prospectivity models shown in Figs. 7-16 and 7-17 are better than the mineral prospectivity models derived via Boolean logic modeling (Fig. 7-5), binary index overlay modeling (Fig. 7-7) and multi-class index overlay modeling (Fig. 7-9). That is

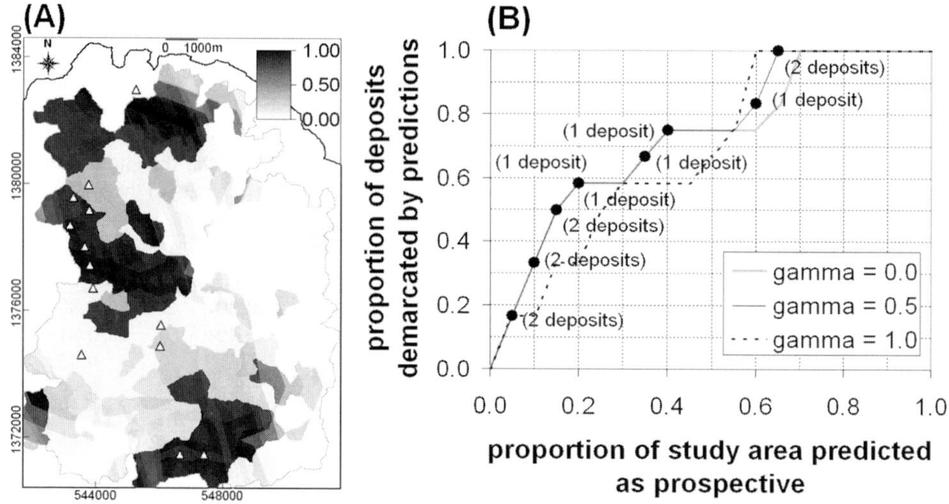

Fig. 7-17. (**A**) An epithermal Au prospectivity map of Aroroy district (Philippines) obtained via fuzzy logic modeling based on evidential maps with fuzzy evidential scores shown in Table 7-VII and on the inference network shown in Fig. 7-15. Triangles are locations of known epithermal Au deposits; whilst polygon outlined in grey is area of stream sediment sample catchment basins (see Fig. 4-11). (**B**) Prediction-rate curves of proportion of deposits demarcated by the predictions versus proportion of study area predicted as prospective. The prediction-rate curve of the map obtained by using γ=0.5 is compared with prediction-rate curves of map obtained by using γ=0 and γ=1 in the final step of the inference network in Fig. 7-15. The prediction-rate curves of the maps obtained by using γ=0.5 and γ=0 mostly overlap each other, meaning that their prediction-rates are mostly equal. The dots, which pertain to the prediction-rate curve of the map derived by using γ=0.5, represent classes of prospectivity values that correspond spatially with a number of cross-validation deposits (indicated in parentheses).

because, if 20% of the case study area is considered prospective, then the former models delineate correctly seven (or about 58%) of the cross-validation deposits (Figs. 7-16B and 7-17B) whilst the latter models delineate correctly six (or 50%) of the cross-validation deposits.

The availability of different fuzzy operators and the ability to modify fuzzy inference networks is an advantage of fuzzy logic modeling compared to binary and multi-class index overlay modeling. The advantage of fuzzy logic modeling compared Boolean logic modeling is mainly in the representation of spatial evidence. However, the assignment of fuzzy evidential scores is as subjective as the assignment of multi-class index scores. No two experts will arrive at the same fuzzy prospectivity scores to classes of the same set of spatial evidence. The disadvantage of fuzzy logic modeling compared to multi-class index overlay modeling is the presumption of equal weights for individual evidential maps. The common disadvantage of all these modeling techniques is the implicit representation of uncertainty of spatial evidence. For example, in fuzzy logic modeling there is no proper and uniform way of assigning fuzzy prospectivity membership scores to locations in a study area with missing data. In many case examples in the literature, locations in a study area without data are assigned either very low fuzzy scores or fuzzy scores equal to zero. It has been demonstrated above that assignment of fuzzy scores equal to zero can produce undesirable and unrealistic results. In this volume, locations in the case study area without stream sediment geochemical data are assigned fuzzy scores of zero (meaning they are not considered in the analysis), although this has the same net effect of using fuzzy scores of zero as exemplified in Fig. 7-16A.

The following section explains a different technique of representing and integrating multi-class evidential maps in order to model mineral prospectivity. This technique – evidential belief modeling – provides for explicit representation of evidential uncertainty.

Evidential belief modeling

Dempster's (1967, 1968) work on the generalisation of Bayesian lower and upper probabilities provided the basis for the theory of evidential belief. Shafer (1976) then defined two evidential belief functions (EBFs), belief and plausibility, to represent the lower and upper probabilities, respectively, that a given body of evidence supports a particular proposition. In the last three decades or so, the Dempster-Shafer theory of evidential belief has attracted considerable attention as a promising method of dealing with some of the basic problems arising in the fusion of data or combination of evidence. Zadeh (1986) provided a simplification of the Dempster-Shafer theory of evidential belief and demonstrated the capability of Dempster's (1968) rule of combination to integrate distinct probability distributions. Walley (1987) suggested, however, that Dempster's (1968) rule of combination is generally neither suitable for combining evidence from independent observations nor appropriate for combining prior beliefs with observational evidence. However, applications of the Dempster-Shafer theory of evidential belief proved its usefulness in combining pieces of evidence from disparate sources (e.g., Cohen, 1985; Lee et al., 1987; Kim and Swain, 1989). Chung and Fabbri (1993) described the representation of geoscience information for data integration based on interpretation of the Dempster-Shafer theory of evidential belief. An et al. (1994b) demonstrated the management or representation of uncertainty in the integration of exploration data using EBFs.

The mathematical formalism of the EBFs is complex (Dempster, 1967; Shafer, 1976). The following explanations for the application of EBFs to mineral prospectivity mapping are simplified and informal. For a piece of spatial evidence that is used in evaluating a proposition (i.e., mineral prospectivity), four values, each in the range of [0,1], are assigned based on evaluation of how much it supports the proposition. These values are *belief* (hereafter denoted as *Bel*), *disbelief* (hereafter denoted as *Dis*), *uncertainty* (hereafter denoted as *Unc*) and *plausibility* (hereafter denoted as *Pls*). The *Bel* and *Pls* represent, respectively, lower and upper degrees of support provided by a given piece of spatial evidence to the proposition. This means, for example, that with a given spatial evidence mineral deposit occurrence is either less (*Bel*) or more (*Pls*) likely. Thus, *Pls* and *Bel* together represent the vague or uncertain '*more-or-less*'

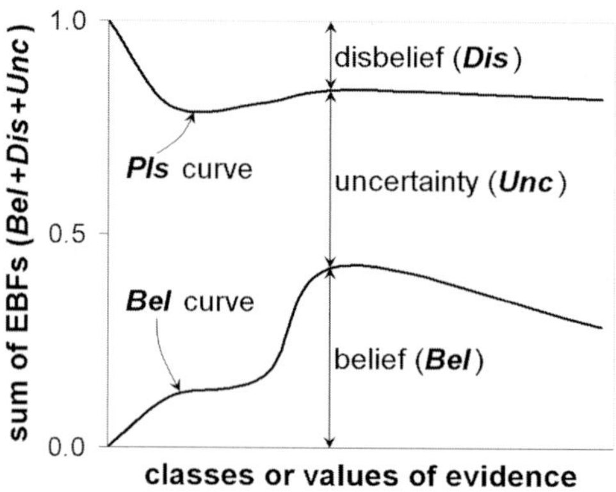

Fig. 7-18. Schematic relationships of evidential belief functions (EBFs). See text for further explanation.

assessment. The *Unc* represents 'ignorance' or 'doubt' that a given piece of spatial evidence supports the proposition. The value of *Unc* is the difference between *Bel* and *Pls*. The *Dis* represents evaluation that a given piece of spatial evidence does not support the proposition.

The four EBFs are inter-related (Fig. 7-18). The sum of *Bel*+*Unc*+*Dis* of a piece of spatial evidence is equal to 1. Likewise, the sum of *Pls*+*Dis* of a piece of spatial evidence is equal to 1. From these two equalities, therefore, *Pls* = *Bel*+*Unc* or *Bel* = *Pls*–*Unc*. The degree of *Unc* influences the relation between *Bel* and *Dis*. If *Unc* = 0 (i.e., there is complete knowledge about a given piece of spatial evidence), then *Bel*+*Dis* = 1 and the relation between *Bel* and *Dis* for a given piece of evidence is binary (i.e., *Bel* = 1–*Dis* or *Dis* = 1–*Bel*), as in the theory of probability. If *Unc* = 1 (i.e., there is complete ignorance or doubt about a given piece of spatial evidence), then *Bel* and *Dis* for a given piece of evidence are both equal to zero. That is, if there is complete uncertainty, then there can be neither belief nor disbelief. Usually, however, *Unc* is neither equal to zero nor equal to one. Therefore, in the case that 0<*Unc*<1, then *Bel* = 1–*Dis*–*Unc* or *Dis* = 1–*Bel*–*Unc*. This means that, because uncertainty is usually present, the relation between *Bel* and *Dis* for a given piece of evidence is usually not binary. This means further that, for a piece of evidence used to evaluate a proposition, one should estimate not only *Bel* but also *Dis* and *Unc*.

In the application of EBFs to knowledge-driven mineral prospectivity mapping, two of the three EBFs – *Bel*, *Dis* and *Unc* – are usually first estimated together in order to represent not only degree of support (or lack of support) to a proposition by a piece of spatial evidence but also the degree of uncertainty about this evidence. The *Pls* is then simply derived from estimates of any of two EBFs and, as shown below, the *Pls* is not used in Dempster's (1968) rule of combination. Estimating *Bel* and *Dis* together is usually the most difficult, because one tends to think of the binary relation between these two EBFs and then neglect *Unc*. Estimating *Dis* and *Unc* together is cumbersome, because of confusion between disbelieving and doubting. So, estimating *Bel* and *Unc* together is usually the more convenient. In the case that 0<*Unc*<1, *Bel* is usually estimated to be less than or equal to 0.5 but never equal to 0.0. Meanwhile, the value of *Unc* is estimated such that (1) the sum *Bel*+*Unc* (i.e., *Pls*) is more than 0.5 but never equal to 1.0, (2) the estimates of *Bel* and *Unc* vary inversely, (3) the derived values of *Dis* vary inversely with the estimates of *Bel* and co-vary with the estimates of *Unc*. These three conditions are important in order to represent the following realistic relations of the EBFs in the case that 0<*Unc*<1. Firstly, the higher the uncertainty, the lower the belief or vice versa. Secondly, the higher the belief, the lower the disbelief or vice versa. Thus, in the usual case that 0<*Unc*<1, the estimates of *Bel* are kept asymptotic to 0.0, whereas the sum *Bel*+*Unc* is kept asymptotic to 1.0. The above-stated conditions for knowledge-driven estimation of *Unc* together with *Bel* do not apply when there is either complete ignorance or doubt (i.e., *Unc*=1) or complete knowledge (i.e., *Unc*=0) about a piece of spatial evidence in relation to a proposition. An example situation of *Unc*=1 in mineral prospectivity mapping is when spatial data are missing. There is no situation of *Unc*=0 in mineral prospectivity mapping because if *Unc* were equal to zero there would be no need for mineral prospectivity mapping.

Once *Bel* and *Unc* have been estimated, for the case of 0<*Unc*<1, the remaining two EBFs (*Dis*, *Pls*) can be easily estimated based on the inter-relations of the EBFs explained above and illustrated in Fig. 7-18. Examples of knowledge-driven estimations of EBFs for mineral prospectivity mapping can be found in Moon (1990, 1993), Chung and Moon (1991), Moon et al. (1991), An (1992), An et al. (1992, 1994a, 1994b), Chung and Fabbri (1993), Wright and Bonham-Carter (1996), Likkason et al. (1997), Carranza (2002), Tangestani and Moore (2002) and Rogge et al. (2006). In practise, knowledge-driven estimates of EBFs are assigned and stored in attribute tables associated with individual maps of spatial data to be used as evidence of mineral prospectivity (see Fig. 7-1). For the present mineral prospectivity case study, Table 7-VIII shows examples of *Bel*, *Unc* and *Dis* estimated for evidential classes of the same sets of evidential maps used in the multi-class index and fuzzy logic modeling (see Tables 7-V to 7-VII). The estimates of *Bel*, like the estimates of the multi-class index scores and the fuzzy membership scores (see Tables 7-V and 7-VI, respectively), are according to the knowledge of spatial associations between the epithermal Au deposit occurrences and the individual sets of spatial data to be used as evidence of epithermal Au prospectivity in the case study area. In accordance with the three conditions given above for estimating *Bel* and *Unc* together, Table 7-VIII shows that the estimates of *Bel* vary inversely with estimates of *Unc* and the derived values of *Dis* vary inversely with the estimates of *Bel* and co-vary with the estimates of *Unc*. The locations without stream sediment geochemical data are assigned *Bel*=0 and *Unc*=1 and, thus, *Dis*=0.

For each spatial evidence map X_i (i=1,2,...,n), three attribute maps representing EBFs Bel_i, Dis_i and Unc_i are then created. The maps of EBFs associated with spatial evidence map X_1 can be combined with the maps of EBFs associated with spatial evidence map X_2 according to Dempster's (1968) rule of combination, which can be implemented by using either an AND or an OR operation (An et al., 1994a).

TABLE 7-VIII

Examples of values of EBFs assigned to evidential classes in individual evidential maps portraying the recognition criteria for epithermal Au prospectivity, Aroroy district (Philippines). Values of *Bel* and *Unc* are first estimated together and then *Dis* is derived as 1–*Bel*–*Unc* (see text for further explanation). Ranges of values in bold include the threshold value of spatial data of optimum positive spatial associations with epithermal Au deposits in the case study area.

	Proximity to NNW[1]				Proximity to FI[2]		
Range (km)	*Bel*	*Unc*	*Dis*	Range (km)	*Bel*	*Unc*	*Dis*
0.00 – 0.08	0.30	0.44	0.26	0.00 – 0.39	0.30	0.44	0.26
0.08 – 0.15	0.35	0.43	0.22	0.39 – 0.58	0.40	0.42	0.18
0.15 – 0.23	0.40	0.42	0.18	0.58 – 0.80	0.45	0.41	0.14
0.23 – 0.32	0.45	0.41	0.14	**0.80 – 1.09**	0.50	0.40	0.10
0.32 – 0.41	0.50	0.40	0.10	1.09 – 1.40	0.40	0.42	0.18
0.41 – 0.52	0.40	0.42	0.18	1.40 – 1.80	0.20	0.46	0.34
0.52 – 0.71	0.30	0.44	0.26	1.80 – 2.32	0.10	0.47	0.43
0.71 – 1.06	0.20	0.46	0.34	2.32 – 2.92	0.05	0.48	0.47
1.06 – 1.73	0.10	0.48	0.42	2.92 – 3.62	0.03	0.49	0.48
1.73 – 3.55	0.05	0.50	0.45	3.62 – 5.92	0.01	0.50	0.49
	Proximity to NW[3]				ANOMALY[4]		
Range (km)	*Bel*	*Unc*	*Dis*	Range	*Bel*	*Unc*	*Dis*
0.00 – 0.18	0.30	0.44	0.26	No data	0.00	1.00	0.00
0.18 – 0.36	0.35	0.43	0.22	0.00 – 0.06	0.05	0.50	0.45
0.36 – 0.54	0.40	0.42	0.18	0.06 – 0.10	0.10	0.48	0.42
0.54 – 0.75	0.45	0.41	0.14	0.10 – 0.16	0.15	0.46	0.39
0.75 – 1.01	0.50	0.40	0.10	0.16 – 0.25	0.25	0.44	0.31
1.01 – 1.29	0.40	0.42	0.18	0.25 – 0.29	0.35	0.43	0.22
1.29 – 1.65	0.30	0.47	0.23	**0.29 – 0.37**	0.50	0.40	0.10
1.65 – 2.24	0.20	0.48	0.32	0.37 – 0.49	0.45	0.41	0.14
2.24 – 3.02	0.10	0.49	0.41	0.49 – 0.78	0.40	0.42	0.18
3.02 – 5.32	0.05	0.50	0.45				

[1]NNW-trending faults/fractures. [2]Intersections of NNW- and NW-trending faults/fractures. [3]NW-trending faults/fractures. [4]Integrated PC2 and PC3 scores obtained from the catchment basin analysis of stream sediment geochemical data (see Chapter 3).

The formulae for combining EBFs of two spatial evidence maps (X_1, X_2) according to an AND operation are defined as (An et al., 1994a):

$$Bel_{X_1X_2} = \frac{Bel_{X_1} Bel_{X_2}}{\beta}, \tag{7.14}$$

$$Dis_{X_1X_2} = \frac{Dis_{X_1} Dis_{X_2}}{\beta}, \text{ and} \tag{7.15}$$

$$Unc_{X_1X_2} = \frac{Unc_{X_1}Unc_{X_2} + Bel_{X_1}Unc_{X_2} + Bel_{X_2}Unc_{X_1} + Dis_{X_1}Unc_{X_2} + Dis_{X_2}Unc_{X_1}}{\beta} \quad (7.16)$$

where $\beta = 1 - Bel_{X_1}Dis_{X_2} - Dis_{X_1}Bel_{X_2}$, which is a normalising factor ensuring that $Bel + Unc + Dis = 1$. Equations (7.14) and (7.15) are multiplicative, so that the application of an AND operation results in a map of integrated *Bel* and integrated *Dis*, respectively, in which the output values represent support and lack of disbelief, respectively, for the proposition being evaluated if pieces of spatial evidence in two input maps coincide (or intersect). In contrast, equation (7.16) is both commutative and associative, so that the application of an AND operation results in a map of integrated *Unc* in which the output values are controlled by pieces of spatial evidence with large uncertainty in either of the two input maps. Therefore, an AND operation is suitable in combining two pieces of complementary spatial evidence (say, X_1 and X_2) in order to support the proposition of mineral prospectivity. In mineral exploration, proximity to faults/fractures and stream sediment geochemical anomalies can represent two sets of complementary spatial evidence of mineral deposit occurrence, because several types of mineral deposits, including epithermal Au, are localised along faults/fractures and, if exposed at the surface, can release metals into the drainage systems and cause anomalous concentrations of metals in stream sediments. (After application of equations (7.14)–(7.16), $Pls_{X_1X_2}$ is derived according to the relationships of the EBFs explained above.)

The formulae for combining EBFs of two spatial evidence maps (X_1, X_2) according to an OR operation are defined as (An et al., 1994a):

$$Bel_{X_1X_2} = \frac{Bel_{X_1}Bel_{X_2} + Bel_{X_1}Unc_{X_2} + Bel_{X_2}Unc_{X_1}}{\beta} \quad (7.17)$$

$$Dis_{X_1X_2} = \frac{Dis_{X_1}Dis_{X_2} + Dis_{X_1}Unc_{X_2} + Dis_{X_2}Unc_{X_1}}{\beta} \quad (7.18)$$

$$Unc_{X_1X_2} = \frac{Unc_{X_1}Unc_{X2}}{\beta} \quad (7.19)$$

where β is the same as in equations (7.14)–(7.16). Equations (7.17) and (7.18) are both commutative and associative, so that the application of OR operation results in a map of integrated *Bel* and integrated *Dis*, respectively, in which the output values are controlled by pieces of spatial evidence with large belief or large disbelief in either of the two input maps. In contrast, equation (7.19) is multiplicative, so that the application of an OR operation results in a map of integrated *Unc* in which the output values is controlled by pieces of spatial evidence with low uncertainty in either of the two input maps. Therefore, an OR operation is suitable in combining two pieces of supplementary (as opposed to complementary) spatial evidence (say, X_1 and X_2) in order to support the proposition of mineral prospectivity. In mineral exploration, proximity to faults/fractures and stream sediment geochemical anomalies can represent two sets of supplementary spatial evidence of the presence of mineral deposits, because not all locations proximal to faults/fractures contain mineral deposits and because not all stream sediment geochemical anomalies necessarily mean the presence of mineral deposits. (After application of equations (7.17)–(7.19), $Pls_{X_1X_2}$ is derived according to the relationships of the EBFs explained above.)

According to Dempster's rule of combination, only EBFs of two spatial evidence maps can be combined each time. The EBFs of maps $X_3,\ldots,X_n$ are combined with already integrated EBFs one after another by re-applying either equations (7.14)–(7.16) or equations (7.17)–(7.19) as deemed appropriate. The final integrated values of *Bel* are considered indices of mineral prospectivity. Furthermore, because equation (7.14) is multiplicative, whilst equation (7.17) is associative and commutative, the output integrated values of *Bel* derived via the former are always less than the corresponding output integrated values of *Bel* derived via the latter. This means that integrated values of EBFs should not be interpreted in absolute terms but in relative terms (i.e., ordinal scale) only and, therefore, in mineral prospectivity modeling integrated values of *Bel* represent relative degrees of likelihood for mineral deposit occurrence.

As in Boolean logic modeling and in fuzzy logic modeling of mineral prospectivity, an inference network is useful in combining logically EBFs of spatial evidence of mineral prospectivity. The inference network used in the Boolean logic modeling (Fig. 7-4), which is more-or-less similar to the inference network used in the fuzzy logic modeling (Fig. 7-15), is applied to logically integrate the EBFs of the spatial evidence maps of epithermal Au prospectivity in the case study area. The geological reasoning behind the integration of the EBFs of the spatial evidence maps of epithermal Au prospectivity in the case study area is, thus, the same as in the earlier application of Boolean logic modeling and similar to the earlier application of the fuzzy logic modeling. The map of integrated *Bel* (Fig. 7-19A) shows a pattern of prospective areas that is more similar to the pattern of prospective areas delineated via multi-class index overlay modeling (Fig. 7-9A) and via fuzzy logic modeling (Figs. 7-16A and 7-17A) than the pattern of prospective areas delineated via Boolean logic modeling (Fig. 7-5A) and via binary index overlay modeling (Fig. 7-7A). However, unlike the earlier mineral prospectivity maps, a prediction-rate curve (Fig. 7-19B) can be constructed for the mineral prospectivity map in Fig. 7-19A with respect to the whole case study area because, for the locations without stream sediment geochemical evidence, there are input EBFs (i.e., *Bel*=0, *Unc*=1 and *Dis*=0) and thus output EBFs. Nevertheless, for proper comparison of predictive performance with the earlier mineral prospectivity maps, a prediction-rate curve with respect only to locations with stream sediment geochemical evidence is also constructed (Fig. 7-19C). Using this curve, if 20% of the case study area is considered prospective, the map of integrated *Bel* delineates correctly seven (or about 58%) of the cross-validation deposits (Fig. 7-19C). This predictive performance of the evidential belief model (Fig. 7-19A) is the same as that of the fuzzy logic model (Fig. 7-

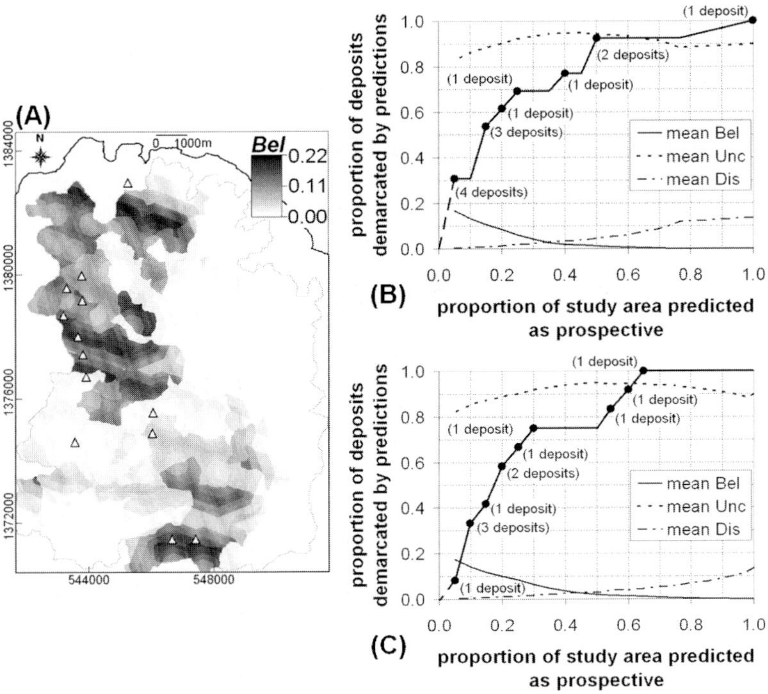

Fig. 7-19. (**A**) A map of integrated *Bel* portraying epithermal Au prospectivity of Aroroy district (Philippines). The inference network in Fig. 7-4 was used in combining input evidential maps with EBFs given in Table 7-VIII. Triangles are locations of known epithermal Au deposits; whilst polygon outlined in grey is area of stream sediment sample catchment basins (see Fig. 4-11). Prediction-rate curves of the map of integrated *Bel* based on (**B**) the whole study area, because the assignment of *Unc* to areas without stream sediment geochemical data (see Table 7-VIII) allows inclusion of those areas in predictive modeling with EBFs, and (**C**) only areas with all input data in order to compare the results with the outputs of those predictive modeling techniques explained earlier. Dots along the prediction-rate curves represent classes of integrated values of *Bel* that correspond spatially with a number of cross-validation deposits (indicated in parentheses). The averages of integrated *Bel*, integrated *Unc* and integrated *Dis* in classes of integrated values of *Bel* are also shown.

17) and is better than those of the Boolean logic model (Fig. 7-5), binary index overlay model (Fig. 7-7) and multi-class index overlay model (Fig. 7-9).

In the evaluation of a mineral prospectivity map derived by evidential belief modeling, unlike in the evaluation of mineral prospectivity maps derived by the other modeling techniques explained earlier, the variations of integrated values of *Unc*, as well as the other integrated EBFs, can be illustrated together with the prediction-rate curve. This allows for additional criteria in evaluating the performance of a predictive model of mineral prospectivity. Thus, based on the prediction-rate curves shown in Fig. 7-19, prospective areas occupying at most 20% of the case study area, inclusive or exclusive

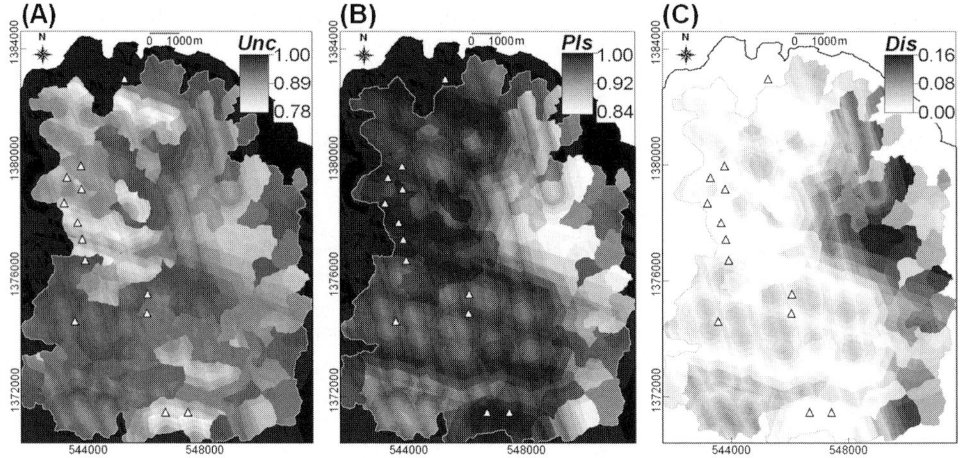

Fig. 7-20. Maps of integrated EBFs [(**A**) *Unc*, (**B**) *Pls* and (**C**) *Dis*] accompanying the map of integrated *Bel* (Fig. 7-19A) for the proposition of epithermal Au prospectivity, Aroroy district (Philippines). The inference network in Fig. 7-4 was used in combining input evidential maps with EBFs given in Table 7-VIII. Triangles are locations of known epithermal Au deposits; whilst polygon outlined in grey is area of stream sediment sample catchment basins (see Fig. 4-11).

of locations without stream sediment geochemical evidence, have lowest degrees of uncertainty in the proposition under examination. The other final maps of integrated EBFs – *Unc*, *Dis* and *Pls* (Fig. 7-20) – also provide geo-information regarding the predictions portrayed in the final map of integrated *Bel* (Fig. 7-19A). The final map of integrated *Unc* (Fig. 7-20A) depicts locations where the input pieces of spatial evidence are insufficient to provide support for the proposition of mineral prospectivity. In the case study area, examples of such locations are obviously those without stream sediment geochemical evidence. The final map of integrated *Pls* (Fig. 7-20B) depicts not only prospective areas but also locations where additional pieces of spatial evidence are required to provide support for the proposition of mineral prospectivity. In the case study area, examples of such locations are in the east-central parts of the area where there are multi-element stream sediment anomalies (see Fig. 5-12) but faults/fractures are scarce (see Fig. 5-13A). The final map of integrated *Dis* (Fig. 7-20C) complements the pieces of spatial geo-information provided by the corresponding final maps of integrated *Bel*, *Unc* and *Pls* in terms of depicting prospective and non-prospective areas as well as locations where the input pieces of spatial evidence are insufficient to provide support for the proposition of mineral prospectivity.

The ability of explicit representation of evidential uncertainty, even in the case of missing evidence, is an advantage of evidential belief modeling compared to the modeling techniques explained earlier. As in fuzzy logic modeling, the availability of different operators and the ability to modify inference networks for combining pieces of evidence is an advantage of evidential belief modeling compared to binary and multi-class index overlay modeling. Perhaps the only disadvantage of evidential belief modeling compared to the modeling techniques explained earlier is that one has to consider not one but two evidential class scores (e.g., *Bel* and *Unc*) simultaneously. However, with four complementary output maps, evidential belief modeling provides better evaluation of predictive model performance especially in terms of determining which input data are problematic; therefore, it provides more pieces of geo-information that are potentially useful in guiding further exploration work.

Calibration of predictive modeling with multi-class evidential maps

There are three approaches by which predictive modeling with multi-class evidential maps can be calibrated: (1) modification of evidential class scores; (2) modification of evidential map weights; and (3) modification of inference networks for combining pieces of spatial evidence.

The first approach to predictive model calibration is relevant to all three techniques for modeling with multi-class evidential maps. The second approach to predictive model calibration is relevant only to multi-class index overlay modeling and fuzzy logic modeling, because evidential belief modeling does not provide ability to incorporate evidential map weights in the modeling process. Porwal et al. (2003b) demonstrate procedures for incorporating evidential map weights in fuzzy logic modeling. For each evidential map, a map weight is assigned based on subjective judgment of relative importance of pieces of spatial evidence. For each class in an evidential map, a class weight is assigned and then the class score is obtained as the product of the map weight and the class weight. The class scores are then transformed into the range [0,1] by applying a fuzzy logistic membership function (see equation (7.21) further below). There are certainly several possible meaningful evidential map weights and evidential class scores that can be assigned and every modeler surely has different opinions about the relative importance or weight of pieces of spatial evidence. Different sets of evidential map weights and evidential class scores result in different mineral prospectivity models, from which the best predictive has to be determined.

The third approach to predictive model calibration is relevant only to fuzzy logic modeling and evidential belief modeling, because multi-class index overlay modeling simply derives the average of weighted class scores. In fuzzy logic modeling and evidential belief modeling, an inference network can be modified by changing operators or by changing the combinations of evidential maps to be integrated by a certain operator. Certainly, one must always evaluate the meaningfulness of an inference network, but the ability to do so depends strongly on quality of available expert knowledge. Different inference networks results in different predictive models, from which the best predictive has to be determined.

Clearly, the generic approach to calibration of knowledge-driven predictive modeling of mineral prospectivity is trial-and-error or comparative analysis to derive an optimum predictive model. By 'optimum', it is meant that a knowledge-driven predictive model of mineral prospectivity is geologically meaningful (i.e., consistent with the conceptual model of mineral prospectivity) and has a high prediction-rate. This quality of an optimum knowledge-driven predictive model of mineral prospectivity can be achieved when there is (a) adequate knowledge of geologic controls on mineral deposits of the type sought and of spatial features indicative of the presence of the same type of mineral deposits and (b) suitable and highly accurate geoscience data sets for representations of spatial evidence of mineral prospectivity. A situation of knowledge-driven mineral prospectivity mapping where the second requirement is available but the first requirement is lacking is more challenging. For this kind of situation (say, in mineral exploration of geologically permissive greenfields areas where mineral deposits of interest are still undiscovered), we turn to a so-called wildcat methodology for knowledge-driven modeling of mineral prospectivity.

WILDCAT MODELING OF MINERAL PROSPECTIVITY

In practise, it is difficult to develop, elicit or model quantitative knowledge of spatial associations between mineral deposits of interest and indicative spatial geological features especially during the early (i.e., reconnaissance) stages of grassroots mineral exploration. The difficulty arises when only a geological map is available for a given greenfields area in which no or very few mineral deposit occurrence are known. In addition, reconnaissance exploration surveys are, in general, more focused on geological 'permissivity' of mineral deposit occurrence rather than on deposit-type prospectivity. Hence, mapping of mineral prospectivity (as opposed to mineral deposit-type prospectivity), which may be used in guiding further exploration, is faced with the problem of how to create and integrate geologically meaningful evidential maps of mineral prospectivity. To solve this problem, a 'wildcat' methodology of predictive mapping of prospective areas can be devised (Carranza, 2002; Carranza and Hale, 2002d). The term 'wildcat', according to Whitten and Brooks (1972), means a "highly speculative exploratory operation". The term also refers to "a borehole (or more rarely a mine) sunk in the hope of finding oil (or ore) in a region where deposits of oil (or metallic ores) have not been recorded" (Whitten and Brooks, 1972). The wildcat methodology is actually a *knowledge-guided data-driven technique* of modeling mineral prospectivity. That is because the evidential class scores are calculated from data, although certain kinds of general knowledge about mineralisation and relative importance of pieces of spatial evidence are applied for meaningful calculation and transformation of evidential class scores and for integration of evidential maps.

The wildcat methodology for modeling mineral prospectivity is built upon the general qualitative knowledge about the characteristics of the geological environments of mineral deposits. For example, hydrothermal mineral deposits generally occur in or near the vicinity of geological features such as igneous intrusions (most often dikes and/or stocks but seldom batholiths) and faults/fractures. In addition, areas containing hydrothermal mineral deposits are usually characterised by surficial geochemical anomalies. In the wildcat methodology, maps of proximity to geological features are first created and integrated in order to represent a spatial evidence of geologic controls. Then, spatial evidence of geologic controls are integrated with spatial evidence of geochemical anomalies.

In order to create maps of spatial indicators of geologic controls, the wildcat methodology makes use of the inverse distance to geological features in the representation or creation of evidential maps of mineral prospectivity. This is based on general knowledge that mineral deposits preferentially occur proximal to rather distal to certain geological features that play certain roles in mineralisation. Thus, for each class of proximity to a set of geological features, an evidential score, S_c (c=1,2,...,n) is defined as:

$$S_c = \frac{1}{\tilde{d}_c} \tag{7.20}$$

where $\tilde{d}_c$ is median distance in each proximity class. Because the types and relative strengths of spatial associations of individual sets of geological features with mineral deposits are (presumably) unknown, scoring bias due to non-uniform classification of data is avoided by using the same number of equal-area or equal-percentile classes of proximity to individual sets of geological features.

For the case study area, Table 7-IX shows values of S_c for 5-percentile intervals of distances to NNW-, NW- and NE-trending faults/fractures and to the mapped units of Nabongsoran Andesite porphyry (Fig. 3-9). The NE-trending faults/fractures and the mapped units of Nabongsoran Andesite porphyry are used here because they are, respectively, plausible structural and heat-source controls on hydrothermal mineralisation, but it is presumed that the case study area is a greenfields exploration area and thus there is lack of knowledge of spatial association between epithermal Au deposits and these geological features. The intersections of NNW- and NW-trending faults/fractures are not used here because, in the reconnaissance stage of exploration, (it is presumed that) there is insufficient a-priori knowledge that these particular types of geological features are associated with hydrothermal mineral deposits in the case study area. Table 7-IX and Fig. 7-21A show that values of S_c decrease exponentially as distance to individual sets of geological features increases. This is a rather pessimistic representation or characterisation of spatial geological evidence of mineral deposit occurrence, especially in the reconnaissance stage of exploration. In addition, the range of values of S_c is different for each set of geological features, suggesting, for example, that the mapped NE-trending faults/fractures are more important structural controls of hydrothermal mineralisation than the mapped NW-trending faults/fractures (Table 7-IX and Fig. 7-21A). Because, in the reconnaissance stage of exploration, (it is presumed that) there is insufficient a-priori knowledge about which sets of geological features are geologic controls on hydrothermal mineralisation in the case study area, then it is reasonable to 'equalise' the range of evidential scores for classes of proximity to individual sets of geological features. This is achieved by applying a fuzzy logistic membership function to a set of values of S_c, thus:

TABLE 7-IX

Evidential scores (S_c) derived as inverse of median distance ($\tilde{d}_c$) [equation (7.20)] of each class of proximity to individual sets of geological features, Aroroy district (Philippines). The ranges of proximity classes are 5-percentile intervals of distances to individual sets of geological features.

Proximity to NNW[1] Distance (km)		S_c	Proximity to NW[2] Distance (km)		S_c	Proximity to NE[3] Distance (km)		S_c	Proximity to NA[4] Distance (km)		S_c
Interval	$\tilde{d}_c$		Interval	$\tilde{d}_c$		Interval	$\tilde{d}_c$		Interval	$\tilde{d}_c$	
0.00–0.04	0.02	**50.0**	0.00–0.01	0.06	**16.7**	0.00–0.03	0.02	**50.0**	0.00–0.36	0.19	**5.3**
0.04–0.08	0.06	**16.7**	0.01–0.18	0.14	**7.1**	0.03–0.06	0.05	**20.0**	0.36–0.74	0.55	**1.8**
0.08–0.11	0.09	**11.1**	0.18–0.27	0.23	**4.3**	0.06–0.09	0.08	**12.5**	0.74–1.10	0.92	**1.1**
0.11–0.15	0.13	**7.7**	0.27–0.36	0.31	**3.2**	0.09–0.12	0.11	**9.1**	1.10–1.45	1.28	**0.8**
0.15–0.19	0.17	**5.9**	0.36–0.44	0.40	**2.5**	0.12–0.15	0.13	**7.7**	1.45–1.79	1.62	**0.6**
0.19–0.23	0.21	**4.8**	0.44–0.53	0.49	**2.0**	0.15–0.17	0.16	**6.3**	1.79–2.19	1.99	**0.5**
0.23–0.27	0.25	**4.0**	0.53–0.64	0.59	**1.7**	0.17–0.20	0.19	**5.3**	2.19–2.60	2.39	**0.4**
0.27–0.32	0.29	**3.4**	0.64–0.75	0.70	**1.4**	0.20–0.24	0.22	**4.5**	2.60–3.05	2.83	**0.4**
0.32–0.36	0.34	**2.9**	0.75–0.88	0.82	**1.2**	0.24–0.27	0.26	**3.8**	3.05–3.51	3.29	**0.3**
0.36–0.41	0.39	**2.6**	0.88–1.01	0.95	**1.1**	0.27–0.30	0.29	**3.4**	3.51–3.96	3.74	**0.3**
0.41–0.46	0.43	**2.3**	1.01–1.15	1.08	**0.9**	0.30–0.33	0.32	**3.1**	3.96–4.39	4.17	**0.2**
0.46–0.52	0.49	**2.0**	1.15–1.29	1.22	**0.8**	0.33–0.37	0.35	**2.9**	4.39–4.80	4.59	**0.2**
0.52–0.60	0.56	**1.8**	1.29–1.44	1.36	**0.7**	0.37–0.41	0.39	**2.6**	4.80–5.21	5.00	**0.2**
0.60–0.71	0.65	**1.5**	1.44–1.65	1.55	**0.6**	0.41–0.45	0.43	**2.3**	5.21–5.68	5.45	**0.2**
0.71–0.84	0.78	**1.3**	1.65–1.92	1.79	**0.6**	0.45–0.50	0.47	**2.1**	5.68–6.22	5.95	**0.2**
0.84–1.06	0.95	**1.1**	1.92–2.24	2.08	**0.5**	0.50–0.55	0.53	**1.9**	6.22–6.80	6.51	**0.2**
1.06–1.35	1.20	**0.8**	2.24–2.60	2.42	**0.4**	0.55–0.64	0.60	**1.7**	6.80–7.40	7.10	**0.1**
1.35–1.73	1.54	**0.6**	2.60–3.02	2.81	**0.4**	0.64–0.80	0.72	**1.4**	7.40–8.01	7.71	**0.1**
1.73–2.23	1.98	**0.5**	3.02–3.58	3.30	**0.3**	0.80–1.06	0.93	**1.1**	8.01–8.62	8.32	**0.1**
2.23–3.55	2.89	**0.3**	3.58–5.32	4.45	**0.2**	1.06–2.04	1.55	**0.6**	8.62–9.96	9.29	**0.1**

[1]NNW-trending faults/fractures. [2]NW-trending faults/fractures. [3]NE-trending faults/fractures. [4]Nabongsoran Andesite porphyry.

$$fS_c = \frac{1}{1+e^{-m(S_c-\tilde{S}_c)}} \tag{7.21}$$

where fS_c is a fuzzified evidential score of proximity class c (=1,2,…,n), S_c is evidential score of class c (equation (7.20)), $\tilde{S}_c$ is median of a set of values of S_c and m is an arbitrary constant that controls the slope and range of values of the fuzzy logistic membership function. The fuzzy logistic membership function is chosen to transform the S_c values because its form is the same as that of the equation for transforming odds (O) to probability (P) [i.e., $P = 1/(1+O)$]. Thus, like probability values, the values of fS_c fall in the range [0,1], although depending on m the calculated values of fS_c may not range from 0 to 1. Suitable values of m are sought so that every set of classes of proximity to individual sets of faults/fractures has a uniform range of fS_c values; that is, from a

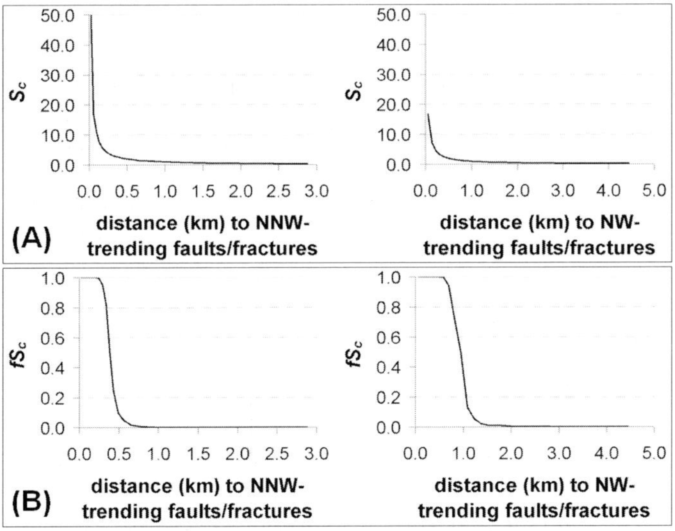

Fig. 7-21. Graphs of distances (km) to NNW- and NW-trending faults/fractures, Aroroy district (Philippines), versus estimates of (**A**) evidential scores [S_c; equation (7.20)] (Table 7-IX) and (**B**) fuzzified evidential scores [fS_c; equation (7.21)] (Table 7-X). See text for further explanation.

smallest possible value closest but not equal to 0 (as demonstrated in the fuzzy logic modeling) to the highest value of 1 (Table 7-X).

Considering that fS_c values greater than 0.5 represent favourability for mineral deposit occurrence, the calculated values of fS_c (Table 7-X) suggest that locations within 0.36 km of NNW-trending faults/fractures are favourable for occurrence of hydrothermal mineral deposits, locations within 0.88 km of NW-trending faults/fractures are favourable for occurrence of hydrothermal mineral deposits and locations within 0.27 km of NE-trending faults/fractures are favourable for occurrence of hydrothermal mineral deposits. Coincidentally, these favourable distances to faults/fractures in the case study area are closely similar to the distances of optimal positive spatial associations between epithermal Au deposits and the individual sets of faults/fractures in the case study area (Chapter 6, Figs. 6-9 and 6-11). In addition, the calculated values of fS_c suggest that locations within 3.05 km of the Nabongsoran Andesite porphyry are favourable for occurrence of hydrothermal mineral deposits. This result is more-or-less consistent with the findings of Carranza and Hale (2002c) that porphyry Cu deposits occur within 3 km of porphyry plutons. Notwithstanding the findings here, the values of fS_c compared to the values of S_c are rather optimistic but realistic representations or characterisations of spatial geological evidence of hydrothermal mineral deposit occurrence, particularly in the reconnaissance stage of exploration. Compared to the variations of S_c with distances to certain geological features (Fig. 7-21A), the variations of fS_c with distances to the same geological features (Fig. 7-21B) are more consistent

TABLE 7-X

Fuzzified evidential scores [fS_c; equation (7.21)] of classes of proximity to individual sets of geological features, Aroroy district (Philippines). The ranges of proximity classes are 5-percentile intervals of distances to individual sets of geological features.

Proximity to NNW[1]		Proximity to NW[2]		Proximity to NE[3]		Proximity to NA[4]	
Range (km)	fS_c	Range (km)	fS_c	Range (km)	fS_c	Range (km)	fS_c
0.00–0.04	**1.0000**	0.00–0.01	**1.0000**	0.00–0.03	**1.0000**	0.00–0.36	**1.0000**
0.04–0.08	**1.0000**	0.01–0.18	**1.0000**	0.03–0.06	**1.0000**	0.36–0.74	**1.0000**
0.08–0.11	**1.0000**	0.18–0.27	**1.0000**	0.06–0.09	**1.0000**	0.74–1.10	**1.0000**
0.11–0.15	**1.0000**	0.27–0.36	**1.0000**	0.09–0.12	**1.0000**	1.10–1.45	**1.0000**
0.15–0.19	**1.0000**	0.36–0.44	**1.0000**	0.12–0.15	**1.0000**	1.45–1.79	**1.0000**
0.19–0.23	**0.9997**	0.44–0.53	**0.9998**	0.15–0.17	**0.9997**	1.79–2.19	**0.9998**
0.23–0.27	**0.9948**	0.53–0.64	**0.9967**	0.17–0.20	**0.9955**	2.19–2.60	**0.9845**
0.27–0.32	**0.9526**	0.64–0.75	**0.9453**	0.20–0.24	**0.9526**	2.60–3.05	**0.9845**
0.32–0.36	**0.8176**	0.75–0.88	**0.7211**	0.24–0.27	**0.7685**	3.05–3.51	**0.5000**
0.36–0.41	**0.5000**	0.88–1.01	**0.5000**	0.27–0.30	**0.5000**	3.51–3.96	**0.5000**
0.41–0.46	**0.2451**	1.01–1.15	**0.1301**	0.30–0.33	**0.2315**	3.96–4.39	**0.0155**
0.46–0.52	**0.0953**	1.15–1.29	**0.0547**	0.33–0.37	**0.1091**	4.39–4.80	**0.0155**
0.52–0.60	**0.0474**	1.29–1.44	**0.0219**	0.37–0.41	**0.0630**	4.80–5.21	**0.0155**
0.60–0.71	**0.0159**	1.44–1.65	**0.0086**	0.41–0.45	**0.0266**	5.21–5.68	**0.0155**
0.71–0.84	**0.0076**	1.65–1.92	**0.0086**	0.45–0.50	**0.0148**	5.68–6.22	**0.0155**
0.84–1.06	**0.0036**	1.92–2.24	**0.0033**	0.50–0.55	**0.0082**	6.22–6.80	**0.0155**
1.06–1.35	**0.0012**	2.24–2.60	**0.0013**	0.55–0.64	**0.0045**	6.80–7.40	**0.0002**
1.35–1.73	**0.0006**	2.60–3.02	**0.0013**	0.64–0.80	**0.0018**	7.40–8.01	**0.0002**
1.73–2.23	**0.0004**	3.02–3.58	**0.0005**	0.80–1.06	**0.0007**	8.01–8.62	**0.0002**
2.23–3.55	**0.0002**	3.58–5.32	**0.0002**	1.06–2.04	**0.0002**	8.62–9.96	**0.0002**

[1]NNW-trending faults/fractures. [2]NW-trending faults/fractures. [3]NE-trending faults/fractures. [4]Nabongsoran Andesite porphyry.

with the conceptual model of multi-class representation of spatial evidence of mineral prospectivity (Fig. 7-8).

The values of S_c and fS_c are calculated and stored in attribute tables associated with maps of classes of proximity to individual sets of geological features. Maps or images of fS_c of individual sets of faults/fractures are then created and input to principal components (PC) analysis (cf. Luo, 1990). The application of PC analysis here is based on the assumption that an integrated spatial evidence of geologic controls on hydrothermal mineralisation can be characterised and can be derived by a quantitative function of linear combinations of proximity to individual sets of geological structures. Interpretation of a particular PC (or eigenvector) as a favourability function representing mineral prospectivity is based on the multivariate association between the input geological variables, as indicated by the magnitude and signs (positive or negative) of the eigenvector loadings. The geological meaning of the multivariate association of the different geological features represented by each of the PCs is also interpreted in terms of general knowledge about mineralisation.

TABLE 7-XI

Principal components of fuzzified evidential scores (fS_c) of classes of proximity to geological features in Aroroy district (Philippines) (see Table 7-X and Fig. 7-21B).

	NNW[1]	NW[2]	NE[3]	NA[4]	% of variance	Cum. % of variance
PC1	0.528	0.592	0.177	0.582	42.70	42.70
PC2	-0.242	-0.207	0.937	0.145	24.47	61.17
PC3	0.743	-0.145	0.253	-0.603	17.76	84.93
PC4	-0.333	0.765	0.164	-0.526	15.07	100.00

[1]NNW-trending faults/fractures. [2]NW-trending faults/fractures. [3]NE-trending faults/fractures. [4]Nabongsoran Andesite porphyry.

For the case study area, PC1 (explaining about 43% of the total variance of fS_c values in the input maps) indicates a strong spatial association between the NNW- and NW-trending faults/fractures and the Nabongsoran Andesite porphyry, which is weakly associated with the NE-trending faults/fractures (Table 7-XI). PC2 (explaining about 24% of the total variance of fS_c values in the input maps) reflects mainly the NE-trending faults/fractures, whilst PC3 (explaining about 18% of the total variance of fS_c values in the input maps) mainly reflects either the NNW-trending faults/fractures or the Nabongsoran Andesite porphyry. PC4 (explaining about 15% of the total variance of fS_c values in the input maps) mainly reflects either the NW-trending faults/fractures or the Nabongsoran Andesite porphyry. Thus, of the four PCs, PC1 is a multivariate association that is the most plausible integrated spatial evidence of heat source and structural controls on hydrothermal mineralisation in the case study area. A map of the PC1 scores can thus be considered a geologically-constrained mineral prospectivity model (Carranza, 2002).

The geologically-constrained wildcat model of hydrothermal mineral deposit prospectivity represented by the map of PC1 scores (Fig. 7-22A), obtained from the PC analysis of fuzzified evidential scores (Table 7-XI), shows intersecting linear patterns (reflecting proximity to faults/fractures), which intersect with circular patterns (reflecting proximity to mapped units of Nabongsoran Andesite porphyry) in the southwestern quadrant of the case study area. The map patterns are different from those of the earlier models of epithermal Au prospectivity in the case study area because the geochemical evidence has not been integrated into the present model. The predictive performance of the map of PC1 scores can be aptly compared only with the map of integrated *Bel* (Fig. 7-19A) because both of these maps have predictions for all locations in the case study area. If 20% of the case study area is considered prospective, then the map of PC1 scores delineates correctly six (or about 46%) of the cross-validation deposits (Fig. 7-22B). This means that, based on 20% predicted prospective zones, the map of PC1 scores is inferior to the map of integrated *Bel* (Fig. 7-19B). If 50% of the case study area is considered prospective, then the map of PC1 scores delineates correctly 12 (or about 92%) of the cross-validation deposits (Fig. 7-22B). This means that, based on 50% predicted prospective zones, the map of PC1 scores performs equally as the map of integrated *Bel* (Fig. 7-19B). If 60% of the case study area is considered prospective, then the map of PC1 scores delineates correctly all of the cross-validation deposits (Fig. 7-22B). This means that, based on 60% predicted prospective zones, the map of PC1 scores is superior to the map of integrated *Bel* (Fig. 7-19B). These comparisons indicate that geologically-constrained wildcat modeling of mineral prospectivity is a potentially useful tool for guiding further exploration in frontier areas.

An integrated geochemical-geological wildcat model of mineral prospectivity can be obtained by either (a) using a spatial evidence of geochemical anomalies together with the pieces of spatial evidence of geologic controls on mineralisation in the PC analysis or (b) taking the product of a spatial evidence of geochemical anomalies and a spatial evidence of geologic controls on mineralisation. In either of these two options, the spatial evidence of geochemical anomalies is transformed, by application of equation (7.21), to the same range of fuzzified evidential scores as the input data for the PC analysis of spatial evidence of geologic controls. The objective of this transformation is to derive values of geochemical evidence that are compatible with the derived values of geological evidence. For the case study area, the integrated PC2 and PC3 scores obtained from the catchment basin analysis of stream sediment geochemical data (Fig. 5-12) are used for the values of S_c in equation (7.21) and the median of these scores are used for

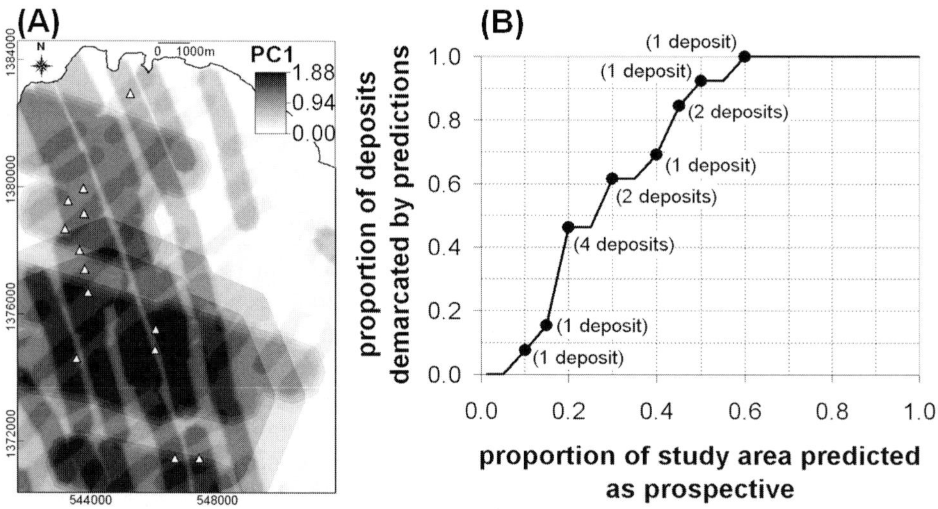

Fig. 7-22. (**A**) A wildcat model of hydrothermal deposit prospectivity, Aroroy district (Philippines), represented by a geologically-meaningful principal component (in this case PC1; Table 7-XI) of fuzzified evidential scores (Table 7-X) obtained as inverse function of distance to geological features (see text for further explanation). Triangles represent locations of known epithermal Au deposit occurrences. (**B**) Prediction-rate curve of proportion of deposits demarcated by the predictions versus proportion of study area predicted as prospective. The dots along the prediction-rate curve represent classes of prospectivity values that correspond spatially with a number of cross-validation deposits (indicated in parentheses).

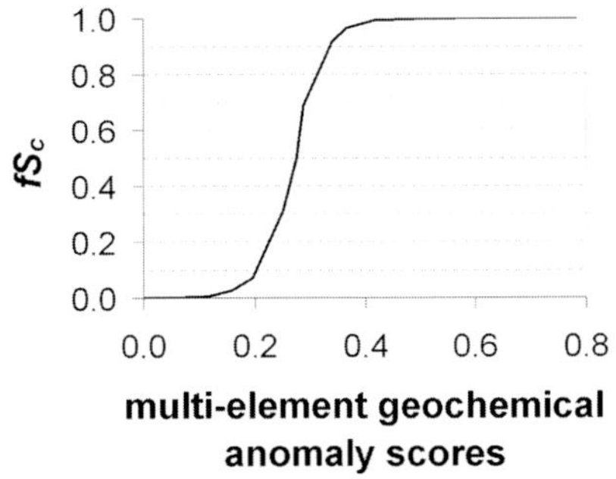

Fig. 7-23. Fuzzified evidential scores (fS_c) of multi-element geochemical anomaly scores (see Fig. 5-12), Aroroy district (Philippines).

the value of $\widetilde{S}_c$ in equation (7.21). The fuzzified evidential scores of spatial evidence of geochemical anomalies in the case study are illustrated in Fig. 7-23. The shape of the fuzzy logic membership function of the multi-element geochemical anomaly scores is similar to the shape of the fuzzy membership function of the same data set (Fig. 7-12B).

Of the two above-stated options for obtaining an integrated geochemical-geological wildcat model of mineral prospectivity, option (a) does not result in a satisfactory model of mineral prospectivity in the case when the spatial evidence of geochemical anomalies does not represent in situ materials as do the pieces of spatial evidence of geologic controls on mineralisation. For example in the case study area, PC1 reflects dissociation between geological and geochemical evidence, PC2 and PC3 both reflect association between geochemical evidence and some structural control evidence and PC4 and PC5 both reflect association between geochemical evidence, heat source control evidence and some structural control evidence (Table 7-XII). Thus, PC4 and PC5 represent equivocal integrated geochemical-geological evidence of mineral prospectivity. Between PC4 and

TABLE 7-XII

Principal components of fuzzified evidential scores (fS_c) of classes of proximity to geological features (see Table 7-X and Fig. 7-21B) and classes of multi-element geochemical anomaly scores (Fig. 7-23) in the Aroroy district (Philippines).

	NNW[1]	NW[2]	NE[3]	NA[4]	ANOM[5]	% of variance	Cum. % of variance
PC1	-0.498	-0.580	-0.182	-0.591	0.181	35.47	35.47
PC2	-0.421	-0.218	0.678	0.199	-0.525	20.86	56.33
PC3	0.258	-0.087	0.685	-0.141	0.661	18.49	74.82
PC4	0.689	-0.335	0.073	-0.422	-0.479	12.92	87.74
PC5	0.181	-0.704	-0.179	0.643	0.159	12.26	100.00

[1]NNW-trending faults/fractures. [2]NW-trending faults/fractures. [3]NE-trending faults/fractures. [4]Nabongsoran Andesite porphyry. [5]Integrated PC2 and PC3 scores obtained from the catchment basin analysis of stream sediment geochemical data (see Chapter 3).

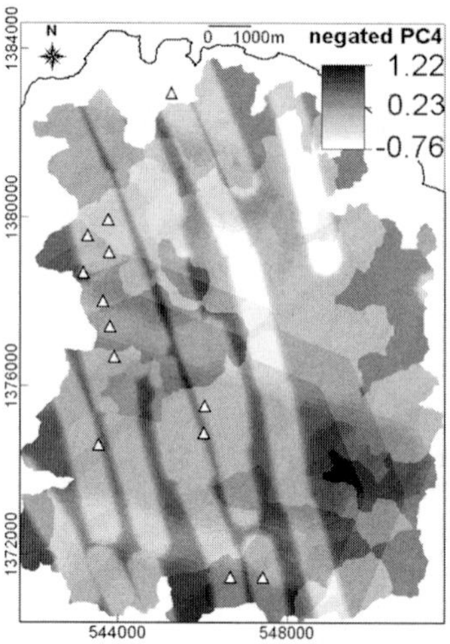

Fig. 7-24. A non-satisfactory integrated geochemical-geological wildcat model of hydrothermal deposit prospectivity, Aroroy district (Philippines). This wildcat predictive model is derived as a principal components (in this case PC4; Table 7-XII) of fuzzified evidential scores of proximity to geological features (Table 7-X) and multi-element geochemical anomaly scores (Fig. 7-23). The PC4 scores are negated (i.e., multiplied by -1) in order to represent mineral prospectivity as high scores. Triangles represent locations of known epithermal Au deposit occurrences.

PC5, the former is the more plausible integrated spatial evidence of hydrothermal deposit prospectivity because the loadings on NW, NA and ANOM are more-or-less similar whereas the latter reflects mainly evidence of heat-source control because the loadings on NA are much higher than the loadings on NW and ANOM. However, the map of PC4 scores (negated by multiplying with -1 because loadings on NW, NA and ANOM are negative) is, just by visual inspection, a non-satisfactory model of hydrothermal deposit prospectivity in the case study area (Fig. 7-24). Therefore, option (b) is offer another method of obtaining an integrated geochemical-geological wildcat model of mineral prospectivity.

An integrated geochemical-geological wildcat model of hydrothermal deposit prospectivity in the case study area (Fig. 7-25A), obtained as a product of the fuzzified evidential scores of multi-element geochemical anomaly scores (Fig. 7-23) and the PC1 scores of geological evidence (Table 7-XI; Fig. 7-22A), is like the earlier models except the model based on EBFs because it does not have predictions in locations without geochemical evidence. However, it shows a pattern of prospective areas that are similar to those of the earlier models of epithermal Au prospectivity in the case study area and therefore it is a much better model than the model shown in Fig. 7-24. If 20-50% of the case study area is considered prospective, then the integrated geochemical-geological wildcat model of hydrothermal deposit prospectivity delineates correctly seven (or about 58%) to nine (or about 75%) of the cross-validation deposits (Fig. 7-25B). This means that, based on 20-50% predicted prospective zones, the integrated geochemical-

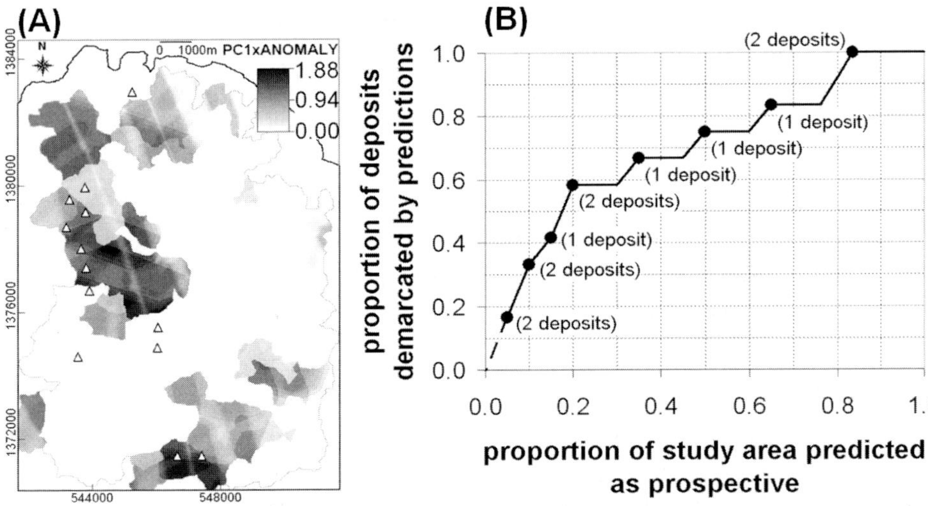

Fig. 7-25. (**A**) An integrated geochemical-geological wildcat model of hydrothermal deposit prospectivity, Aroroy district (Philippines), derived as product of fuzzified evidential scores of multi-element geochemical anomaly scores (Fig. 7-23) and the PC1 scores (Table 7-XI; Fig. 7-22A) of fuzzified evidential scores of proximity to geological features (Table 7-X). Triangles represent locations of known epithermal Au deposit occurrences. (**B**) Prediction-rate curve of proportion of deposits demarcated by the predictions versus proportion of study area predicted as prospective. The dots along the prediction-rate curve represent classes of prospectivity values that correspond spatially with a number of cross-validation deposits (indicated in parentheses).

geological wildcat model performs equally as well as the fuzzy logic model (Fig. 7-17B) and the evidential belief model (Fig. 7-19C). However, if more 50% of the case study area is considered prospective, then the integrated geochemical-geological wildcat model is inferior to the fuzzy logic model and evidential belief model. These results, nonetheless, indicate that wildcat modeling of mineral prospectivity is a potentially useful tool for guiding further exploration in greenfields frontier areas.

The wildcat methodology bears out the usefulness of reconnaissance (or small-scale) geological maps in first-pass assessment of mineral prospectivity of greenfields geologically permissive areas. The methodology, which is a knowledge-guided data-driven technique for modeling of mineral prospectivity, is sensitive to the widths of classes of proximity to geological features and to the types of geological features used in the analysis. As demonstrated by Carranza and Hale (2002d), narrower classes of proximity to geological features results in predictive models of mineral prospectivity with higher prediction-rates. In the case study, classes of proximity narrower than the 5-percentile intervals of map distances are likely to result in higher prediction-rates although proving this hypothesis is not an objective here. The choice of which type of geological features to include in modeling mineral prospectivity depends on the general knowledge of which general types of mineral deposits are likely to occur in certain greenfields geologically permissive areas. A good knowledge of general characteristics of mineral deposits (e.g., Cox and Singer, 1986; Roberts et al., 1988; Berger and Drew, 2002) and their tectonic environments (e.g., Mitchell and Garson, 1981; Sawkins, 1989; Pirajno, 1992; Robb, 2004) is therefore essential in the wildcat methodology of modeling mineral prospectivity as well as in the other techniques for knowledge-driven modeling of mineral prospectivity.

CONCLUSIONS

Several techniques for knowledge-driven modeling of mineral prospectivity exist. All the techniques demonstrated here are aided by a GIS in terms of creating and integrating evidential maps and evaluating the performance of mineral prospectivity maps. It is better to use multi-class evidential maps than binary evidential maps in modeling of mineral prospectivity. Representation and incorporation of evidential uncertainty, as in evidential belief modeling, result in better predictive models of mineral prospectivity, at least in the case study area. The performance of knowledge-driven mineral prospectivity maps depends chiefly on the subjective nature of expert judgments that are applied in creating evidential maps and/or in integrating evidential maps. Therefore, depending on the technique applied, deriving an optimum knowledge-driven predictive model of mineral prospectivity entails trial-and-error or comparative analysis by (a) adjustment of evidential scores of classes in evidential maps, (b) adjustment of evidential map weights and/or (c) adjustment of inference networks for combining evidential maps.

Despite the subjectivity of 'expert' knowledge applied in knowledge-driven modeling of mineral prospectivity, the techniques demonstrated here are useful in first-pass assessments of mineral prospectivity of greenfields geologically permissive areas where no or very few mineral deposits of interest are known to occur. This is further demonstrated by the results of the application of the knowledge-guided data-driven wildcat technique of mineral prospectivity, which is based on generic knowledge of geological environments of mineral deposits rather than on knowledge of empirical spatial associations between mineral deposits of the type sought and indicative geological features. However, because of the subjective nature of 'expert' knowledge applied in knowledge-driven or knowledge-guided data-driven modeling of mineral prospectivity, it is obligatory to evaluate performance of derived mineral prospectivity maps. It is possible to do so in cases where a few occurrences of mineral deposits of interest are known, but it is difficult, if not impossible, to do so in cases where there are no known occurrences of mineral deposits of interest. In any case, the quality and quantity of exploration data that are available in greenfields geologically permissive areas also influence the quality of a knowledge-driven model of mineral prospectivity. Presumably, more and better data allow better expert judgments in creating and integrating evidential maps of mineral prospectivity.

Knowledge-driven modeling of mineral prospectivity is also applicable in relatively well-explored areas where the objective is to find new exploration targets for mineral deposits of the type sought in the presence of several discovered mineral deposits of the type sought. However, in order to achieve objectivity in creating evidential maps, it is then preferable to apply data-driven modeling of mineral prospectivity.

Geochemical Anomaly and Mineral Prospectivity Mapping in GIS
by E.J.M. Carranza
Handbook of Exploration and Environmental Geochemistry, Vol. 11 (M. Hale, Editor)

Chapter 8

DATA-DRIVEN MODELING OF MINERAL PROSPECTIVITY

INTRODUCTION

Data-driven mineral prospectivity mapping is appropriate in areas representing moderately- to well-sampled (or so-called 'brownfields') mineralised landscapes, where the objective is to demarcate new targets for further exploration of undiscovered deposit-type locations based on the following suppositions. Known deposit-type locations are a sample set of locations with high likelihood of mineral deposit occurrence in a mineralised landscape. This sample set embodies or provides a collection of geological knowledge that there are certain combinations of spatial evidential features (e.g., proximity to faults/fractures, geochemical anomalies, etc.) associated with every deposit-type location in a mineralised landscape. This collection of geological knowledge, which constitutes a conceptual model of prospectivity recognition criteria for undiscovered deposit-type locations, indicates that mineral deposit occurrence is a function of the degrees of presence and relative importance of individual pieces of spatial evidential features (see Fig. 1-2). Thus, if more important evidential features are present in one location than in another location in a mineralised landscape, then the former location has higher mineral prospectivity than the latter location. The conceptual model of mineral prospectivity recognition criteria provides the framework for establishing or quantifying empirical spatial associations between a set of known deposit-type locations and individual sets of spatial evidential data in most, if not all, locations in a mineralised landscape under investigation (see Fig. 1-3). The quantified empirical spatial associations between known deposit-type locations and individual sets of spatial evidential data depict relative indices of likelihood of deposit-type occurrence portrayed in predictor maps. These predictor maps are then integrated in order to delineate, in a mineralised landscape, new targets for further exploration of undiscovered deposit-type locations. The quantified empirical spatial associations of known deposit-type locations and individual sets of spatial evidential data could also be used in re-defining previously established conceptual models of prospectivity recognition criteria for mineral deposits of the type sought in moderately- to well-sampled mineralised landscapes.

In a study area, the relative index of prospectivity (P_D) for mineral deposits (D, the target variable) of the type sought, based on known D deposit-type locations, can be defined as a function (f) of combinations of a number of X_i (i=1,2,…,n) spatial evidential features (i.e., explanatory/predictor variables), thus:

$$P_D = f(X_i, \ldots, X_n) \,. \tag{8.1}$$

For a study area, the map of D (i.e., the target variable) and each of the individual maps of explanatory/predictor variables are usually partitioned into equal-sized unit cells, which are usually squares. That is, relative indices of prospectivity (P_D) are expressed as degrees of likelihood of deposit-type occurrence per unit cell. Each of the individual maps of X_i spatial evidential features are discretised further into a number of C_{ji} (j=1,2,…,m) classes of evidence. For each unit cell in the study area, P_D can be defined further as:

$$P_D = f(wC_{ji}, \cdots, wC_{mn}) \,, \tag{8.2}$$

where wC_{ji} represents evidential weights (i.e., degree of spatial associations) of C_{ji} classes of individual X_i spatial evidential features with respect to D. The evidential weights relate to the degree of spatial coincidence or association between D and every C_{ji} in X_i maps. Unit cells characterised by high values of wC_{ji} in most, if not all, maps of X_i, therefore, have spatial geological attributes that are similar to (in terms of spatial association with) the unit cells containing the known locations of D. Equations (8.1) and (8.2), thus, simultaneously involve interpolative and extrapolative analyses of unit cells that likely contain unknown (or undiscovered) locations of D based on the spatial associations of C_{ji} in X_i maps with unit cells containing the known locations of D.

In Chapter 6, the methods for quantifying spatial associations between a map of deposit-type locations and individual maps of relevant geoscience spatial data do not lead directly to the derivation of values of wC_{ji} for C_{ji} classes in X_i maps of spatial evidential features. However, there are different data-driven techniques for deriving directly values of wC_{ji} for C_{ji} classes in X_i maps of spatial evidential features with respect to D (i.e., creating predictor maps) and then combining these predictor maps, via an integration function f (equation (8.2)), in order to obtain a predictive model of prospectivity (i.e., a map of P_D) for mineral deposits of the type sought. Chung and Fabbri (1993) proposed a unified mathematical framework for spatial predictive modeling of geo-objects (e.g., geohazard-prone areas, prospective areas, etc.).

There are two types of mathematical techniques for GIS-based data-driven predictive modeling of mineral prospectivity: bivariate and multivariate. Bivariate techniques (Table 8-I) involve pairwise analysis of spatial association between a map of D and a map of X_i with C_{ji} classes. In the applications of bivariate techniques, predictor maps are explicitly created and then integrated after the pairwise analyses of spatial associations. Weights-of-evidence modeling is the most commonly used bivariate technique. Multivariate techniques (Table 8-II) involve simultaneous analysis of spatial associations between a map of D and maps of X_i with C_{ji} classes. In the applications of multivariate techniques, predictor maps are usually created automatically and then integrated 'on-the-fly' or dynamically whilst the spatial associations between a map of D and maps of X_i

TABLE 8-I

Bivariate mathematical models/methods used for data-driven mapping of mineral prospectivity.

Model/method	References to examples
Weights-of-evidence modeling	Bonham-Carter et al. (1988, 1989), Agterberg et al. (1990, 1993a), Bonham-Carter and Agterberg (1990), Agterberg (1992), Bonham-Carter (1991, 1994), Cheng and Agterberg (1999), Mihalasky (1999), Raines (1999), Singer and Kouda (1999), Pan and Harris (2000), Mihalasky and Bonham-Carter (2001), Harris et al. (2001b), Agterberg and Cheng (2002), Harris et al. (2003), Carranza (2004b), Porwal et al. (2001, 2006a), Coolbaugh and Bedell (2006), Harris and Sanborn-Barrie (2006), Porwal (2006)
Evidential belief modeling	Chung and Fabbri (1993), An et al. (1994b), Carranza (2002), Carranza and Hale (2003), Carranza et al. (2005, 2008a, 2008b; 2008c), this volume

with C_{ji} classes are being quantified. It seems that logistic regression and artificial neural networks are the most commonly used multivariate techniques for data-driven predictive modeling of mineral prospectivity. The multivariate techniques outnumber the bivariate techniques for creating and then integrating predictor maps in mineral prospectivity modeling. This indicates that, because of the highly complex nature of spatial associations between mineral deposits and geological features, it is in most cases more desirable to develop and/or apply multivariate rather than bivariate techniques for data-driven modeling of mineral prospectivity. In some cases it is even more desirable to develop and/or apply hybrid methods for data-driven modeling of mineral prospectivity, like fuzzy weights-of-evidence modeling (Cheng and Agterberg, 1999; Porwal, 2006; Porwal et al., 2006a), data-driven fuzzy modeling (Luo and Dimitrakopoulos, 2003; Porwal et al., 2003b) and neuro-fuzzy modeling (Porwal et al., 2004; Porwal, 2006).

The different methods of GIS-based data-driven modeling of mineral prospectivity are well-documented in the literature and are now mostly well-established. In Tables 8-I and 8-II, the researches described in the references cited for each method range from seminal studies in developing a method, to innovative or adaptive studies providing improvements of a method, to instructive studies in various cases demonstrating or addressing further certain aspects that are vital in the application of a method. Moreover, some of the references cited in Tables 8-I and 8-II compare and contrast some of the techniques for data-driven modeling of mineral prospectivity. Therefore, this chapter does not attempt to explain and demonstrate each of the different methods of GIS-based data-driven techniques for modeling mineral prospectivity. However, one bivariate technique (evidential belief modeling) and one multivariate technique (discriminant analysis) are explained and demonstrated here in a case study of mapping epithermal Au prospectivity in the Aroroy district (Philippines).

TABLE 8-II

Multivariate mathematical models/methods for data-driven mapping of mineral prospectivity.

Model/method	References to examples
Discriminant analysis	Chung (1977), Prelat (1977), Bonham-Carter and Chung (1983), Harris and Pan (1999), Pan and Harris (2000), Harris et al. (2003), this volume
Characteristic analysis	Botbol et al. (1977, 1978), McCammon et al. (1983, 1984), Harris (1984), Pan and Harris (1992a, 2000)
Logistic regression analysis	Chung (1978, 1983), Chung and Agterberg (1980, 1988), Bonham-Carter and Chung (1983), Agterberg (1988, 1992, 1993), Agterberg et al. (1993a), Harris and Pan (1991, 1999), Sahoo and Pandalai (1999), Pan and Harris (2000), Harris et al. (2001b, 2006), Carranza and Hale (2001b), Carranza (2002), Harris et al. (2003)
Favourability analysis	Pan (1989, 1993a, 1993b, 1993c), Pan and Portefield (1995), Pan and Harris (1992b, 2000)
Likelihood ratio analysis	Chung and Fabbri (1993), Chung et al. (2002), Chung and Keating (2002), Chung (2003), Harris and Sanborn-Barrie (2006)
Artificial neural networks	Singer and Kouda (1996, 1997, 1999), Harris and Pan (1999), Pan and Harris (2000), Brown et al. (2000, 2003), Bougrain et al. (2003), Harris et al. (2003), Porwal et al. (2003a, 2004), Rigol-Sanchez et al. (2003), Harris and Sanborn-Barrie (2006), Porwal (2006), Skabar (2005, 2007a, 2007b), Nykänen (2008)
Bayesian network classifiers	Porwal (2006), Porwal et al. (2006b), Porwal and Carranza (2008)

In the case study area (covering roughly 130 km^2; see Fig. 3-9), there are 13 known locations of epithermal Au deposits and the epithermal Au prospectivity recognition criteria (defined in Chapter 6) are as follows.

- Proximity to NNW-trending faults/fractures (representing structural controls).
- Proximity to NW-trending faults/fractures (representing structural controls).
- Proximity to intersections of NNW- and NW-trending faults/fractures (representing structural controls as well as proxies for heat source controls).
- Presence of multi-element stream sediment geochemical anomalies (representing surficial evidence).

The spatial data sets used in the case study are: (a) distance to NNW-trending faults/fractures; (b) distance to NW-trending faults/fractures; (c) distance to intersections of NNW- and NW-trending faults/fractures; and (d) integrated PC2 and PC3 scores obtained from the catchment basin analysis of stream sediment geochemical data (see Chapter 5 and Fig. 5-12).

Before explaining and demonstrating evidential belief modeling and discriminant analysis of mineral prospectivity, this chapter first discusses techniques or strategies addressing the issues of (a) objective selection of a suitable unit cell size for data-driven modeling mineral prospectivity, (b) selection of coherent deposit-type locations for data-driven modeling of mineral prospectivity and (c) cross-validation of data-driven mineral prospectivity models. These three issues are vital to the application of any bivariate or multivariate technique of GIS-based data-driven modeling of mineral prospectivity, so it is of foremost importance that they are given prior considerations.

SELECTION OF SUITABLE UNIT CELL SIZE FOR MODELING

Because known deposit-type locations are *samples* of a mineralised landscape under investigation, it is essential to consider and determine a suitable *sampling density* in data-driven modeling of mineral prospectivity. By doing so makes data-driven modeling of mineral prospectivity consistent with the general principles of the statistical theory, sampling theory and information theory. Mineral prospectivity is one type of (spatial) information, which is derived systematically via (statistical) analysis of multiple sets of various geo-exploration data (i.e., explanatory/predictor variables) at known deposit-type locations and at non-deposit locations. Prospective areas, as we recall from Chapter 2, are not only *definition-limited* but also *sampling-limited* geo-objects.

Sampling density in data-driven modeling of mineral prospectivity is defined by a *unit cell size* that is used for representing known deposit-type locations and in discretising spatial data of explanatory/predictor variables. A unit cell is equivalent to a *grid cell* in sampling. It is also equivalent to a *pixel* in GIS (see Fig. 2-5). The size of a unit cell is described by its length or width, if it is a rectangle, or by its diameter or radius, if it is a circle. The size of a unit cell defines the *spatial resolution* (grid or pixel resolution) of models of spatial data (see Chapter 2). The size of a unit cell, therefore, defines the spatial accuracy of the derived pieces of geo-information portrayed in maps or images. Fine unit cells represent high sampling density and thus good spatial accuracy of information, whereas coarse unit cells represent low sampling density and thus poor spatial accuracy of information. This comparison of unit cell sizes and, thus, sampling densities, in the context of mineral prospectivity mapping, is pertinent only to a particular areas, because the spatial distributions of mineral deposits and their spatial associations with geological features vary from one area to another. The size of a unit cell also determines the scale of a cartographic map. The finer the unit cell, the larger the map scale; the coarser the unit cell, the smaller the map scale. However, map scale does not symbolise and, thus, should not be confused with size of a study area.

Most, if not all, studies of GIS-based data-driven modeling of mineral prospectivity make use of equal-sized (usually square) unit cells. The choice of a suitable unit cell size in GIS-based data-driven modeling of mineral prospectivity is based on (a) the knowledge that every mineral deposit is unique, even if they are classified to certain types of mineral deposits, and (b) the assumption that each unit cell must contain just one of each of the known locations of mineral deposits of the type sought. However, the choice of a suitable unit cell size in most studies of GIS-based data-driven modeling of mineral prospectivity is usually subjective. The following discussions explain and demonstrate analytical tools that can aid in the objective selection of a suitable square unit cell size for GIS-based data-driven modeling of mineral prospectivity.

The choice of a suitable unit cell size must be based on the spatial configuration or pattern of locations of a-priori samples, i.e., the known locations of mineral deposits of the type sought. In sampling theory, this strategy is referred to as *retrospective sampling*, which is applied to previously sampled areas, as opposed to *prospective sampling*, which is applied to unsampled areas. Because known locations of mineral deposits are usually depicted as points in data-driven modeling of mineral prospectivity, especially in regional- to district-scales of mapping, algorithms of point pattern analysis for measures of dispersion (Boots and Getis, 1988; Rowlingson and Diggle, 1993), which are independent of the size of a study area, can be used to determine distances from every deposit-type location and the corresponding probabilities associated with these distances that there is one neighbour deposit-type location situated next to another deposit-type location. The range of distances in which there is zero probability of one neighbour deposit-type location situated next to another deposit-type location is a set of choices for a suitable unit cell size. Fig. 8-1 shows the results of the application of measures of dispersion via point pattern analysis to different types of mineral deposits and to geothermal prospects in four different areas. For each area, the results suggest different ranges of suitable unit cell sizes for data-driven modeling of prospectivity for the types of Earth resources under examination. For data-driven modeling of prospectivity for epithermal Au deposits in the Aroroy district (Philippines), a suitable unit cell size is at most 560 m. For data-driven modeling of prospectivity for epithermal Au deposits in the Cabo de Gata area (Spain), a suitable unit cell size is at most 160 m. For data-driven modeling of geothermal prospectivity in West Java (Indonesia), a suitable unit cell size is at most 750 m. For data-driven modeling of prospectivity for alkalic porphyry Cu-Au deposits in British Columbia, a suitable cell size is at most 330 m. The results of the application of measures of dispersion via point pattern analysis (e.g., Fig. 8-1) are considered further in a graphical analysis, which is described below, to aid the objective selection of a suitable unit cell size for GIS-based data-driven modeling of mineral prospectivity.

Based on a unit cell size [denoted hereafter as $N(\bullet)$], a study area T has $N(T)$ total number of unit cells, has $N(D)$ number of unit cells each containing just one D deposit-type location and has $N(T)$–$N(D)$ number of unit cells not containing D. An a-priori estimate of prospectivity (i.e., in the absence of spatial evidence) of mineral deposits of the type sought in a study area is the ratio $[N(D)] : [N(T)–N(D)]$. This ratio represents the spatial contrast between mineralised cells and barren cells. Empirically, the ratio $[N(D)] : [N(T)–N(D)]$ must be a very small value because mineralisation is a relatively rare geological phenomenon, meaning that $N(\bullet)$ must be suitably fine. All possible $N(\bullet)$ within the range of distances with zero probability of one neighbour D situated next to another D (Fig. 8-1) result in very small values of the ratio $[N(D)] : [N(T)–N(D)]$. However, Fig. 8-2 shows that the ratio $[N(D)] : [N(T)–N(D)]$ increases exponentially as

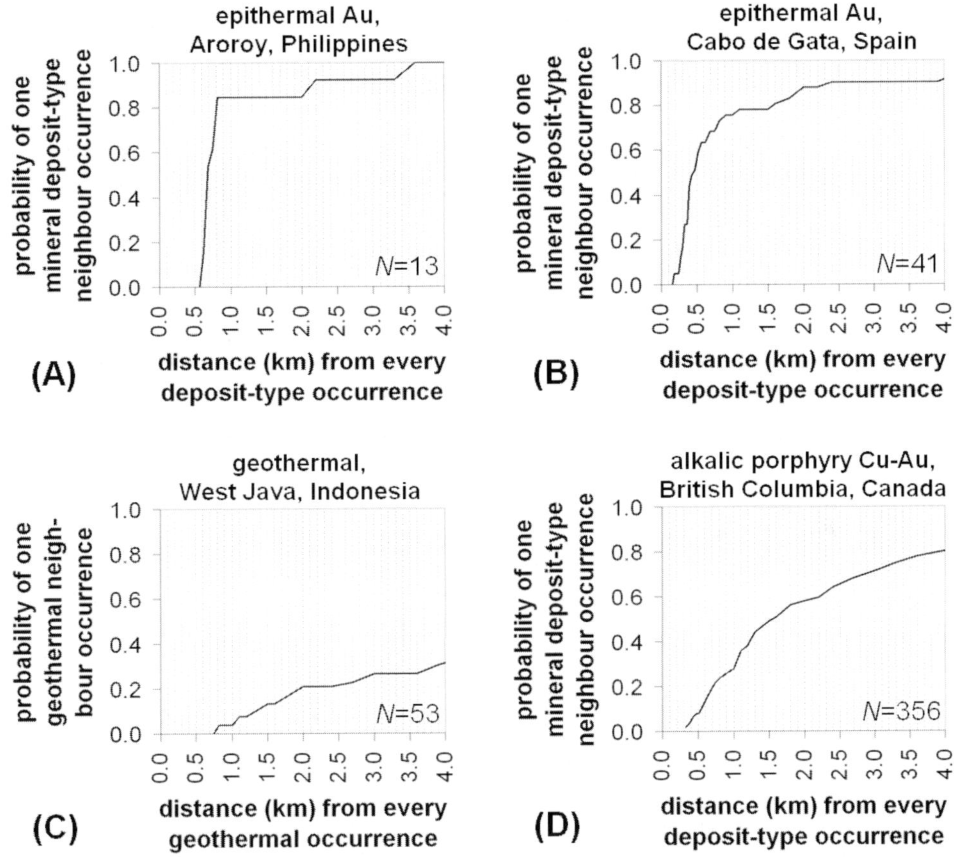

Fig. 8-1. Distances and corresponding probabilities that one resource-type location is situated next to another resource-type location in a study area. Results of application of measures of dispersion via point pattern analysis (Boots and Getis, 1988; Rowlingson and Diggle, 1993) to the locations of (**A**) epithermal Au deposits in Aroroy, Philippines (see Fig. 3-9), (**B**) epithermal Au deposits in Cabo de Gata, Spain (see Carranza et al., 2008a), (**C**) geothermal occurrences in West Java, Indonesia (see Carranza et al., 2008c) and (**D**) alkalic porphyry Cu-Au deposits in British Columbia, Canada (see Carranza et al., 2008b). N denotes number of resource-type locations.

the $N(\bullet)$ increases linearly. Because $N(D)$ is constant when each D is contained in only one cell (except in Fig. 8-2D when $N(\bullet) \geq 330$ m), the exponential increase in the ratio $[N(D)] : [N(T)–N(D)]$ is due to the exponential decrease in $[N(T)–N(D)]$ as $N(\bullet)$ linearly increases.

The graphs in Fig 8-2 do not readily indicate, however, which $N(\bullet)$ is the most suitable for data-driven modeling of prospectivity of the deposit-type of interest in each of the areas under examination. Nevertheless, the individual data sets plotted in Fig. 8-2

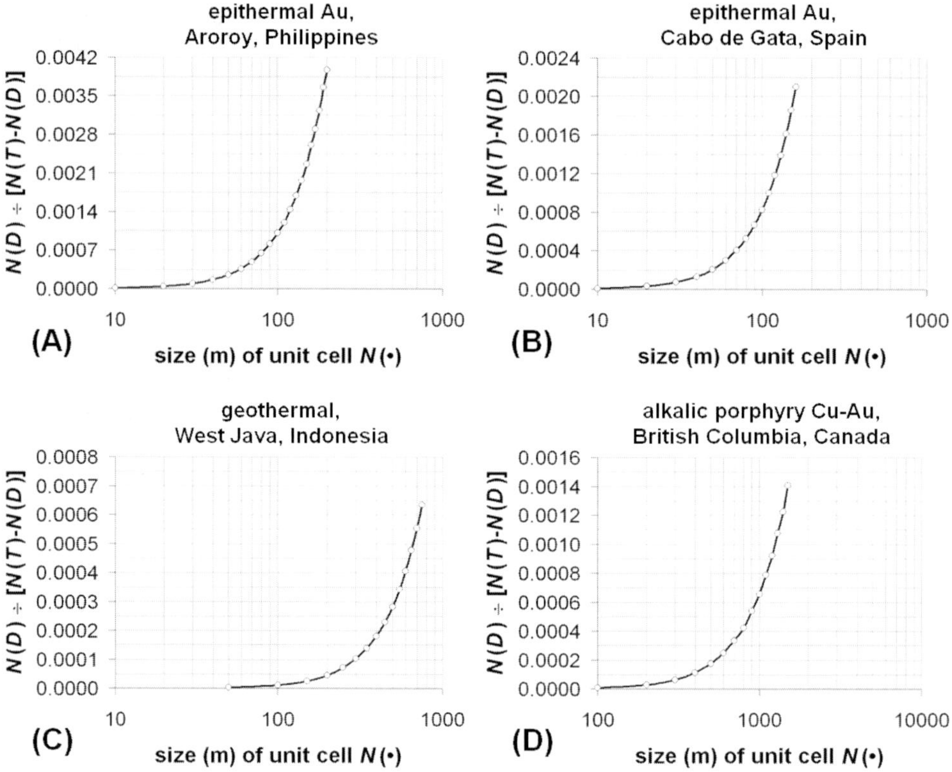

Fig. 8-2. Variations of the ratio [*N*(*D*)] : [*N*(*T*)–*N*(*D*)] as a function of unit cell size *N*(•). The results shown are mostly for the ranges of zero-probability distances derived from the point pattern analyses (Fig. 8-1) of (**A**) epithermal Au deposits, Aroroy (Philippines), (**B**) epithermal Au deposits, Cabo de Gata (Spain), (**C**) geothermal occurrences, West Java (Indonesia) and (**D**) alkalic porphyry Cu-Au deposits, British Columbia (Canada).

can be investigated further to find the most suitable *N*(•) per area. One procedure of doing so is to determine the amount (expressed in %) of increase in the ratio [*N*(*D*)] : [*N*(*T*)–*N*(*D*)] from a unit cell size $N(\bullet)_i$ to the next coarser unit cell size $N(\bullet)_{i+1}$ (where 1 denotes the unit cell size interval used to create the graphs in Fig. 8-2). An alternative procedure is to determine the amount (expressed in %) of decrease in the ratio [*N*(*D*)] : [*N*(*T*)–*N*(*D*)] from a unit cell size $N(\bullet)_i$ to the next finer unit cell size $N(\bullet)_{i-1}$ (where 1 denotes the unit cell size interval used to create the graphs in Fig. 8-2). Based on the first procedure, the graphs in Fig. 8-3 indicate that the rate of increase in the ratio [*N*(*D*)] : [*N*(*T*)–*N*(*D*)] is much higher when using finer unit cells than when using coarser unit cells. The results shown in Fig. 8-3 are consistent with the knowledge that, in raster-based GIS (cf. Stein et al., 2001; Hengl, 2006; Nykänen and Raines, 2006), the overall spatial information content in a map (in this case a map of *D*) decreases and increases as

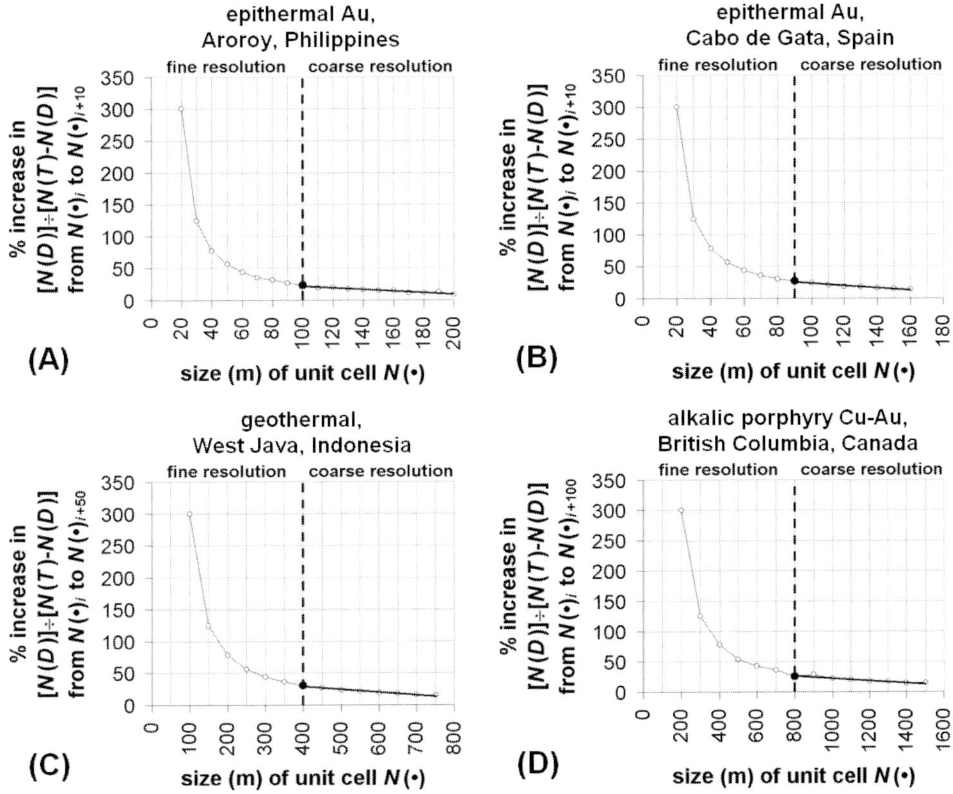

Fig. 8-3. Variations in the rate of percent increase in the ratio $[N(D)] : [N(T)–N(D)]$ as function of linear increase in unit cell size $N(\bullet)$ for representation of deposit-type locations: (**A**) epithermal Au deposits in Aroroy, Philippines (see Fig. 3-9); (**B**) epithermal Au deposits in Cabo de Gata, Spain (see Carranza et al., 2008c); (**C**) geothermal occurrences in West Java, Indonesia (see Carranza et al., 2008a); and (**D**) alkalic porphyry Cu-Au deposits in British Columbia, Canada (see Carranza et al., 2008b). Changes in fine resolution $N(\bullet)$ result in exponentially decreasing rates (thin curves) of increase in the ratio $[N(D)] : [N(T)–N(D)]$, whilst changes in coarse resolution $N(\bullet)$ result in weak linearly decreasing rates (thick curves) of increase in the ratio $[N(D)] : [N(T)–N(D)]$. See text for further explanation.

the spatial resolution becomes coarser and finer, respectively. The results of the second procedure lead to identical interpretations as the results of the first procedure.

The graphs in Fig. 8-3 allow distinction between $N(\bullet)$ that can be considered 'fine' resolution and $N(\bullet)$ that can be considered 'coarse' resolution. Gradual (i.e., equal interval) changes in the sizes of fine resolution $N(\bullet)$ are associated with exponential rates of increase in the spatial information content in a map of D, whilst gradual changes in the sizes of coarse resolution $N(\bullet)$ are associated with linear rates of increase in the spatial information content in a map D. A $N(\bullet)$ about the transition from fine resolution $N(\bullet)$ to coarse resolution $N(\bullet)$ represents a threshold $N(\bullet)$ that can be considered the most suitable $N(\bullet)$. A $N(\bullet)$ that is either much finer or much coarser than the most suitable $N(\bullet)$ is likely an impractical representation of D. Thus, according to the results shown in Fig 8-3, the most suitable $N(\bullet)$ is either the coarsest fine resolution $N(\bullet)$ or the finest coarse resolution $N(\bullet)$. For data-driven modeling of prospectivity for epithermal Au deposits in the Aroroy district (Philippines), the results suggest that the most suitable $N(\bullet)$ is 100 m (Fig. 8-3A). For data-driven modeling of prospectivity for epithermal Au deposits in the Cabo de Gata area (Spain), the results suggest that the most suitable $N(\bullet)$ is 90 m (Fig. 8-3B). For data-driven modeling of geothermal prospectivity in West Java (Indonesia), the result suggest that the most suitable $N(\bullet)$ is 400 m (Fig. 8-3C). For data-driven modeling of prospectivity for alkalic porphyry Cu-Au deposits in British Columbia, the result suggest that the most suitable $N(\bullet)$ is 800 m (Fig. 8-3D). Based on the analyses of the graphs in Fig. 8-3, it seems that the most suitable $N(\bullet)$ is approximately an inflection point in each of the curves of $[N(D)] : [N(T)–N(D)]$ versus $N(\bullet)$ shown in Fig. 8-2. Provided that it is so, visual inspection of an inflection point in a curve of $[N(D)] : [N(T)–N(D)]$ versus $N(\bullet)$ is, however, difficult because such a curve is very smooth (Fig. 8-2). The technique of deriving the curves shown in Fig. 8-3 aids, therefore, in identification of an inflection point in a curve of $[N(D)] : [N(T)–N(D)]$ versus $N(\bullet)$ and in selection of a most suitable $N(\bullet)$.

In contrast to and notwithstanding of the results of the analyses illustrated in Figs. 8-2 and 8-3, the following previous works of GIS-based data-driven modeling of mineral prospectivity each used a $N(\bullet)$ based on subjective judgment in view of the distance-probability relation. In data-driven modeling of prospectivity for epithermal Au deposits in the Cabo de Gata area (Spain), Carranza et al. (2008a) chose and used a $N(\bullet)$ of 100 m, which is within the range of distances in which there is zero probability of one neighbour epithermal Au deposit location situated next to another epithermal Au deposit location (Fig. 8-1B) and which is slightly coarser than the most suitable $N(\bullet)$ of 90 m suggested by the results presented in Fig. 8-3B. In data-driven modeling of geothermal prospectivity in West Java (Indonesia), Carranza et al. (2008c) selected and used a $N(\bullet)$ of 500, which is within the range of distances in which there is zero probability of one neighbour geothermal location situated next to another geothermal location (Fig. 8-1C) and which is slightly coarser than the most suitable $N(\bullet)$ of 400 m suggested by the results displayed in Fig. 8-3C. In data-driven modeling of prospectivity for alkalic porphyry Cu-Au deposits in British Columbia, Carranza et al. (2008b) used a $N(\bullet)$ of 1 km, which is outside the range of distances in which there is zero probability of one neighbour alkalic porphyry Cu-Au deposit location situated next to another alkalic porphyry Cu-Au deposit location (Fig. 8-1A) and which is slightly coarser than the most suitable $N(\bullet)$ of 800 m suggested by the results shown in Fig. 8-3D. In these three case studies, the common reason for selecting and using a $N(\bullet)$ that is slightly coarser than the most suitable $N(\bullet)$ suggested by the results shown in Figs. 8-3B to 8-3D is simplicity of area calculations (i.e., as multiples of number of unit cells or pixels) in the application of a raster-based or pixel-based GIS.

The application of point pattern analysis (Boots and Getis, 1988; Rowlingson and Diggle, 1993), as shown in Fig. 8-1, can therefore be useful in deriving a preliminary set of choices for a suitable $N(\bullet)$. However, the final choice of a suitable unit cell size must also consider (a) the scale of the field geological observations used in constructing a mineral deposit occurrence database, (b) the scales of the input maps or images of spatial data of explanatory/predictor variables and (c) the scale of the desired output mineral prospectivity map(s). These considerations were taken by Carranza et al. (2008b) in data-driven modeling of prospectivity for alkalic porphyry Cu-Au deposits in British Columbia. The current British Columbia MINFILE mineral inventory database (BCGS, 2007) contains records of prospect- to mine-camp-scale (usually larger than 1:10,000) data for 356 locations of alkalic porphyry Cu-Au deposits. In contrast, the scale of the geologic map is 1:250,000 (Massey et al., 2005), whereas the airborne magnetic and gravity data were captured in 1-km and 2-km grids, respectively (Geoscience Data Repository, 2006a, 2006b), which translate to map scales of about 1:400,000 and 1:800,000, respectively (see Hengl, 2006). Therefore, in view the different scales or spatial resolutions of input spatial data of the explanatory/predictor variables, Carranza et al. (2008b) used an 'average' unit cell size of 1 km. The use of a larger unit cell size than indicated by the distance-probability relation can mean that more than one deposit-type location is covered by a unit cell, and this was the case for some unit cells in the British Columbia study. Here, however, some of the alkalic porphyry Cu-Au deposits (e.g., Axe (Adit Zone), Axe (South Zone) and Axe (West Zone)) described in the MINFILE database probably represent one large alkalic porphyry Cu-Au deposit so, if that is the case, using a larger unit cell size than indicated by the distance-probability relation is justifiable.

Whether a preliminary set of choices for a suitable $N(\bullet)$ indicated by the distance-probability relation is adopted or adapted, the analysis of the rate of increase in the ratio $[N(D)] : [N(T)–N(D)]$ as function of equal-interval change in $N(\bullet)$, as illustrated in Fig. 8-3, is robust regardless of the number of deposit-type locations and the size of a study area. Nevertheless, it is also imperative to verify if the most suitable $N(\bullet)$ suggested by results of analyses depicted in Figs. 8-2 and 8-3 is reasonably consistent with the average lateral extents (at prospect- to mine-camp-scales) of known occurrences of mineral deposits of the type sought. Data of lateral extents of known occurrences of mineral deposits of the type sought are, unfortunately, not available in many cases. If such is the case, then the sorts of analyses demonstrated here, although mainly graphical, provide an objective way of selecting the most suitable $N(\bullet)$ for GIS-based data-driven modeling of mineral prospectivity. After having made a final objective choice of a suitable $N(\bullet)$, one must next determine which of the known locations of mineral deposits of the type sought are suitable in data-driven modeling of mineral prospectivity.

SELECTION OF COHERENT DEPOSIT-TYPE LOCATIONS FOR MODELING

Every mineral deposit, even if classified into a deposit-type, is unique and has characteristics that are, to a certain extent, dissimilar to other mineral deposits of the same type. It follows that multivariate spatial data signatures of deposit-type locations are, to a certain extent, dissimilar or non-coherent. Because modeling of mineral prospectivity involves 'fitting' (i.e., establishing spatial associations between) a map of D with several maps of spatial data of X_i evidential features, dissimilarity (i.e., heterogeneity or non-coherence) of multivariate spatial data signatures of deposit-type locations can undermine the quality of a data-driven model of mineral prospectivity. Carranza et al. (2008b) have shown that uncertainties of a data-driven model of mineral prospectivity can be reduced and that fitting- and prediction-rates of a data-driven model of model prospectivity can be improved by using a set of coherent deposit-type locations (i.e., with similar multivariate spatial data signatures).

A two-stage methodology for selection of coherent deposit-type locations is explained and demonstrated here (after Carranza et al., 2008b): (1) analysis of mineral occurrence favourability scores of individual spatial data sets with respect to deposit-type and non-deposit locations; and (2) analysis of deposit-type locations with similar multivariate spatial data signatures. This two-stage methodology is demonstrated in the case study area (Aroroy district, Philippines). Before doing so, let us address first the issues of (a) increasing the number of locations of the target variable in a study area if it is considered and/or found insufficient (e.g., 13 as in the case study area here) to derive, depending on the method (Tables 8-I and 8-II), a proper (e.g., statistically significant) data-driven model of mineral prospectivity and (b) selecting non-deposit locations required in the analysis of coherent deposit-type locations and in the application of multivariate methods of data-driven modeling of mineral prospectivity (Table 8-II).

The issue of increasing the number of locations of the target variable can be addressed by considering locations immediately around each of the known deposit-type locations as proxy deposit-type locations. This consideration is based on the assumption that locations immediately around known deposit-type locations are also probably (albeit weakly) mineralised. Thus, given a map of D partitioned into equal-sized unit cells, the eight unit cells surrounding the unit cell representing a deposit-type location can be considered proxy deposit-type locations. (For the case study area, there are 104 (i.e., 13×8) proxy deposit-type locations.) Alternatively, an optimum distance buffer around each deposit-type location can be sought via point pattern analysis (Boots and Getis, 1988; Rowlingson and Diggle, 1993) such that (a) there is zero probability of a neighbour deposit-type location within a buffered deposit-type location and (b) based on knowledge that mineralisation is a very rare geological phenomenon, the total number of unit cells representing deposit-type and proxy deposit-type locations forms a very small percentage (say, 1%) of the total number of unit cells in a study area (Carranza and Hale, 2001b, 2003; Carranza, 2002). The application of proxy deposit-type locations reduces artificial spatial associations between evidential data and deposit-type locations (Stensgaard et al., 2006), which usually occur when the number of the latter is small (e.g., Carranza, 2004b). However, like the multivariate spatial data signatures of deposit-type locations, the multivariate spatial data signatures of proxy deposit-type locations are, to a certain extent, dissimilar or non-coherent. So, the two-stage methodology for selection of coherent deposit-type locations is also demonstrated in selecting coherent proxy deposit-type locations.

The issue of selecting non-deposit locations can be addressed by considering the following three selection criteria (Carranza et al., 2008b). Firstly, in contrast to deposit-type locations, which exhibit non-random spatial patterns (see Chapter 6), non-deposit locations must be random (or randomly selected) so that their multivariate spatial data signatures are likely non-coherent. Point pattern analysis (Diggle, 1983; Boots and Getis, 1988) can be applied to evaluate degrees of spatial randomness of selected non-deposit locations (see Chapter 6 for application to deposit-type locations). Secondly, random (or randomly-selected) non-deposit locations must be distal to (or located far away from) deposit-type locations under study because locations proximal to deposit-type locations likely have similar multivariate spatial data signatures to deposit-type locations and thus probably do not qualify as non-deposit locations. Point pattern analysis (Diggle, 1983; Boots and Getis, 1988) can be applied to determine the minimum distance from every deposit-type location within which there is 100% probability of a neighbour deposit-type location. In some cases, the criterion of 'minimum distance with 100% probability' may leave insufficient locations for random selection of non-deposit locations, so one may consider a lower probability distance. (For the case study area, non-deposit locations are randomly selected beyond 2.2 km of any deposit-type location; within this distance from any deposit-type location there is 90% probability of a deposit-type location (Fig. 8-1A).) Thirdly, the number of distal and random non-deposit locations must be equal to the number of deposit-type locations, because the latter locations are rare. This criterion applies especially when logistic regression is used, as in the analysis of coherent deposit-type locations (see below). The use of equal number of 'zeros' (e.g., non-deposit locations) and 'ones' (e.g., deposit-type locations) in logistic regression is optimal when the latter is rare (Breslow and Cain, 1988; Schill et al., 1993). In cases of rare 'ones', King and Zeng (2001) aver that the information content contributed by the independent variables used in logistic regression starts to diminish as the number of 'zeros' exceeds the number of 'ones'. (For the case study, 117 distal and random non-deposit locations are selected in order to match the total number of deposit-type and proxy deposit-type locations. Two sets of distal and random non-deposit locations are generated (Fig. 8-4) in order to demonstrate the reproducibility and robustness of the methodology of selecting coherent deposit-type locations.)

We now turn to the two-stage analysis of coherent deposit-type locations.

Analysis of mineral occurrence favourability scores at deposit-type locations

This section describes and discusses the first stage in selecting coherent deposit-type locations and coherent proxy deposit-type locations. This involves deriving mineral occurrence favourability scores (*MOFS*) of spatial data with respect to deposit-type and

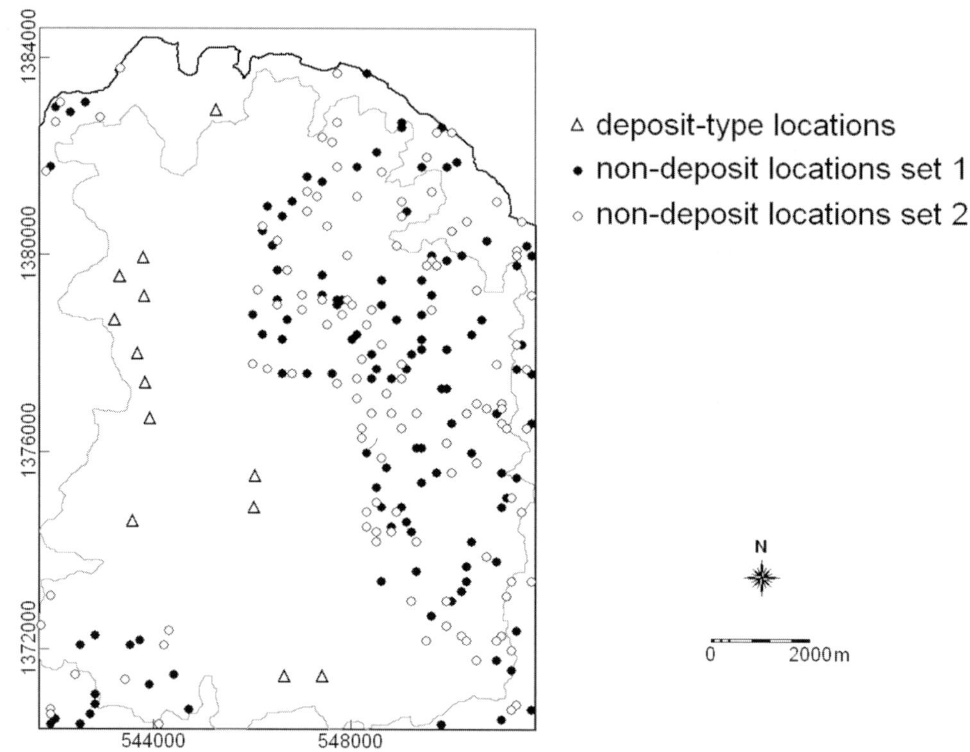

Fig. 8-4. Two sets of randomly-selected non-deposit locations situated distal to [deposit-type] locations of epithermal Au deposits in Aroroy district (Philippines). Polygon outlined in grey is area of stream sample catchment basins (see Fig. 4-11).

non-deposit locations. The *MOFS* represent likelihood of mineral occurrence, in the range [0,1], as a function of spatial data representing the presence of indicative geological features. Deriving the *MOFS* involves establishing spatial associations between maps of individual spatial data sets and a map of deposit-type locations. For this purpose, either the distance distribution method or the distance correlation method, which are explained and demonstrated in Chapter 6, can be used. Only spatial data sets exhibiting positive spatial associations with the deposit-type locations are used further in the analysis. Thus, for the case study area (see results of analysis in Chapter 6), the spatial data sets used are (1) distance to NNW-trending faults/fractures, (b) distance to NW-trending faults/fractures, (3) distance to intersections of NNW- and NW-trending faults/fractures and (4) integrated PC2 and PC3 scores obtained from the catchment basin analysis of stream sediment geochemical data (see Chapter 5 and Fig. 5-12). For spatial data representing distances to geological features (e.g., faults/fractures), distances equal to or less than the distance of optimum positive spatial association with the

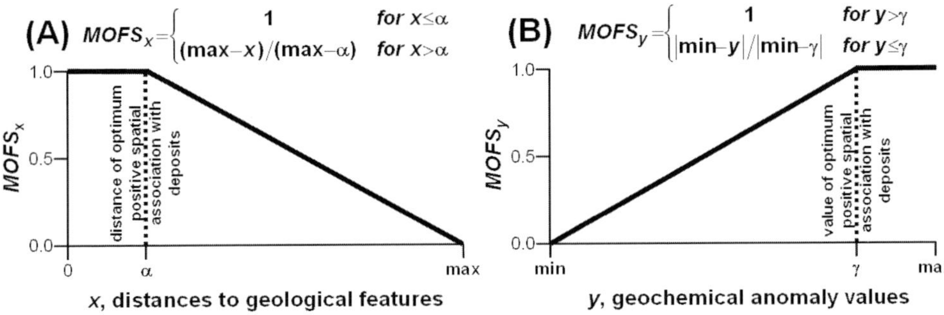

Fig. 8-5. Scheme of assigning or calculating mineral occurrence favourability scores (*MOFS*) to spatial data, such as (**A**) distances to geological features (e.g., faults/fractures) and (**B**) geochemical anomaly values, based on their spatial association with deposit-type locations.

deposit-type locations are assigned *MOFS* of [1], whilst distances greater than the distance of optimum spatial association with the deposit-type locations are assigned linearly decreasing *MOFS* from [1] to [0] (Fig. 8-5A). For spatial data representing geochemical anomaly values, values greater than the value of optimum positive spatial association with the deposit-type locations are assigned *MOFS* of [1], whilst values equal to or less than the value of optimum spatial association with the deposit-type locations are assigned linearly decreasing *MOFS* from [1] to [0] (Fig. 8-5B).

A database of *MOFS* of spatial data is then created for the deposit-type locations, the proxy deposit-type locations and randomly-selected distal non-deposit locations. In a GIS, creating this database involves an overlay operation (plus table operation) between a map of *MOFS* and a map of point locations under study (see, for example, Figs. 3-7 and/or 5-6). Then, it is instructive to create boxplots of *MOFS* of spatial data at the deposit-type locations, the proxy deposit-type locations and randomly selected non-deposit locations in order to visualise the overall dissimilarities in the spatial characteristics of these locations. For example, in the case study area, the locations of epithermal Au deposits and their immediate surroundings are strongly dissimilar to non-deposit locations in terms of proximity to faults/fractures (Figs. 8-6A to 8-6C) and are moderately to strongly dissimilar to non-deposit locations in terms of geochemical anomalies (Fig. 8-6D). However, Fig. 8-6 also shows that even individual deposit-type locations and individual proxy deposit-type locations exhibit dissimilarities in terms of proximity to faults/fractures and geochemical anomalies, which indicate that multivariate spatial data signatures of deposit-type locations and of proxy deposit-type locations are, to a certain extent, dissimilar or non-coherent.

Analysis of coherent deposit-type locations

This section describes and discusses the second stage in selecting coherent deposit-type locations and coherent proxy deposit-type locations. The concept of the analysis in

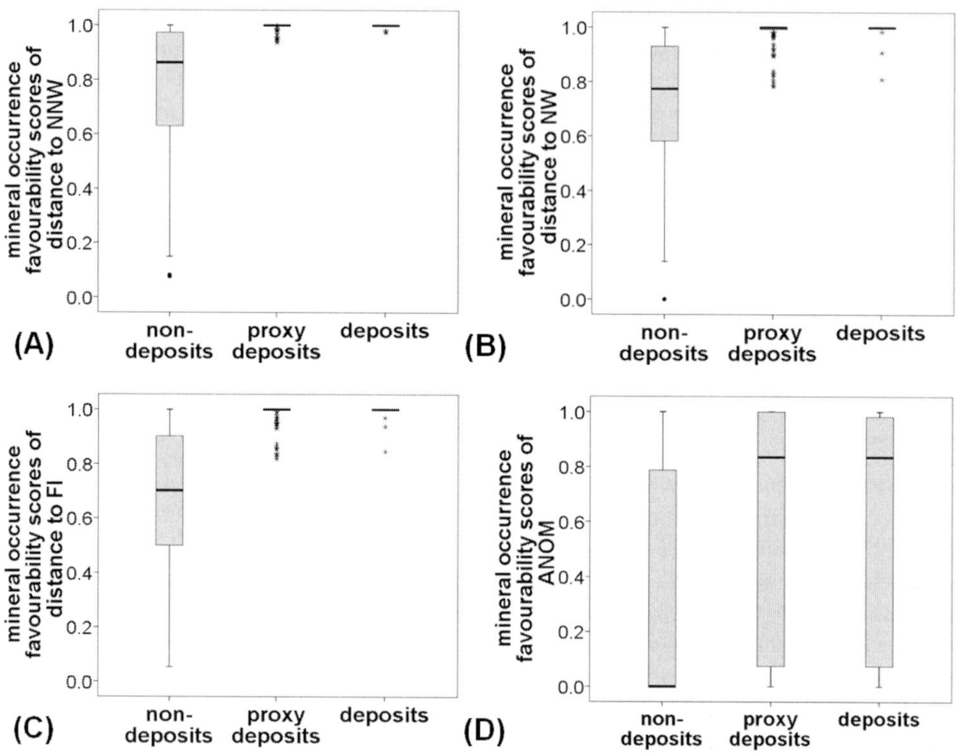

Fig. 8-6. Boxplots of mineral occurrence favourability scores (*MOFS*) of spatial data at deposit-type, proxy deposit-type and non-deposit locations in the Aroroy district (Philippines): (**A**) distance to NNW-trending faults/fractures; (**B**) distance to NW-trending faults/fractures; (**C**) distance to intersections of NNW- and NW-trending faults/fractures (FI); and (**D**) integrated PC2 and PC3 scores (ANOM) obtained from the catchment basin analysis of stream sediment geochemical data (see Fig. 5-12). See Fig. 3-4 for explanation of features of a boxplot.

this stage is the following. Deposit-type locations and proxy deposit-type locations can each be given a mineral occurrence (Y) score of [1], whereas non-deposit locations can each be given a Y score of [0]. However, based on the *MOFS* of certain spatial data (Fig. 8-6), the likelihood of mineral occurrence at deposit-type and proxy deposit-type locations, given certain spatial evidence, is not always maximum (or 1) and the likelihood of mineral occurrence at non-deposit locations is not always minimum (or 0). By modeling a mathematical relationship between mineral occurrence scores, Y_i, at i $(=1,2,\ldots,n)$ deposit-type, proxy deposit-type and non-deposit locations and a number of $(j=1,2,\ldots,m)$ sets of $MOFS_{ji}$ of spatial data at the same i $(=1,2,\ldots,n)$ deposit-type, proxy deposit-type and non-deposit locations, a predicted mineral occurrence score, $\hat{Y}_i$, can be derived for the individual deposit-type, proxy deposit-type and non-deposit locations. A predicted mineral occurrence score ($\hat{Y}_i$) represents a multivariate spatial data signature at every deposit-type, proxy deposit-type and non-deposit location. By modeling a mathematical relationship between a set of mineral occurrence scores (Y_i) and multiple sets of $MOFS_{ji}$ of spatial data at deposit-type, proxy deposit-type and non-deposit locations, weak dissimilarities in multivariate spatial data signatures of deposit-type locations and weak to moderate dissimilarities in multivariate spatial data signatures of proxy deposit-type locations, as indicated in Fig. 8-6, can be enhanced. This allows distinction between coherent and non-coherent deposit-type locations and between coherent and non-coherent proxy deposit-type locations. Then a threshold $\hat{Y}_i$ can be sought to distinguish between coherent and non-coherent deposit-type locations and between coherent and non-coherent proxy deposit-type locations.

Because the mineral occurrence score, Y_i, is a dichotomous variable, logistic regression is appropriate in modeling the relationship between Y_i and $MOFS_{ji}$ in order to derive $\hat{Y}_i$ in the unit range [0,1], viz. (Rock, 1988a; Hosmer and Lemeshow, 2000):

$$\hat{Y}_i = 1\Big/\left[1 + e^{-(b_0 + b_j MOFS_{ji} + \cdots + b_m MOFS_{mn})}\right] \tag{8.3}$$

where b_0 is a constant and b_j is the coefficient of the jth (j=1,2,…,m) $MOFS_{ji}$ independent variable. In logistic regression, the relationship between the dependent and independent variables is not a linear function. Data of independent variables used in logistic regression can be of any form; they can be dichotomous, nominal, interval or ratio variables (Hosmer and Lemeshow, 2000). Logistic regression makes no assumption about the distribution of data of independent variables; they do not have to be normally distributed, linearly related or of equal variance. However, for any of the ith (i=1,2,…,n) cases (e.g., deposit-type or non-deposit locations) with missing values for at least one of the jth (j=1,2,…,m) independent variables (in this case $MOFS_{ji}$ for the geochemical data; see Fig. 5-12), it is very difficult, if not impossible, to estimate $\hat{Y}_i$. Current solutions to the problem of missing data of independent variables in logistic regression are still somewhat controversial and not yet routine (Rubin, 1996; Allison, 2002; Paul et al., 2003). For the case study, deposit-type and non-deposit locations without geochemical data are simply assigned a *MOFS* of [0].

The logistic regression coefficients (b_j) of the jth (j=1,2,…,m) $MOFS_{ji}$ independent variable are determined via the maximum likelihood method (Cox and Snell, 1989), whereby the square of the difference between Y_i and $\hat{Y}_i$ is minimised and tested for goodness-of-fit (e.g., via the Hosmer-Lemeshow test (Hosmer and Lemeshow, 2000)). Because the relationship between independent and dependent variables is not a linear function in logistic regression, the coefficients b_j may not have straightforward interpretations as they do in ordinary linear regression (Rock, 1988a). Thus, it is imperative to test the statistical significance of logistic regression coefficients (e.g., using the Wald statistic (Menard (2001)). In addition, a backward stepwise logistic regression is instructive in eliminating independent variables that do not contribute significantly (e.g., at the 90% level) to the logistic regression.

TABLE 8-III

A logistic regression model of relationship between the dichotomous dependent variable mineral occurrence score (Y_i) and the independent variables $MOFS_j$ at i (=1,2,...n) deposit-type, proxy deposit-type and non-deposit locations (Aroroy district, Philippines). The model is based on set 1 of non-deposit locations (Fig. 8-4).

Independent variable ($MOFS_j$)	Coefficient (b_j, b_0)	Wald statistic	Significance (α)
Distance to NNW[1]	41.086	16.812	0.000
Distance to NW[2]	-25.490	5.075	0.024
Distance to FI[3]	38.550	7.922	0.005
ANOM[4]	1.918	9.272	0.002
Constant	-52.877	21.834	0.000

[1]NNW-trending faults/fractures. [2]NW-trending faults/fractures. [3]Intersections of NNW- and NW-trending faults/fractures. [4]Integrated PC2 and PC3 scores obtained from the catchment basin analysis of stream sediment geochemical data (Chapter 3, Fig. 5-12).

Based on set 1 of 117 non-deposit locations (Y_i = 0) (Fig. 8-4) and the set of 117 deposit-type and proxy deposit-type locations (Y_i = 1) in the case study area, a final logistic regression model indicates that the $MOFS_{ji}$ of all X_i sets of spatial data of indicative geological features at locations of epithermal-Au deposits and their immediate surroundings are statistically dissimilar (at 95% significance level) from the *MOFS* of the same sets of spatial data at non-deposit locations (Table 8-III). The magnitudes of the coefficients of the jth $MOFS_j$ reflect the degree of dissimilarity of the deposit-type and proxy deposit-type locations from the non-deposit location. (Fig. 8-6). For example, the small coefficient of the $MOFS_j$ of the geochemical anomaly (Table 8-III) reflects the weak to moderate dissimilarity of the $MOFS_{ji}$ of the geochemical anomaly values at the deposit-type and proxy deposit-type locations from the $MOFS_{ji}$ of the geochemical anomaly values at non-deposit locations (Fig. 8-6D). However, the different magnitudes and signs of the coefficients of the jth $MOFS_j$ of distances to structural features are consistent with the results of analyses of spatial associations and are meaningful in terms of geologic controls on epithermal Au mineralisation in the case study area (see Chapter 6, Table 6-IX, Fig. 6-16). For example, the coefficients of the jth $MOFS_j$ of distances to geological structures suggest that NNW-trending faults/fractures and intersections between NNW- and NW-trending faults/fractures are more important than NW-trending faults/fractures as structural controls on epithermal Au mineralisation in the case study area. Therefore, the final logistic regression model in Table 8-III is considered meaningful and useful for selecting deposit-type and proxy deposit-type locations with similar or coherent multivariate spatial data signatures.

A one-dimensional scatter plot of predicted mineral occurrence scores ($\hat{Y}_i$) versus ID numbers of deposit-type, proxy deposit-type and non-deposit locations allows visualisation and distinction between coherent and non-coherent deposit-type and proxy deposit-type locations (Fig. 8-7). Fig. 8-7A is the result of the logistic regression analysis

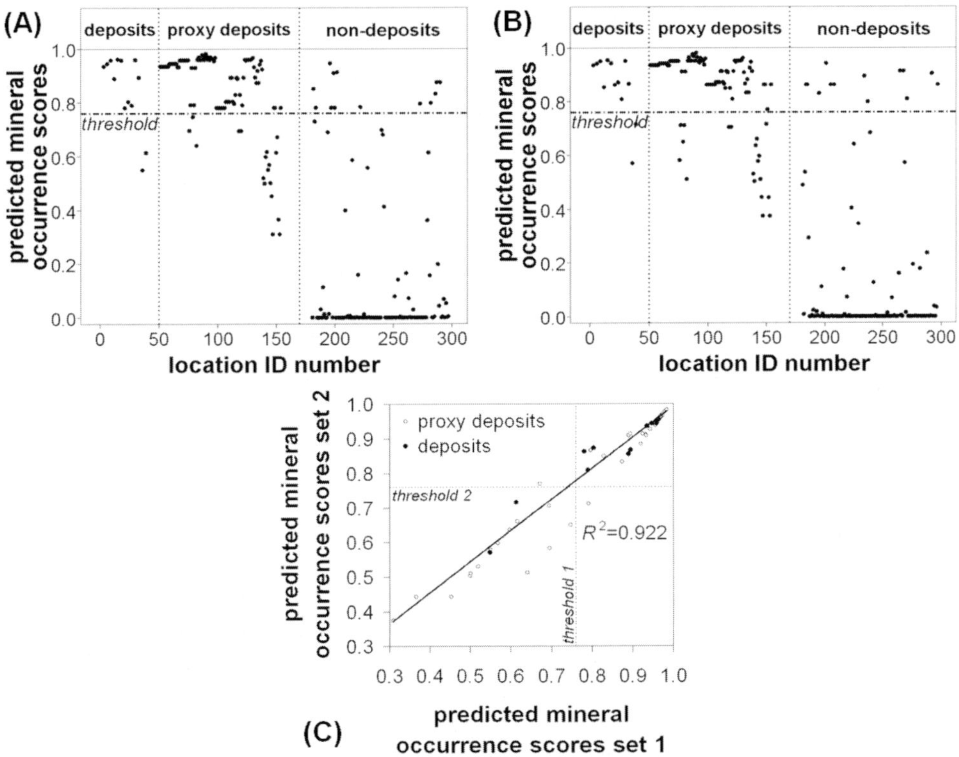

Fig. 8-7. (**A**) Scatter plot of predicted mineral occurrence scores (derived by logistic regression analysis) versus ID numbers of deposit-type, proxy deposit-type and randomly selected (set 1) non-deposit locations (see Fig. 8-4). (**B**) Scatter plot of predicted mineral occurrence scores (derived by logistic regression analysis) versus ID numbers of deposit-type, proxy deposit-type and randomly selected (set 2) non-deposit locations (see Fig. 8-4). In both scatter plots, the threshold predicted mineral occurrence score is 0.76, above which deposit-type and proxy deposit-type locations are considered coherent. (**C**) Scatter plot of the two sets of predicted mineral occurrence scores at deposit-type and proxy deposit-type locations, indicating the reproducibility and robustness of the logistic regression technique to distinguish between coherent and non-coherent deposit-type (as well as proxy deposit-type) locations.

described in Table 8-III, which is based on set 1 of 117 non-deposit locations (Fig. 8-4); whereas Fig. 8-7B is the result of a replicate logistic regression analysis based on set 2 of 117 non-deposit locations (Fig. 8-4). The results clearly distinguish the deposit-type and proxy deposit-type locations from the non-deposit locations in both set 1 and set 2. For all 13 epithermal Au deposit locations and most of the 104 proxy deposit-type locations $\hat{Y}_i$ is greater than 0.5, whilst for most of the non-deposit locations $\hat{Y}_i$ is less than 0.5. The results suggest that a threshold $\hat{Y}_i$ of 0.76 is suitable to differentiate between coherent deposit-type and proxy deposit-type locations from non-coherent deposit-type and proxy deposit-type locations. For most non-deposit locations $\hat{Y}_i$ is less than 0.76; however, the results show that a few non-deposit locations have multivariate spatial data signatures similar to the coherent deposit-type and proxy deposit-type location locations and these non-deposit locations are thus plausible prospective targets.

The two sets of logistic regression analyses indicate that, in the case study area, there are 11 coherent locations of epithermal Au deposits having strongly similar multivariate spatial data signatures, which are dissimilar from the multivariate spatial data signatures of two locations of epithermal Au deposits. The first and second logistic regression analyses indicate, respectively, that there are 86 and 85 proxy deposit-type locations with strongly similar and thus coherent multivariate spatial data signatures, which are similar and coherent to the multivariate spatial data signatures of the 11 coherent locations of epithermal Au deposits. Between the results of the two logistic regression experiments, all deposit-type locations have the same classifications whilst only three (~3% of 104) proxy deposit-type locations have different classifications. The values of $\hat{Y}_i$ scores of the coherent deposit-type and proxy deposit-type locations derived from the two logistic regression analyses have nearly perfect correlation (Fig. 8-7C). The results shown in Fig. 8-7 indicate, therefore, the reproducibility and robustness of the proposed technique for objective selection of coherent deposit-type (as well as proxy deposit-type) locations.

Fig. 8-8 shows the 11 coherent locations (represented as unit cells of 100×100 m) of epithermal Au deposits in the case study area. Fig. 8-8 also shows 86 proxy locations of epithermal Au deposits having multivariate data signatures that are coherent with the deposit-type locations per analysis using set 1 of non-deposit locations (Fig. 8-4). The location of epithermal Au deposit #12 and the locations immediately surrounding it are non-coherent with the 11 coherent locations of epithermal Au deposits because they are situated in the area without stream sediment geochemical data. The location of epithermal Au deposit #13 and most of the locations immediately surrounding it are non-coherent with the 11 coherent locations of epithermal Au deposits because these locations are characterised by background integrated PC2 and PC3 scores of the multi-element geochemical data based on catchment basin analysis (see Fig. 5-12). These results indicate that, given the same spatial evidential data sets used in deriving the *MOFS*, using deposits #12 and #13 in data-driven modeling of epithermal Au prospectivity in the study area is likely to result in a predictive model that is poorer than a predictive model derived by not using them. An indirect proof of this proposition is shown in Fig. 8-9, which compares the spatial associations (quantified via the distance distribution method; see Chapter 6) of the locations of 13 epithermal Au deposits and the 11 coherent locations of epithermal Au deposits with intersections of NNW- and NW-trending faults/fractures in the case study area. The D curves indicate that, in the case study area, all the deposit-type locations have weaker spatial associations with indicative geological features compared to the coherent deposit-type locations (see also explanations for Fig. 6-10A because Fig. 8-9A is identical to it). This goes to show that using all deposit-type locations in data-driven modeling of mineral prospectivity is likely

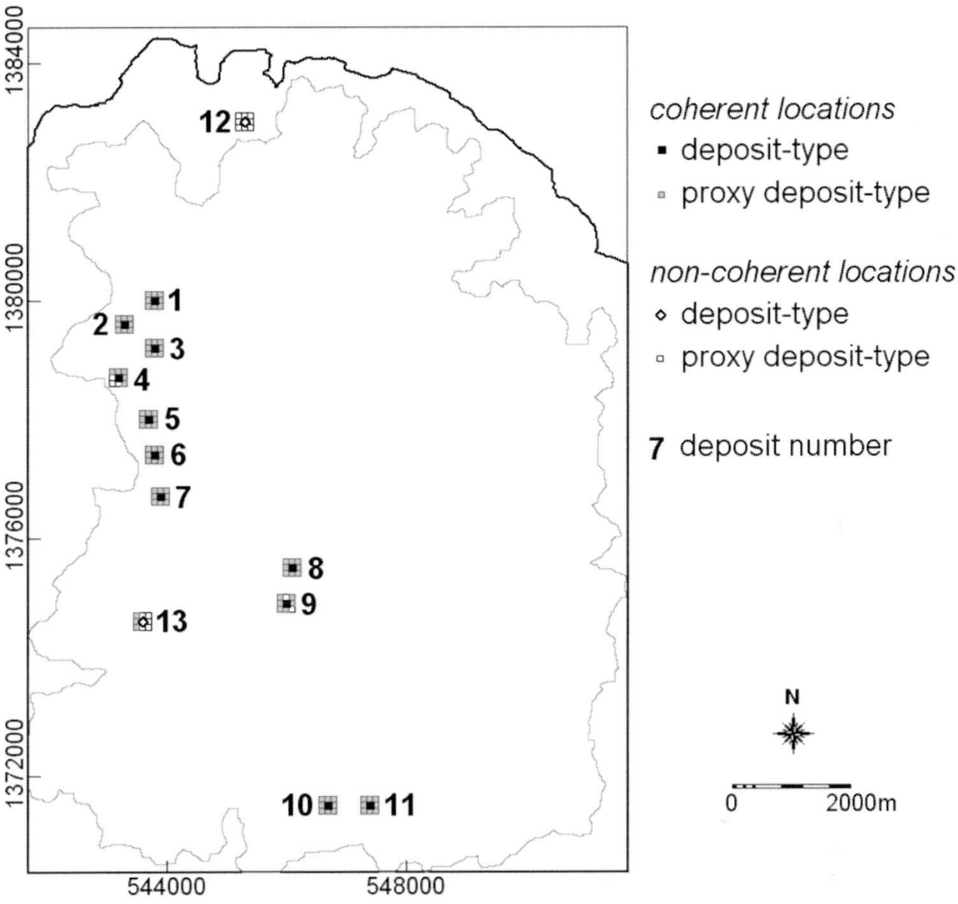

Fig. 8-8. Coherent and non-coherent locations of epithermal Au deposits and their immediate surroundings (proxy deposit-type locations), Aroroy district (Philippines) per analysis using set 1 of randomly-selected non-deposit locations (Figs. 8-4 and 8-7A). Polygon outlined in grey is area of stream sample catchment basins (see Fig. 4-11). See text for further explanation.

to result in predictor maps with weaker predictive strengths than using only coherent deposit-type locations.

The usefulness of coherent deposit-type locations (Carranza et al., 2008b) and the advantage of coherent proxy deposit-type locations over all proxy deposit-type locations (Stensgaard et al., 2006) in GIS-based data-driven modeling of mineral prospectivity are further demonstrated in this volume (see below). We now turn to discuss strategies for the cross-validation of data-driven models of mineral prospectivity.

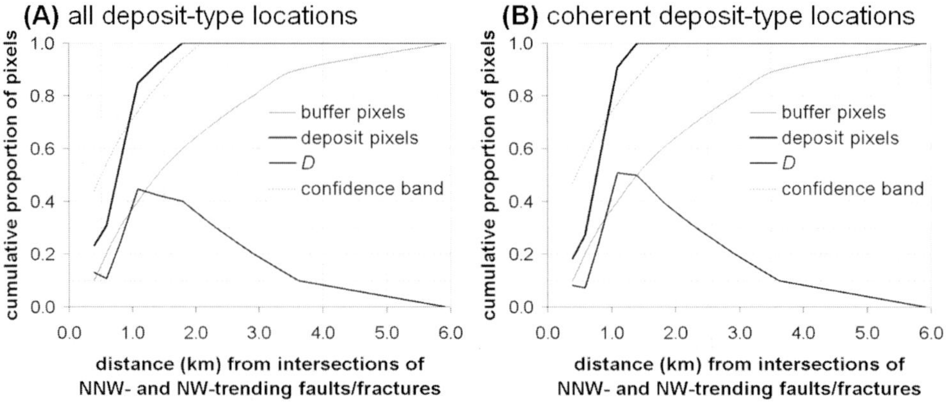

Fig. 8-9. Comparison of cumulative proportions of distance buffer and deposit pixels around intersection of NNW- and NW-trending faults/fractures, Aroroy district (Philippines): (**A**) all 13 locations of epithermal Au deposits; (**B**) only 11 coherent locations of epithermal Au deposits. Coherent deposit-type locations have stronger spatial association with indicative geological features compared to all deposit-type locations. See text for further explanation.

CROSS-VALIDATION OF DATA-DRIVEN MODELS OF PROSPECTIVITY

From Chapter 1, we recall the two fundamental assumptions in modeling of mineral prospectivity (see Fig. 1-2): (1) a specific location is prospective if it is characterised by the same or similar evidential features as known locations of mineral deposits of the type sought and (2) if more important evidential features are present in one location than in another location in a mineralised landscape, then the former has higher mineral prospectivity than the latter. The first assumption relates to the degree of fit (i.e., degree of spatial association) between evidential features and training (or prediction) deposits-type locations used in data-driven modeling of mineral prospectivity. The first assumption is validated by quantifying a *fitting-rate* of a data-driven model of mineral prospectivity against the training deposit-type locations. The second assumption is related not only to the degree of fit between evidential features and training deposits-type locations but also to the 'degree of fit' between a data-driven model of mineral prospectivity and undiscovered deposit-type locations. This second 'degree of fit' is validated by quantifying a *prediction-rate*, which can only be actually determined by waiting (endlessly) for new discoveries of mineral deposits of the type sought in a study area. An empirical prediction-rate can be quantified, however, by subdividing the set of known deposit-type locations into a training subset and a testing (or cross-validation) subset. The deposit-type locations in the test subset are presumed undiscovered in order to derive a prediction-rate curve.

In the preceding chapter, Fig. 7-2 shows the schematic GIS-based procedures for creating a prediction-rate curve associated with a mineral prospectivity map. The same schematic procedures can be applied for creating a fitting-rate curve, but using a map of the subset of prediction (or training) deposit-type locations instead of the subset of cross-validation (or testing) deposit-type locations. By switching the roles of the training and testing subsets, at least two data-driven models of mineral prospectivity are thus derived, which provide the opportunity to answer the two model validation questions posed in Chapter 1. There are various strategies of cross-validation in data-driven modeling of mineral prospectivity (cf. Agterberg and Bonham-Carter, 2005; Chung and Fabbri, 2005; Fabbri and Chung, 2008), the common objective of which is to establish an optimum predictive model of mineral prospectivity having the best fitting- and prediction-rates.

N–n strategies

Given a set of a number, N, of known deposit-type locations, n (≤50% of N) deposit-type locations can be used for testing and the remaining N–n deposit-type locations are used for training. In cases where N is relatively small, one deposit-type location is used for testing and the remaining N–1 deposit-type locations are used for training so that data-driven modeling of mineral prospectivity is thus performed N times, each time with a different N-1 training subset and a different N^{th} testing subset. In cases where N is large (say 321 as in the case of modeling of prospectivity for alkalic porphyry Cu-Au deposits in British Columbia (see Carranza et al., 2008b), the N-1 (or *jack-knife*) strategy can be impractical. In such a case, n>1 deposit-type locations out of the N deposit-type locations can be used for testing and the remaining N–n deposit-type locations are used for training so that it is not necessary to perform data-driven modeling of mineral prospectivity N times. Still, several (although less than N) iterations of mineral prospectivity modeling with different N–n subsets are necessary to establish an optimum predictive model. In each of these iterations, the n deposit-type locations are usually chosen randomly.

The N–(n>1) strategy is probably the most commonly used strategy of cross-validation in data-driven modeling of mineral prospectivity. Skabar (2005) presented a sound version of this strategy by replicating an original set of deposit-type locations four times. From each replicate set, ¾ and ¼ of the deposit-type locations were used for training and testing, respectively, and each of the four testing subsets did not contain common deposit-type locations.

Deposit-type classification strategies

Because mineral exploration endeavours to find mineral deposits, especially those with commercially viable concentrations of minerals or metals for mining purposes, it is instructive to derive mineral prospectivity models that provide the opportunity for discovery of high-grade and large tonnage mineral deposits likely to be commercially viable. Because high-grade and/or large-tonnage mineral deposits of the type sought are usually rare and thus few in number, the validation strategy when using these mineral deposits in data-driven prospectivity modeling is usually the *N*-1 strategy. In many cases, however, data of grade and/or tonnage of known deposit-types of interest are not available in mineral inventory databases of national geological survey organisations, although data about their status (e.g., mine, past producer, prospect, showing, etc.) are usually available in these databases. The mineral deposit status attributes can be used as basis for cross-validation in data-driven modeling of mineral prospectivity. For example, Carranza and Hale (2000) and Carranza (2002) used locations of large-scale gold deposits (i.e., mines and prospects of private mining companies) and small-scale gold deposits (i.e., artisanal workings by local people) as training and testing subsets and vice versa for predictive modeling of prospectivity for epithermal Au in the Baguio district (Philippines). Similarly, Carranza et al. (2008c) used showings/indications of geothermal activity and developed/explored geothermal prospects as training and testing subsets, respectively, for predictive modeling of geothermal prospectivity in West Java (Indonesia).

Spatial subdivision strategies

Within a study area, a representative portion (e.g., the most prospective region) of a mineralised landscape containing an adequate number of samples (i.e., deposit-type locations) can be chosen as a training region. The weights (wC_{ji}) for C_{ji} classes in X_i maps of spatial evidential features (see equation (8.2)) derived in the training region are then applied for data-driven modeling of mineral prospectivity in the whole study area. Alternatively, a study area may be subdivided into, say, four equal regions, each of which is used as a training region thereby creating four mineral prospectivity models. In both of these cross-validation strategies, the deposit-type locations outside the training region form a testing subset. It can be argued, however, that these cross-validation strategies form a sort of biased sampling because the geological features and especially the geologic controls on mineralisation vary from one region to another. It follows that using different and spatially (or geologically) non-coherent training regions may result in predictive models of mineral prospectivity that are dissimilar not only in terms of empirical spatial associations but also in terms of genetic associations between deposit-type occurrences and geological features. Nevertheless, the application of spatially coherent training and testing regions to cross-validation allow recognition of different regions of a mineralised landscape having similar geology and thus mineral prospectivity compared to the known most prospective region(s) (Agterberg and Bonham-Carter, 2005). GIS-based shape-analytical tools can be useful in determining spatially coherent prospective regions in a mineralised landscape (see Gardoll et al. (2000) for details).

Other strategies of cross-validation

It is also useful to perform experiments by varying not only the compositions of the training and testing subsets but also (a) the combinations of X_i evidential maps (e.g.,

Harris and Sanborn-Barrie, 2006; Woldai et al., 2006), (b) the widths or intervals of C_{ji} classes in some of the X_i evidential maps (e.g., Harris and Sanborn-Barrie, 2006) and, perhaps, (c) the unit cell size. These and the foregoing cross-validation experiments aim at predictive model calibration.

In the demonstration of evidential belief modeling and discriminant analysis of mineral prospectivity in the case study (see below), the following combinations of N–n and deposit-type classification strategies for cross-validation are applied.

- In order to illustrate the utility of coherent deposit-type locations,
  (a) coherent deposit-type locations comprise one training subset (n=11 out of N=13; Fig. 8-8) and
  (b) all deposit-type locations comprise another training set (N=13; Fig. 8-8).
  The derived data-driven models of mineral prospectivity are then cross-validated against a testing set consisting of all proxy deposit-type locations (N=104; Fig. 8-8).
- In order to illustrate the utility of coherent proxy-deposit type locations,
  (a) coherent proxy deposit-type locations are used for training (n=86 out of N=104; Fig. 8-8) and
  (b) randomly-selected proxy deposit-type locations are also used for training (n=86 out of N=104).
  The derived data-driven models of mineral prospectivity are then cross-validated against all deposit-type locations (N=13).

EVIDENTIAL BELIEF MODELING OF MINERAL PROSPECTIVITY

Brief and informal explanations of the concept of evidential belief functions (denoted hereafter as EBFs), which are based on the theory of evidential belief (Dempster 1967, 1968; Shafer, 1976), as applied to knowledge-driven modeling of mineral prospectivity are given in Chapter 7. Those explanations are adapted here for data-driven modeling of mineral prospectivity.

Estimates of EBFs relate to the proposition that "*this location is prospective for mineral deposits of the type sought*". That is, estimates of EBFs represent, according to the general definition of index of prospectivity in equation (8.2), values of wC_{ji} for each of the C_{ji} classes in X_i maps of spatial evidential features with respect to D known locations of mineral deposits of interest in a study area. For the jth C_{ji} class in the ith X_i evidential map, four EBFs, each in the range [0,1], are estimated to evaluate the proposition of mineral prospectivity. These four EBFs are *belief* (or *Bel*), *disbelief* (or *Dis*), *uncertainty* (or *Unc*) and *plausibility* (or *Pls*). The *Bel* and *Pls*, respectively, represent lower and upper degrees of support to the proposition given a jth C_{ji} class of ith X_i spatial evidence. The *Unc* represents a measure of 'doubt' that the given j^{th} C_{ji} class of ith X_i spatial evidence supports the proposition. The *Dis* represents a degree of opposition to the proposition given the jth C_{ji} class of ith X_i spatial evidence.

The four EBFs are inter-related (see Fig. 7-18). The sum of *Bel*+*Unc*+*Dis* for the jth C_{ji} class in the ith X_i evidential map is equal to 1. Likewise, the sum of *Pls*+*Dis* for the jth

C_{ji} class in the ith X_i evidential map is equal to 1. From these two equalities, therefore, *Pls* = *Bel*+*Unc* or *Bel* = *Pls*–*Unc*. The degree of *Unc* influences the relation between *Bel* and *Dis*. If *Unc* = 0 (i.e., any j^{th} C_{ji} class in the ith X_i evidential map is 'totally accurate and precise' with respect to *D*), then *Bel*+*Dis* = 1 and the relation between *Bel* and *Dis* for the jth C_{ji} class in the ith X_i evidential map is binary (i.e., *Bel* = 1–*Dis* or *Dis* = 1–*Bel*), as in the theory of probability. If *Unc* = 1 (i.e., any j^{th} C_{ji} class in the ith X_i evidential map is 'totally inaccurate and imprecise' with respect to *D*), then *Bel* and *Dis* for the jth C_{ji} class in the ith X_i evidential map are both equal to zero. That is, if there is complete uncertainty, then there can be neither belief nor disbelief. Usually, however, *Unc* is neither equal to 0 nor equal to 1 (i.e., any j^{th} C_{ji} class in the ith X_i evidential map is neither 'totally accurate and precise' nor 'totally inaccurate and imprecise' with respect to *D*). Therefore, in the usual case that 0<*Unc*<1, then *Bel* = 1–*Dis*–*Unc* or *Dis* = 1–*Bel*–*Unc*. This means that, because uncertainty is usually present, the relation between *Bel* and *Dis* for a given piece of evidence is usually not binary. This further means that, for any j^{th} C_{ji} class in the ith X_i evidential map that is used to evaluate the proposition of mineral prospectivity, not only *Bel* and *Dis* but also *Unc* must be modeled.

Most of the published applications of EBFs to mineral prospectivity mapping are knowledge-driven (Moon 1990, 1993; Chung and Moon, 1991; Moon et al., 1991; An, 1992; An et al., 1992, 1994a, 1994b; Chung and Fabbri, 1993; Wright and Bonham-Carter, 1996; Likkason et al., 1997; Carranza, 2002; Tangestani and Moore, 2002; Chapter 7 of this volume). Knowledge-driven estimation of EBFs is suitable for modeling of mineral prospectivity in frontier or less-explored mineralised landscapes where there are no or very few known locations of mineral deposits of the type sought. Data-driven estimation of EBFs, however, can be performed in modeling of mineral prospectivity in moderately- to well-explored mineralised landscapes where there are several known locations of mineral deposits of the type sought (see references cited in Table 8-1).

The minimum number of deposit-type locations used in data-driven estimation of EBFs depends on the size of a study area, because data-driven estimates of EBFs, like estimates of wC_{ji} for C_{ji} classes in X_i evidential maps via application of other data-driven techniques, are based on size of study area. However, a minimum deposit density (e.g., ratio of the number of deposit-type pixels (or unit cells) to the number of 'study area' pixels) that results in geologically meaningful data-driven estimates of wC_{ji} for C_{ji} classes in X_i evidential maps has not yet been established. Data-driven estimates of EBFs are meaningful if they represent geologically sound empirical spatial associations between mineral deposits of the type sought and certain geological features (see Chapter 7). Nevertheless, Carranza (2002) showed geologically meaningful results of applications of data-driven EBFs to mineral prospectivity mapping based on (a) 12 locations of porphyry Cu deposits in an area of roughly 920 km^2 and (b) 17 locations of vein-type Cu-Au deposits in an area of roughly 1,450 km^2. These imply that application of data-driven EBFs to model mineral prospectivity in the present case study area, where there are 13 locations of epithermal Au deposits in about 130 km^2, is likely to yield geologically meaningful results.

GIS-based data-driven estimation of EBFs

Suppose that in a study area T, comprising $N(T)$ total number of unit cells or pixels, there are a number of known mineral deposits, D, of the type sought occurring in $N(D)$ number of pixels (Fig. 8-10). Suppose further that X_i (i=1,2,…,n) spatial evidence maps, each with a number of C_{ji} (j=1,2,…,m) classes of spatial evidence, have been created to represent certain prospectivity recognition criteria. Each of the jth C_{ji} class of the ith X_i spatial evidence map has $N(C_{ji})$ number of pixels (Fig. 8-10). The sum of $N(C_{ji})$ number of pixels in any i^{th} X_i spatial evidence map is equal to $N(T)$. By overlaying a binary map of D on each multi-class evidential map, the number of C_{ji} pixels overlapping with D pixels [i.e., $N(C_{ji} \cap D)$] is determined. From this, the number of C_{ji} pixels not overlapping with D pixels [i.e., $N(C_{iji}) - N(C_{ji} \cap D)$] can be derived. The values of $N(T)$, $N(D)$, $N(C_{ji})$ and $N(C_{ji} \cap D)$ are the ones used in data-driven estimations of the EBFs.

The degree of belief ($Bel_{C_{ji}}$) for the jth C_{ji} (j=1,2,…,m) class of the ith X_i (i=1,2,…,n) spatial evidence map with respect to D is, according to An et al. (1994b), estimated as

$$Bel_{C_{ji}} = \frac{N(C_{ji} \cap D)}{N(C_{ji})} . \tag{8.4}$$

The degree of disbelief ($Dis_{C_{ji}}$) for the jth C_{ji} (j=1,2,…,m) class of the ith X_i (i=1,2,…,n) spatial evidence map with respect to D is, according to An et al. (1994b), estimated as

$$Dis_{C_{ji}} = \frac{N(C_{ji}) - N(C_{ji} \cap D)}{N(C_{ji})} . \tag{8.5}$$

Then, the degree of uncertainty ($Unc_{C_{ji}}$) for the jth C_{ji} (j=1,2,…,m) class of the ith X_i (i=1,2,…,n) spatial evidence map with respect to D is, according to the relationships of the EBFs (see Chapter 7, Fig. 7-18), estimated as

$$Unc_{C_{ji}} = 1 - \frac{N(C_{ji} \cap D)}{N(C_{ji})} - \frac{N(C_{ji}) - N(C_{ji} \cap D)}{N(C_{ji})} = 1 - Bel_{C_{ji}} - Dis_{C_{ji}} . \tag{8.6}$$

The degree of plausibility ($Pls_{C_{ji}}$) for the jth C_{ji} (j=1,2,…,m) class of the ith X_i (i=1,2,…,n) spatial evidence map with respect to D is, according to the relationships of the EBFs (see Chapter 7, Fig. 7-18), estimated as

$$Pls_{C_{ji}} = Bel_{C_{ji}} + Unc_{C_{ji}} . \tag{8.7}$$

There are two problems associated with the application of equations (8.4) to (8.6). Firstly, because $N(C_{ji}) = N(C_{ji} \cap D) + [N(C_{ji}) - N(C_{ji} \cap D)]$, it follows that $Unc_{C_{ji}}$ in equation (8.6) is equal to [0], whereas there is always uncertainty. Secondly, equations (8.4) and (8.5) represent conditional probability that a mineral deposit of the type sought exists and does not exist, respectively, given C_{ij}. The estimates of both $Bel_{C_{ji}}$ and $Dis_{C_{ji}}$ via equations (8.4) and (8.5), respectively, thus represent the relationship of the jth C_{ij} class in the ith X_i spatial evidence map with D only but do not represent the relationship of the jth C_{ij} class with the other m^{th} C_{mi} classes in the ith X_i spatial evidence map. Chung and Fabbri (1993) aver that the relationships among the classes in an evidential map, aside from their spatial relationships to D, must be considered and represented in a mathematical function f (see equations (8.1) and (8.2)) for combining predictor maps of mineral prospectivity. The following modified equations (Carranza, 2002; Carranza and Hale, 2003) have been proposed to overcome the problems associated with equations (8.4) to (8.6).

The $Bel_{C_{ji}}$ for the jth C_{ji} (j=1,2,…,m) class of the ith X_i (i=1,2,…,n) spatial evidence map with respect to D is re-defined and estimated as

$$Bel_{C_{ji}} = \frac{W_{C_{ji}D}}{\sum\limits_{j=1}^{m} W_{C_{ji}D}} , \tag{8.8a}$$

$$\text{where } W_{C_{ji}D} = \frac{\dfrac{N(C_{ji} \cap D)}{N(C_{ij})}}{\dfrac{N(D) - N(C_{ji} \cap D)}{N(T) - N(C_{ji})}} . \tag{8.8b}$$

The numerator in equation (8.8b) is the conditional probability that D exists given the presence of C_{ji}. It means simply that a target mineral deposit occurs in C_{ji}. The denominator in equation (8.8b) is the conditional probability that D exists given the absence of C_{ij}. It means simply that a target mineral deposit occurs outside C_{ji}. Thus, the $W_{C_{ji}D}$ is the relative weight of every C_{ji} in terms of D being more likely present instead of being more likely absent in C_{ji} than would be expected due to chance. Accordingly, the $Bel_{C_{ji}}$ in equation (8.8a) is a the relative strength of the spatial association of the jth C_{mi} class in the ith X_i spatial evidence map with D compared to the strengths of spatial associations of the other m^{th} C_{mi} classes in the ith X_i spatial evidence map with D.

The $Dis_{C_{ji}}$ for the jth C_{ji} (j=1,2,…,m) class of the ith X_i (i=1,2,…,n) spatial evidence map with respect to D is re-defined and estimated as

$$Dis_{C_{ji}} = \frac{W_{C_{ji}\overline{D}}}{\sum_{j=1}^{m} W_{C_{ji}\overline{D}}}, \tag{8.9a}$$

where
$$W_{C_{ji}\overline{D}} = \frac{\dfrac{N(C_{ji}) - N(C_{ji} \cap D)}{N(C_{ji})}}{\dfrac{N(T) - N(D) - [N(C_{ji}) - N(C_{ji} \cap D)]}{N(T) - N(C_{ji})}} \tag{8.9b}$$

and $\overline{D}$ means pixels or unit cells not known to contain D. The numerator in equation (8.8b) is the conditional probability that D does not exist given the presence of C_{ji}. It means simply that a target mineral deposit occurs outside C_{ji}. The denominator in equation (8.8b) is the conditional probability that D does not exist given the absence of C_{ij}. It means simply that a target mineral deposit occurs in C_{ji}. Thus, $W_{C_{ji}\overline{D}}$ is the relative weight of C_{ji} in terms of D being more likely absent instead of being more likely present in C_{ji} than would be expected due to chance. Conversely, $W_{C_{ji}\overline{D}}$ is the relative weight of C_{ji} in terms of $\overline{D}$ being more likely present instead of being more likely absent in C_{ji} than would be expected due to chance. Accordingly, the $Dis_{C_{ji}}$ in equation (8.9a) is the relative strength of the spatial association of the jth C_{mi} class in the ith X_i spatial evidence map with $\overline{D}$ (but not with D) compared to the strengths of spatial associations of the other m^{th} C_{mi} classes in the ith X_i spatial evidence map with $\overline{D}$.

There are two effects when the relative strengths of spatial associations of jth C_{mi} classes in the ith X_i spatial evidence map with D and $\overline{D}$ are considered. First, for every j^{th} C_{ji} class in the ith X_i spatial evidence map, the values of $Bel_{C_{ji}}$ and $Dis_{C_{ji}}$ estimated according to equations (8.8a) and (8.9a), respectively, are constrained to the unit interval [0,1], which is consistent with the concept of probability. Second, for every j^{th} C_{ji} class in the ith X_i spatial evidence map, the sum of the values of $Bel_{C_{ji}}$ and $Dis_{C_{ji}}$ is less than [1], which allows modeling of $Unc_{C_{ji}}$, thus

$$Unc_{C_{ji}} = 1 - \frac{W_{C_{ji}D}}{\sum\limits_{j=1}^{m} W_{C_{ji}D}} - \frac{W_{C_{ji}\overline{D}}}{\sum\limits_{j=1}^{m} W_{C_{ji}\overline{D}}} = 1 - Bel_{C_{ji}} - Dis_{C_{ji}}. \quad (8.10)$$

The re-defined equations for GIS-based data-driven estimations of *Bel*, *Dis* and *Unc* for mineral prospectivity mapping thus take into account (a) the relative proportions of known deposit-type locations in every class of spatial evidence (see the numerators of equations (8.8b) and (8.9)), which define the spatial associations between the deposit-type locations and classes of spatial evidence in individual evidential maps and (b) the relative weights and thus relationships among the classes of evidence in each evidential map with respect to D and $\overline{D}$.

Applications of equations (8.8a), (8.9a) and (8.10) to hypothetical data shown in Fig. 8-10 indicate that classes of spatial evidence with the highest density of deposit-type locations [i.e., $N(C_j \cap D) \div N(C_j)$] have the highest Bel_{C_j}, lowest Dis_{C_j} and lowest Unc_{C_j}. Conversely, the hypothetical example in Fig. 8-10 indicates that classes of spatial evidence with the lowest density of deposit-type locations have the lowest Bel_{C_j}, highest Dis_{C_j} and highest Unc_{C_j}. The data-driven estimates of EBFs based on the artificial data (Fig. 8-10) indicate that (a) the *Bel* is a measure of relative strengths of spatial associations between geo-objects of interest and classes of spatial evidence and (b) both the *Dis* and *Unc* are 'inverse' measures of relative strengths of spatial association between geo-objects of interest and classes of spatial evidence. It is further shown in the case study that the applications of the re-defined equations for GIS-based data-driven estimations of EBFs for mineral prospectivity mapping result in values of *Bel* portraying empirical spatial associations between deposit-type locations and geological features that are comparable to and interpretable as the results of applications of the methods for quantifying spatial associations described and explained in Chapter 7.

In the application of equations (8.8) and (8.9) for GIS-based data-driven estimations of EBFs for mineral prospectivity mapping, one must take note of the result that $W_{C_{ji}D} = 0$ (equation (8.8b)). This means that $N(C_{ji} \cap D) = 0$, which results in $Bel_{C_{ji}} = 0$ (equation (8.8a)). If $W_{C_{ji}D} = 0$, then the corresponding estimate of $W_{C_{ji}\overline{D}}$ (which actually is not equal to zero since $N(C_{ji}) - [N(C_{ji} \cap D)] \neq 0$; equation (8.9b)) must be discarded and instead the $W_{C_{ji}\overline{D}}$ is replaced with or re-set to [0] (Fig. 8-11). By doing this, the corresponding $Dis_{C_{ji}} = 0$ (equation (8.9a)) and the corresponding $Unc_{C_{ji}} = 1$ (equation (8.10)). The logic of this is that if there is no belief then there is also no disbelief but there is only uncertainty.

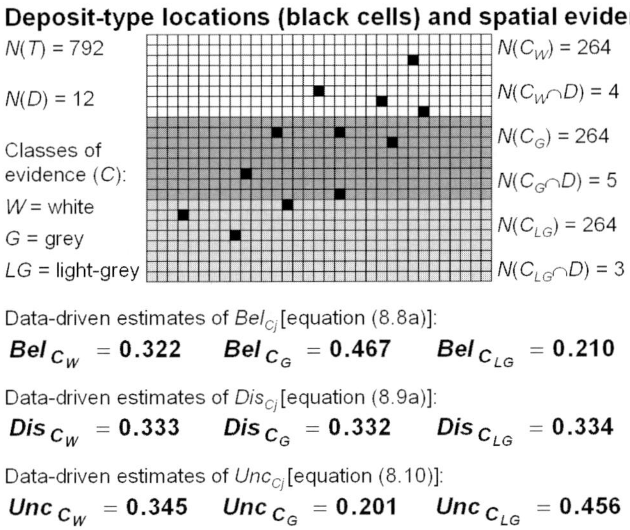

Fig. 8-10. Hypothetical data of deposit-type locations and a map of spatial evidence with three classes (C_j) (j = grey scale) for illustration of data-driven estimates of Bel_{Cj} (equation (8.8a)), Dis_{Cj} (equation (8.9a)) and Unc_{Cj} (equation (8.10)). Based on a suitable unit cell size $N(\bullet)$, a study region is discretised into equal-area unit cells $N(T)$. The number of unit cells corresponding with a deposit-type location $N(D)$ is determined. The areas occupied by each evidential class correspond to a number of unit cells $N(C_j)$. The numbers of class evidence unit cells coinciding with deposit-type locations $N(C_j \cap D)$ are determined via map overlay operation. The values of $N(T)$, $N(D)$, $N(C_j)$ and $N(C_j \cap D)$ are then used in data-driven estimations of the EBFs (see equations (8.8a), (8.9a), (8.10) and text for further explanation).

Unlike in the applications of most multivariate methods to data-driven modeling of mineral prospectivity (Table 8-II), parts of a study area with missing data are considered and included in the application of the bivariate methods to data-driven modeling of mineral prospectivity (Table 8-I). In the ith X_i spatial evidence map, locations with missing data comprise a jth C_{ji} class labeled as, say, "no data". In data-driven estimation of EBFs, regardless of whether the jth C_{ji} class of "no data" coincides with some known deposit-type locations, the derived values of both $W_{C_{ji}D}$ and $W_{C_{ji}\bar{D}}$ are discarded and re-set to [0] so that the resulting estimates of both $Bel_{C_{ji}}$ and $Dis_{C_{ji}}$ are [0] and thus $Unc_{C_{ji}} = 1$ (Fig. 8-11). These are logical representations of locations with missing spatial evidence.

Calibration of data-driven estimation of EBFs

The classification of data of continuous fields (e.g., distance to geological features, geochemical anomalies) can introduce artifacts in the variations of spatial associations between deposit-type locations and geological features as depicted by variations in values of wC_{ji} (see equation (8.2)) of C_{ji} classes of individual X_i spatial evidential features with respect to D. These artifacts, which represent *systemic* (or procedural) *errors* in predictive modeling of geo-objects, are manifested as abrupt (or 'noisy') rather than gradual variations in the values wC_{ji} because some data classes coincide with many deposit-type locations, some data classes coincide with few deposit-type locations and some data classes do not coincide with any deposit-type locations. This problem is intrinsic not only in the application of data-driven modeling with EBFs but also in the application of weights-of-evidence (WofE) modeling (see Bonham-Carter, 1994, pp. 319-321). In WofE modeling the problem is overcome by calculating values representing wC_{ji} based on cumulative classes of data. However, in accordance with the theory of evidential belief (Dempster 1967, 1968; Shafer, 1976), classes of data are treated as discrete and independent bodies of evidence so that equations (8.8) and (8.9) apply to data-driven estimation of EBFs of non-cumulative classes of data. Nevertheless, non-noisy variations of data-driven EBFs can be achieved by performing model calibration experiments with different class intervals of data. Let us illustrate this with some examples in the case study area.

In one example, the integrated PC2 and PC3 scores obtained from the catchment basin analysis of stream sediment geochemical data (see Chapter 5, Fig. 5-12) are classified into a map of classes of integrated PC2 and PC3 scores, which is then crossed with (or overlaid on) the map of locations of epithermal Au deposits (Fig. 8-11A). In one calibration experiment, the integrated PC2 and PC3 scores are classified into more-or-less 10-percentile intervals (Fig. 8-11B). Only one class does not coincide with any epithermal Au deposit location. However, because the class intervals (i.e., number of class pixels) are not constant (which is common in raster-based analysis with a somewhat coarse pixel size), the graph of the values of *Bel* versus the upper limits of classes of integrated PC2 and PC3 scores (Fig. 8-12A) is somewhat noisy, indicating that the data-driven EBFs are somewhat improperly calibrated. In a second experiment, the eight classes of more-or-less 10-percentile intervals in the first experiment are re-classified into four classes of more-or-less 20-percentile intervals (Fig. 8-11C). The new curve of the values of *Bel* versus the upper limits of classes of integrated PC2 and PC3 scores is smooth (Fig. 8-12A), which indicates that the data-driven EBFs in the second experiment are properly calibrated. This is so because the new curve of the values of *Bel* versus the upper limits of classes of integrated PC2 and PC3 scores shows that the epithermal Au deposits are associated spatially with high integrated PC2 and PC3 scores and that the threshold score separating anomalous and background integrated PC2 and PC3 scores is about 0.3. This result is consistent with the result of the distance distribution analysis of the spatial association between the integrated PC2 and PC3 scores obtained from the catchment basin analysis of stream sediment geochemical data (see Chapter 5, Fig. 5-12) and the known locations of epithermal Au deposits in the study area (see Chapter 6, Figs. 6-12E and 6-12F).

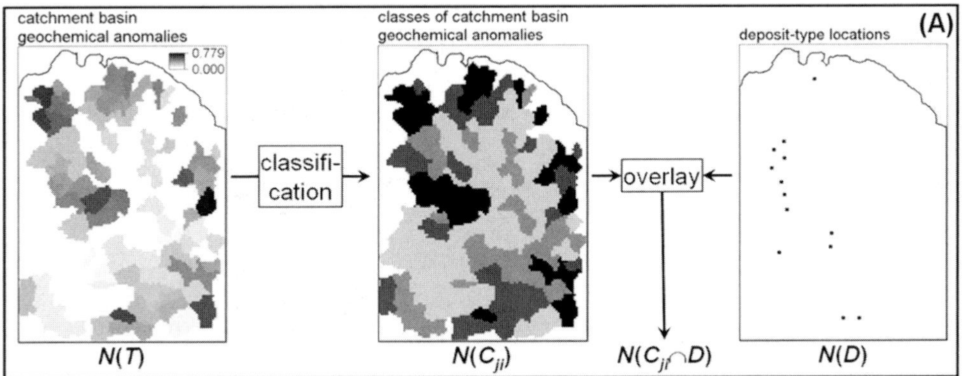

(B)

anomaly	npixcl	npixt	npixclmo	npixmo	wcmo	twcmo	bel	wcmon	wcmon2	twcmon	dis	unc
no data	2844	12914	1	13	0.000	10.461	0.000	1.001	0.000	6.997	0.000	1.000
0.058	2931	12914	3	13	1.022	10.461	0.098	1.000	1.000	6.997	0.143	0.759
0.099	1058	12914	0	13	0.000	10.461	0.000	1.001	0.000	6.997	0.000	1.000
0.154	946	12914	1	13	1.054	10.461	0.101	1.000	1.000	6.997	0.143	0.756
0.251	1104	12914	1	13	0.892	10.461	0.085	1.000	1.000	6.997	0.143	0.772
0.289	1026	12914	2	13	2.107	10.461	0.201	0.999	0.999	6.997	0.143	0.656
0.368	977	12914	2	13	2.222	10.461	0.212	0.999	0.999	6.997	0.143	0.645
0.489	976	12914	2	13	2.224	10.461	0.213	0.999	0.999	6.997	0.143	0.644
0.781	1052	12914	1	13	0.940	10.461	0.090	1.000	1.000	6.997	0.143	0.767

(C)

upper limits of C_{ji}	$N(C_{ji})$	$N(T)$	$N(C_{ji} \cap D)$	$N(D)$	$W_{C_{ji}D}$	$\sum_{j=1}^{m} W_{C_{ji}D}$		$W_{C_{ji}\bar{D}}$	re-set $W_{C_{ji}\bar{D}}$	$\sum_{j=1}^{m} W_{C_{ji}\bar{D}}$		
anomaly	npixcl	npixt	npixclmo	npixmo	wcmo	twcmo	bel	wcmon	wcmon2	twcmon	dis	unc
no data	2844	12914	1	13	0.000	5.666	0.000	1.001	0.000	3.998	0.000	1.000
0.099	3989	12914	3	13	0.671	5.666	0.118	1.000	1.000	3.998	0.250	0.632
0.251	2050	12914	2	13	0.964	5.666	0.170	1.000	1.000	3.998	0.250	0.580
0.368	2003	12914	4	13	2.421	5.666	0.427	0.999	0.999	3.998	0.250	0.323
0.781	2028	12914	3	13	1.610	5.666	0.284	0.999	0.999	3.998	0.250	0.466

Fig. 8-11. (**A**) GIS operations involved in deriving map variables required for data-driven estimation of *Bel*, *Dis* and *Unc* (see equations (8.8), (8.9) and (8.10), respectively) for classes of catchment basin geochemical anomaly values with respect to locations of epithermal Au deposits in the Aroroy district (Philippines). In order to demonstrate the effect of data classification in data-driven estimation of EBFs, the catchment basin geochemical anomaly values are first (**B**) classified into more-or-less 10-percentile distance intervals and then (**C**) then some of the 10-percentile class intervals of the geochemical anomaly values are merged. In (**C**) the names of columns in the tables are annotated with the variables used in equations (8.8) to (8.10).

In another example, distances to NNW-trending faults/fractures are classified in order to derive a map of proximity classes to NNW-trending faults/fractures, which is then crossed with (or overlaid on) the map of locations of epithermal Au deposits (Fig. 8-13A). In one calibration experiment, the distances to NNW-trending faults/fractures are classified into more-or-less 5-percentile intervals (Fig. 8-13B). Many classes do not coincide with locations of epithermal Au deposits and some classes that coincide with locations of epithermal Au deposits are non-contiguous. The graph of the values of *Bel*

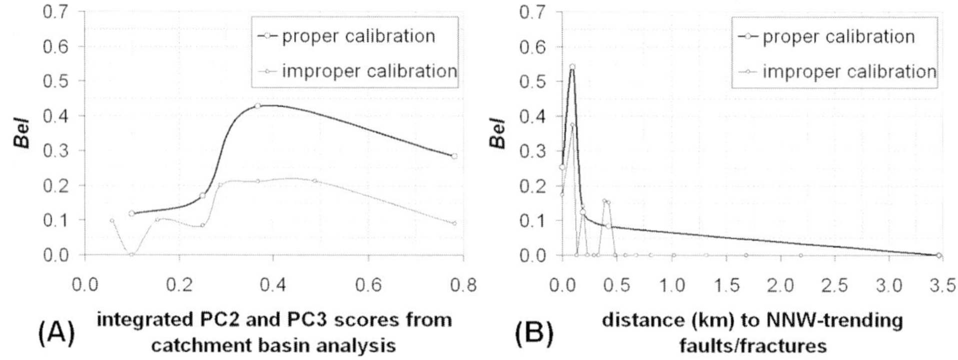

Fig. 8-12. Variations of spatial associations between locations of epithermal Au deposits and geological features in Aroroy district (Philippines) as depicted by the plots of data-driven estimates of *Bel* versus upper limits of classes of (**A**) integrated PC2 and PC3 scores obtained from the catchment basin analysis of stream sediment geochemical data (see Chapter 5, Fig. 5-12) and (**B**) distances to NNW-trending faults/fractures. Smooth curves represent properly calibrated map classes, whilst rough or noisy curves represent improperly calibrated map classes.

versus the upper class limits (Fig. 8-12B) is noisy, indicating that the data-driven EBFs are improperly calibrated. In another calibration experiment, some of the classes (i.e., 96.9-193.6 m, 193.7-425.9 m and 426.0-3465.4 m) in the first experiment are merged (Fig. 8-13C) whilst some of the classes are retained (0.0 m and 0.1-96.8 m). The new curve of the values of *Bel* versus the upper class limits of proximity to NNW-trending faults/fractures is not noisy (Fig. 8-12B), which indicates that the data-driven EBFs in the second experiment are properly calibrated. This is so because the new curve of the values of *Bel* versus the upper class limits of proximity to NNW-trending faults/fractures shows that epithermal Au deposits are associated spatially with NNW-trending faults/fractures and that the spatial association is optimal within 250 m of NNW-trending faults/fractures. This is consistent with the result of the distance correlation analysis of the spatial association between NNW-trending faults/faults and the known locations of epithermal Au deposits in the study area (see Chapter 6, Table 6-IX).

One may wonder why the first two classes of distances to NNW-trending faults/fractures (Figs. 8-13B and 8-13C) are not merged like the other classes. If the first two classes of distances to NNW-trending faults/fractures are merged, the resulting class coincides with eight of the 13 locations of epithermal Au deposits (Fig. 8-14). If a class of data of a continuous field (e.g., distances to geological features), C_{ji}, coincides with more than 50% of the locations of mineral deposits of the type sought and if $N(C_{ji})$ is less than 50% of $N(T)$, then applications of equations (8.8) to (8.10) could result in a negative value of *Unc* (Fig. 8-14). A negative value of *Unc* is certainly incorrect and it demonstrates a caveat of the equations (8.8) and (8.9) for data-driven estimation of EBFs. When a negative value of *Unc* occurs, it is imperative to examine the logic of combining or separating data classes. For example, the two locations of epithermal Au

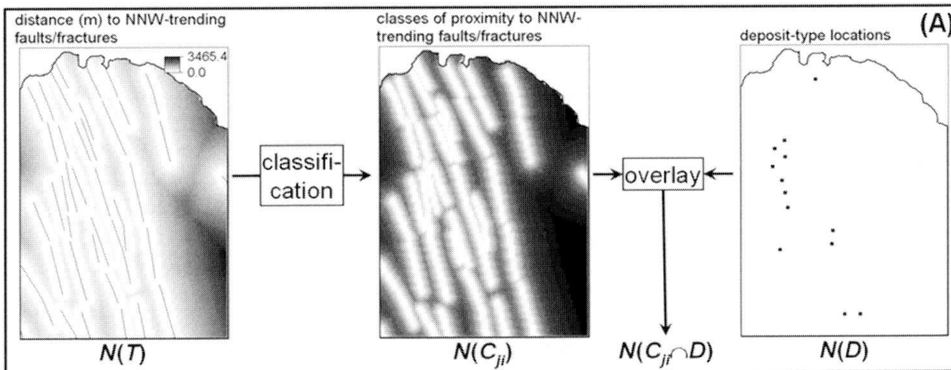

(B)

distance	npixcl	npixt	npixclmo	npixmo	wcmo	twcmo	bel	wcmon	wcmon2	twcmon	dis	unc
0.0	782	12914	2	13	2.821	16.151	0.175	0.998	0.998	4.992	0.200	0.625
96.8	1605	12914	6	13	6.040	16.151	0.374	0.997	0.997	4.992	0.200	0.426
135.5	553	12914	0	13	0.000	16.151	0.000	1.001	0.000	4.992	0.000	1.000
193.6	950	12914	2	13	2.290	16.151	0.142	0.999	0.999	4.992	0.200	0.658
232.3	537	12914	0	13	0.000	16.151	0.000	1.001	0.000	4.992	0.000	1.000
290.4	883	12914	0	13	0.000	16.151	0.000	1.001	0.000	4.992	0.000	1.000
329.1	458	12914	0	13	0.000	16.151	0.000	1.001	0.000	4.992	0.000	1.000
387.2	863	12914	2	13	2.539	16.151	0.157	0.999	0.999	4.992	0.200	0.643
425.9	423	12914	1	13	2.461	16.151	0.152	0.999	0.999	4.992	0.200	0.648
484.0	637	12914	0	13	0.000	16.151	0.000	1.001	0.000	4.992	0.000	1.000
580.8	776	12914	0	13	0.000	16.151	0.000	1.001	0.000	4.992	0.000	1.000
677.6	628	12914	0	13	0.000	16.151	0.000	1.001	0.000	4.992	0.000	1.000
813.1	604	12914	0	13	0.000	16.151	0.000	1.001	0.000	4.992	0.000	1.000
1026.1	614	12914	0	13	0.000	16.151	0.000	1.001	0.000	4.992	0.000	1.000
1316.5	663	12914	0	13	0.000	16.151	0.000	1.001	0.000	4.992	0.000	1.000
1684.3	653	12914	0	13	0.000	16.151	0.000	1.001	0.000	4.992	0.000	1.000
2187.7	641	12914	0	13	0.000	16.151	0.000	1.001	0.000	4.992	0.000	1.000
3465.5	644	12914	0	13	0.000	16.151	0.000	1.001	0.000	4.992	0.000	1.000

(C)

upper limits of C_{ji}	$N(C_{ji})$	$N(T)$	$N(C_{ji} \cap D)$	$N(D)$	$W_{C_{ji}D}$	$\sum_{j=1}^{m} W_{C_{ji}D}$		$W_{C_{ji}\bar{D}}$	re-set $W_{C_{ji}\bar{D}}$	$\sum_{j=1}^{m} W_{C_{ji}\bar{D}}$		
distance	npixcl	npixt	npixclmo	npixmo	wcmo	twcmo	bel	wcmon	wcmon2	twcmon	dis	unc
0.0	782	12914	2	13	2.821	11.165	0.253	0.998	0.998	3.995	0.250	0.497
96.8	1605	12914	6	13	6.040	11.165	0.541	0.997	0.997	3.995	0.250	0.209
193.6	1503	12914	2	13	1.380	11.165	0.124	1.000	1.000	3.995	0.250	0.626
425.9	3164	12914	3	13	0.924	11.165	0.083	1.000	1.000	3.995	0.250	0.667
3465.4	5860	12914	0	13	0.000	11.165	0.000	1.002	0.000	3.995	0.000	1.000

Fig. 8-13. (**A**) GIS operations involved in deriving map variables required for data-driven estimation of *Bel*, *Dis* and *Unc* (see equations (8.8), (8.9) and (8.10), respectively) for classes of proximity to NNW-trending faults/fractures with respect to locations of epithermal Au deposits in the Aroroy district (Philippines). In order to demonstrate the effect of data classification in data-driven estimation of EBFs, the distances (m) to NNW-trending faults/fractures are first (**B**) classified into more-or-less 5-percentile distance intervals and then (**C**) some of the 5-percentile distance class intervals are merged. In (**C**) the names of columns in the tables are annotated with the variables used in equations (8.8) to (8.10).

deposits situated on NNW-trending faults/fractures (i.e., distance=0, Figs. 8-13B and 8-13C) plausibly represent (nearly) exposed deposits, whereas the six locations of

distance	npixcl	npixt	npixclmo	npixmo	wcmo	twcmo	bel	wcmon	wcmon2	twcmon	dis	unc
96.8	2387	12914	8	13	7.056	9.360	0.754	0.997	0.997	2.997	0.333	-0.087
193.6	1503	12914	2	13	1.380	9.360	0.147	1.000	1.000	2.997	0.334	0.519
425.9	3164	12914	3	13	0.924	9.360	0.099	1.000	1.000	2.997	0.334	0.567
3465.4	5860	12914	0	13	0.000	9.360	0.000	1.002	0.000	2.997	0.000	1.000

Fig. 8-14. Illustration of occurrence of a negative value for *Unc* by application of equations (8.8), (8.9) and (8.10) for data-driven estimation of EBFs of classes of data with respect to deposit-type locations. The classified data are distances (m) to NNW-trending faults/fractures, which are considered in predictive modeling of epithermal Au prospectivity in Aroroy district (Philippines). See text for further explanations and Fig. 8-13C for the names of columns.

epithermal Au deposits situated between 0 and 100 m from NNW-trending faults/fractures plausibly represent concealed deposits. If so, then merging the first two classes of distances to NNW-trending faults/fractures (Figs. 8-13B and 8-13C) into one class (Fig. 8-14) is geologically inappropriate because the contained locations of epithermal Au deposits plausibly represent different levels of erosion. This is a generic problem in 2-D modeling of mineral prospectivity.

A negative value of *Unc* could also occur in the applications of equations (8.8) to (8.9) if a class of data of a discrete field (e.g., lithologic units as a prospectivity recognition criterion of favourable host rocks) coincides with more than 50% of the locations of mineral deposits of the type sought and if $N(C_{ji})$ is less than 25% of $N(T)$. In such a case, it is imperative to re-examine the map of the data of a discrete field in terms of (a) accuracy of class boundaries (e.g., lithologic contacts), (b) compatibility of its level of data attribute classification (e.g., rock type versus lithologic formation) and thus scale to the scale of the mineral prospectivity mapping being performed (Raines et al., 2007), (c) compatibility of its scale to the scale of map of mineral deposit occurrences and (d) its relevance to the conceptual model of mineral prospectivity under consideration. In addition, it also imperative to re-examine (a) accuracy of deposit-type locations and (b) suitability of the unit cell size used in modeling. Addressing these issues could, more often than not, overcome the occurrence of a negative value of *Unc* in data-driven estimation of EBFs of classes of data of a discrete field.

The occurrence of a negative value of *Unc* highlights not only the limitations of equations (8.8) and (8.9) for data-driven EBF modeling of mineral prospectivity but also the general limitations of 2-D modeling of mineral prospectivity. Unfortunately, in 2-D (as well as in 3-D) modeling of mineral prospectivity, there is no rule-of-thumb for the correct interval or number of classes into which evidential data of continuous fields must be discretised. Thus, the occurrence of a negative of *Unc* provides the opportunity to re-examine if the data sets are accurate and if classes of evidential data are sound in the context of mineral deposit occurrence and the scale of mineral prospectivity mapping being performed. The caveats of data-driven estimation of EBFs thus provide for a knowledge-guided data-driven modeling of mineral prospectivity (Carranza et al., 2008a) and they preclude application of the technique as a 'black-box' method in which the geological significance of the modeling procedures is overlooked. The examples discussed above demonstrate that performing model calibration experiments with different classes of data is imperative in knowledge-guided data-driven creation of predictor maps (cf. Roy et al., 2006), not only via application of the data-driven EBFs but also via application of the other data-driven methods listed in Tables 8-I and 8-II. Nevertheless, the examples discussed above demonstrate that, provided that the classes of evidential data are prudently examined and properly calibrated, the applications of equations (8.8) to (8.10) for data-driven estimation of EBFs result in geologically meaningful empirical spatial associations between deposit-type locations and indicative geological features and, thus, are useful in the creation of predictor maps for mineral prospectivity mapping.

Integration of data-driven EBFs

Data-driven estimates of EBFs are calculated and then stored usually in attribute tables associated with the individual X_i spatial evidence maps (Figs. 8-11, 8-13 and 8-14). Attribute maps of EBFs (i.e., predictor maps) for each of the X_i spatial evidence maps are then created. Only attribute maps of Bel_i, Dis_i and Unc_i are used for integration of predictor maps according to the application of Dempster's (1968) rule of combination. We recall from the introduction to EBFs in Chapter 7 that, according to Walley (1987), Dempster's (1968) rule of combination is generally neither suitable for combining evidence from independent observations nor appropriate for combining prior beliefs with observational evidence. This means that Dempster's (1968) rule of combining EBFs is suitable in modeling of mineral prospectivity because predictor maps used in most, if not all, cases are conditionally dependent with respect to locations of mineral deposits of the type sought for at least two reasons. Firstly, many predictor maps of mineral prospectivity are derived from a common data set (e.g., maps of proximity to individual sets of faults/fractures are derived from a geological map), which means that they are to some extents 'observationally' dependent on each other. Secondly, predictor maps each represent Earth processes that, at some periods in the geologic time scale and at some environments in the Earth's crust, interacted simultaneously with each other and caused the formation of mineral deposits. Inferences about the inter-play of geological processes involved in mineralisation can be represented in the logical (or sequential) integration of predictor maps portrayed as EBFs.

The formulae for combining maps of EBFs via either an AND or an OR operation (An et al., 1994a), according to Dempster's (1968) rule of combination, are given in Chapter 7 (equations (7.14)-(7.16) and (7.17)-(7.19)) and are not repeated here. An AND or an OR operation represents a function f in equation (8.2). An inference network is applied to logically combine predictor maps representing EBFs of two sets of spatial evidence at a time. An inference network is a series of logical steps, each of which represents a hypothesis of inter-relationship between two sets of processes (portrayed in predictor maps) that represent (a) controls on the occurrence of a geo-object (e.g., mineral deposits) and/or (b) spatial features that indicate the presence of the geo-object. The inference network applied in the knowledge-driven Boolean logic modeling (see Chapter 7, Fig. 7-4) and in the knowledge-driven evidential belief modeling (Chapter 7) is also applied to logically integrate the data-driven estimates EBFs of the spatial evidence maps of epithermal Au prospectivity in the case study area.

Case study application of data-driven EBFs

The objectives of the case study are to illustrate the utility, in mineral prospectivity mapping, of (a) coherent deposit-type locations and (b) coherent proxy deposit-type locations. The spatial evidential data sets used in the case study are given in the introduction section of this chapter. The strategies applied for cross-validation of the predictive models, portrayed as integrated values of *Bel*, of epithermal Au prospectivity in the case study area are given the preceding section.

Calibration experiments were performed not only in deriving data-driven estimates of EBFs for classes of catchment basin analysis anomaly values (Fig. 8-11) and classes of proximity to NNW-trending faults/fractures (Fig. 8-13) but also in deriving data-driven estimates of EBFs for classes of proximity to NW-trending faults/fractures and proximity to intersections of NNW- and NW-trending faults/fractures. These calibration experiments resulted in common properly calibrated classes of each of the data sets (Table 8-IV) with respect to the training sets of 13 known and 11 coherent (out of the 13 known) locations of epithermal Au deposits and 86 randomly-selected and 86 coherent proxy locations of epithermal Au deposits (Fig. 8-8). The data classes in Table 8-IV are considered properly calibrated because the plots of the data-driven estimates of *Bel* versus the upper limits of the properly calibrated classes with respect to the different training sets are not noisy (Fig. 8-15) and facilitate recognition of threshold data values that are associated spatially with the training sets of deposit and proxy deposit locations. The graphs in Fig. 8-15 indicate that the training deposit and proxy deposit locations are mostly within 0.3 km of NNW-trending faults/fractures and within 1.5 km of NW-trending faults/fractures and intersections of NNW- and NW-trending faults/fractures and that the training deposit and proxy deposit locations coincide mostly with integrated PC2 and PC3 multi-element scores greater than 0.3. These results are generally consistent with the results of analyses of spatial associations between the same spatial data sets and the locations of epithermal Au deposits in the study area (see Chapter 6, Table 6-IX).

The graphs in Fig. 8-15 show that, for the range of data values that are associated spatially with the deposit and proxy deposit locations, the applications of the training sets of coherent deposit and proxy deposit locations result in higher values of *Bel* than the applications of the training sets of all deposit locations and randomly-selected proxy deposit locations. Conversely, the graphs in Fig. 8-15 show that, for the range of data values that lack spatial association with the deposit and proxy deposit locations, the applications of the training sets of all deposit locations and randomly-selected proxy deposit locations result in higher values of *Bel* than the application of the training sets of coherent deposit and proxy deposit locations. These results imply that, compared to the applications of all deposit locations and randomly-selected proxy deposit locations, the

TABLE 8-IV

Properly calibrated classes of spatial data sets for evidential belief modeling of epithermal Au prospectivity, Aroroy district (Philippines). Ranges of data values in bold are associated spatially with the epithermal Au deposits in the case study area (see Chapter 6, Table 6-IX). The plots of values of *Bel* versus the upper limits of classes of data are shown in Fig. 8-15.

Classes of proximity to NNW[1a]		Classes of proximity to NW[2]	
Class code	Range[1b] (km)	Class code	Range (km)
NNW1	**0.00**	NW1	**0.00 – 0.33**
NNW2	**0.01 – 0.10**	NW2	**0.34 – 0.74**
NNW3	**0.11 – 0.19**	NW3	**0.75 – 1.26**
NNW4	**0.20 – 0.43**	NW4	1.27 – 2.19
NNW5	0.44 – 3.47	NW5	2.20 – 5.23
Classes of proximity to FI[3]		**Classes of ANOMALY[4a]**	
Class code	Range (km)	Class code	Range[4b]
FI1	**0.00 – 0.54**	ANOM1	**0.369 – 0.781**
FI2	**0.55 – 0.99**	ANOM2	**0.252 – 0.368**
FI3	**1.00 – 1.59**	ANOM3	0.100 – 0.251
FI4	1.60 – 2.75	ANOM4	≤0.099
FI5	2.76 – 5.87	ANOM5	'no data'

[1a]NNW-trending faults/fractures. [1b]Upper limits of proximity range are equivalent to the distances (m) in the first column of table in Fig. 8-13C. [2]NW-trending faults/fractures. [3]Intersections of NNW- and NW-trending faults/fractures. [4a]Integrated PC2 and PC3 scores obtained from the catchment basin analysis of stream sediment geochemical data (see Chapter 5, Fig. 5-12). [4b]Upper limits of ANOMALY range are the same as values of anomaly in the first column of table in Fig. 8-11C, but in decreasing order.

applications of coherent deposit and proxy deposit locations result in better predictive models of epithermal Au prospectivity in the study area.

The map of integrated values of *Bel* based on the training set of 13 known locations of epithermal Au deposits (Fig. 8-16A) and the map of integrated values of *Bel* based on the training set of 11 coherent locations of epithermal Au deposits (Fig. 8-17A) both show circular patterns of intermediate and high values of integrated *Bel* reflecting the spatial evidence of proximity to intersections of NNW- and NW-trending faults/fractures. The values of integrated *Bel* are highest mostly where the circular patterns intersect with the patterns reflecting the spatial evidence of proximity to NNW-trending faults/fractures, which are more conspicuous in Fig. 8-16A than in Fig. 8-17A. The pattern of intermediate and high integrated values of *Bel* in both maps (Figs. 8-16A and 8-17A) seem odd but are consistent with the conceptual model of epithermal Au mineralisation in dilational or extensional settings as depicted in Fig. 6-16.

The fitting-rates of the map of integrated values *Bel* based on the training set of 11 coherent locations of epithermal Au deposits (Fig. 8-17B) are better than the fitting-rates of the map of integrated values *Bel* based on the training set of 13 known locations of

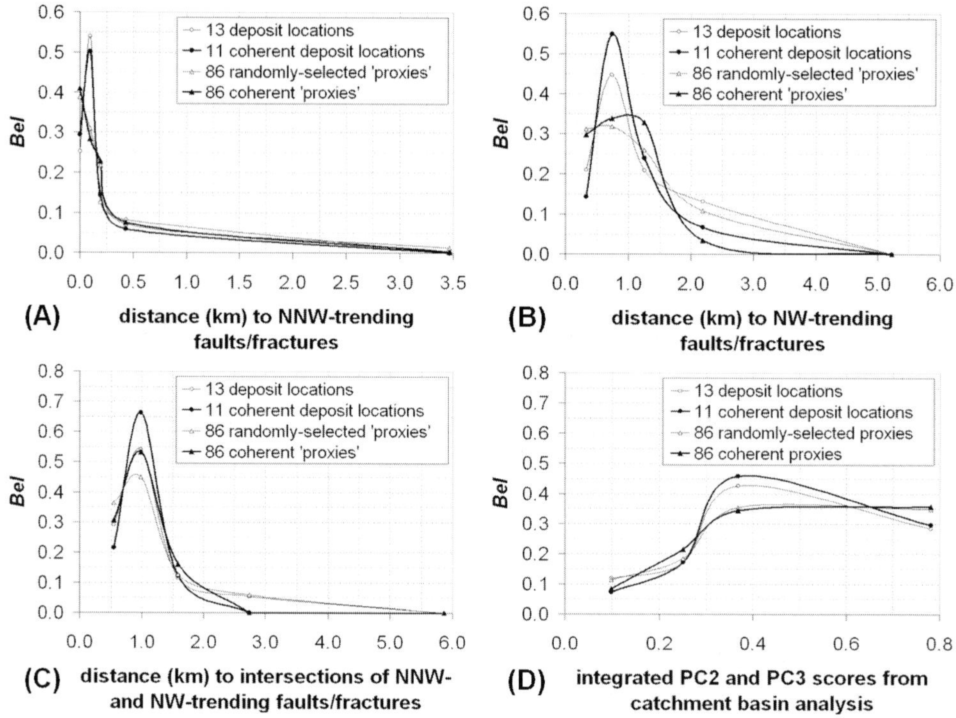

Fig. 8-15. Variations of spatial associations of individual sets of geological features with all and only coherent locations of epithermal Au deposits and with randomly-selected and coherent proxy locations of epithermal Au deposits in Aroroy district (Philippines) as depicted by the plots of data-driven estimates of *Bel* versus upper limits of classes of (**A**) distances to NNW-trending faults/fractures, (**B**) distances to NW-trending faults/fractures, (**C**) distances to intersections of NNW- and NW-trending faults/fractures and (**D**) integrated PC2 and PC3 scores obtained from the catchment basin analysis of stream sediment geochemical data (see Chapter 5, Fig. 5-12).

epithermal Au deposits (Fig. 8-16B). This is why the former map is less 'noisy' than the latter map. The prediction-rates of the map of integrated values of *Bel* based on the training set of 11 coherent locations of epithermal Au deposits (Fig. 8-17B) are also better than the prediction-rates of the map of integrated values *Bel* based on the training set of 13 known locations of epithermal Au deposits (Fig. 8-16B). For example, if 10% and 30% of the study area are considered prospective, then the map in Fig. 8-17A has prediction-rates of 38% and 75% (Fig. 8-17B), respectively, whereas the map in Fig. 8-16A has prediction-rates of 29% and 70% (Fig. 8-1B), respectively. In addition, the map of integrated values of *Bel* based on the training set of 11 coherent locations of epithermal Au deposits has lower values of integrated *Unc* (Fig. 8-17B) than the map of integrated values *Bel* based on the training set of 13 known locations of epithermal Au

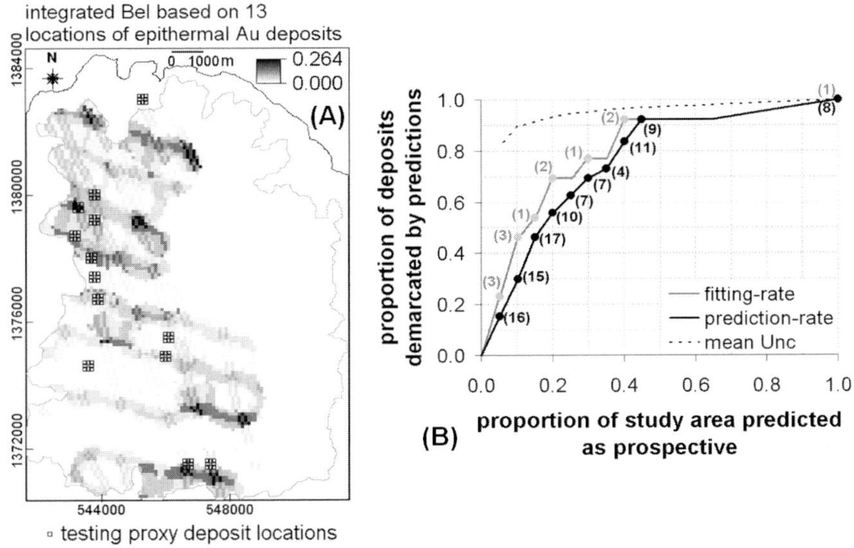

Fig. 8-16. (**A**) Epithermal Au prospectivity map of Aroroy district (Philippines) portrayed as integrated values of *Bel* of spatial evidence layers with respect to a training set of 13 locations of epithermal Au deposits. Polygon outlined in grey is area of stream sediment sample catchment basins (see Fig. 4-11). The testing set of 104 proxy deposit locations immediately surrounding each of the known locations epithermal Au deposits (see Fig. 8-8) is shown as reference to the prediction-rate. (**B**) Fitting and prediction-rate curves of, respectively, proportions of training deposits and testing proxy deposits demarcated by the predictions versus proportion of the study area predicted as prospective based on the integrated values of *Bel*. The grey and black dots represent classes of integrated values of *Bel* that correspond spatially with certain numbers of training deposit locations (in grey) and certain numbers of testing proxy locations (in black), respectively.

deposits (Fig. 8-16B). These results illustrate the advantage of using coherent deposit-type locations in predictive modeling of mineral prospectivity.

The map of integrated values *Bel* based on a training set of 86 randomly-selected proxy locations of epithermal Au deposits (Fig. 8-18A) and the map of integrated values *Bel* based on the training set of 86 coherent proxy locations of epithermal Au deposits (Fig. 8-19A) also show circular patterns of intermediate and high values of integrated *Bel* reflecting the spatial evidence of proximity to intersections of NNW- and NW-trending faults/fractures. Compared to the former map, the latter map is more similar to the maps in Fig. 8-16A and 8-17A. The maps in Figs. 8-18A and 8-19A both display intermediate and high values of integrated *Bel* reflecting the patterns of anomalous sample catchment basins in the western parts of the study area (see Fig. 5-12), although these patterns are more conspicuous in Fig. 8-18A than in Fig. 8-19A. These observations indicate that the map of integrated values of *Bel* based on the training set of

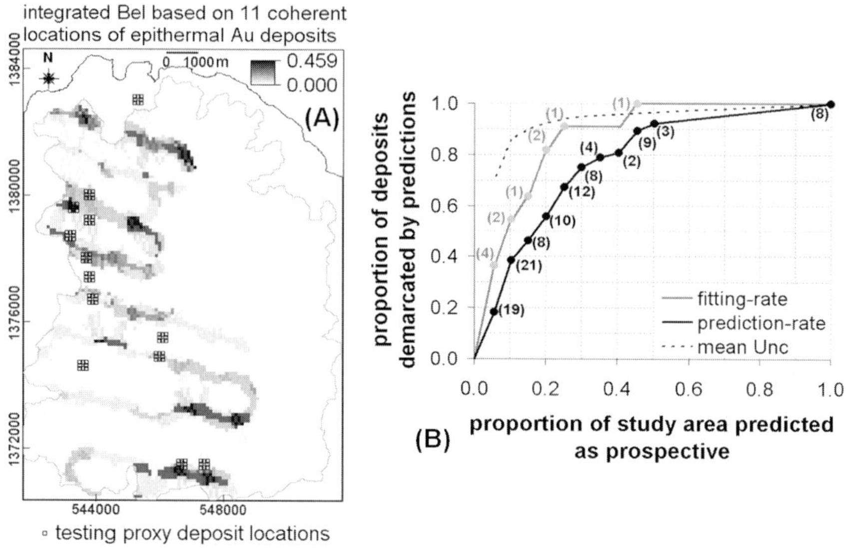

Fig. 8-17. (**A**) Epithermal Au prospectivity map of Aroroy district (Philippines) portrayed as integrated values of *Bel* of spatial evidence layers with respect to a training set of coherent locations of epithermal Au deposits. Polygon outlined in grey is area of stream sediment sample catchment basins (see Fig. 4-11). The testing set of 104 proxy deposit locations immediately surrounding each of the known locations epithermal Au deposits (see Fig. 8-8) is shown as reference to the prediction-rate. (**B**) Fitting and prediction-rate curves of, respectively, proportions of training deposit locations and testing proxy deposit locations demarcated by the predictions versus proportion of the study area predicted as prospective based on the integrated values of *Bel*. The grey and black dots represent classes of integrated values of *Bel* that correspond spatially with certain numbers of training deposit locations (in grey) and certain numbers of testing locations (in black), respectively.

86 coherent proxy locations of epithermal Au deposits (Fig. 8-19A) portrays patterns derived from each of the four input predictor maps whilst the map of integrated values *Bel* based on a training set of 86 randomly-selected proxy locations of epithermal Au deposits (Fig. 8-18A) portrays patterns derived mostly from at least two of the four predictor maps. This is why the fitting-rates of the former map (Fig. 8-19B) are better than the fitting-rates of the latter map (Fig. 8-18B).

In terms of prediction-rate, the map of integrated values of *Bel* based on the training set of 86 coherent proxy locations of epithermal Au deposits (Fig. 8-19A) and the map of integrated values of *Bel* based on a training set of 86 randomly-selected proxy locations of epithermal Au deposits (Fig. 8-18A) both delineate the same percentages (38-54%) of testing deposit locations if 10-20% of the study area is considered prospective (Figs. 8-18B and 8-19B). The map of integrated values of *Bel* based on a training set of 86 randomly-selected proxy locations of epithermal Au deposits (Fig. 8-18A) is better than the map of integrated values of *Bel* based on the training set of 86 coherent proxy

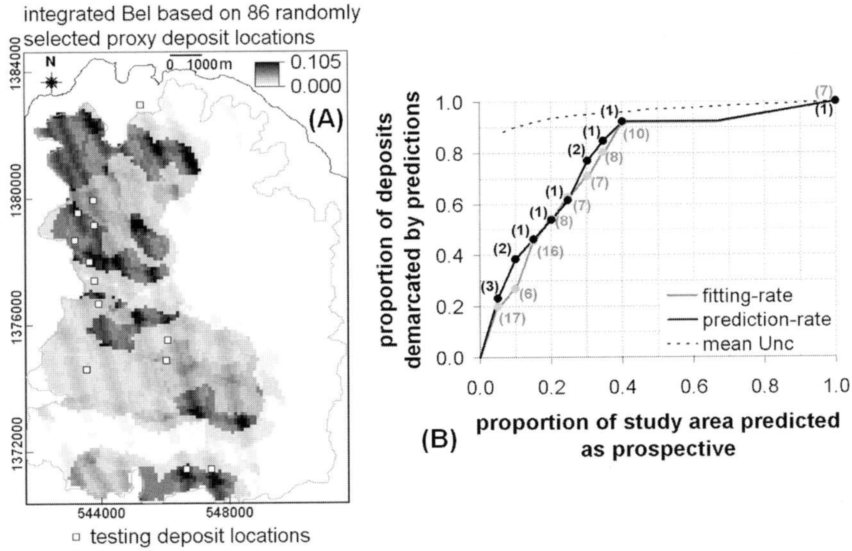

Fig. 8-18. (**A**) Epithermal Au prospectivity map of Aroroy district (Philippines) portrayed as integrated values of *Bel* of spatial evidence layers with respect to a training set of 86 randomly-selected (from 104) proxy locations of epithermal Au deposits (Fig. 8-8). Polygon outlined in grey is area of stream sediment sample catchment basins (see Fig. 4-11). The testing set of locations of 13 epithermal Au deposits is shown as reference to the prediction-rate. (**B**) Fitting and prediction-rate curves of, respectively, proportions of training proxy deposits and testing deposits demarcated by the predictions versus proportion of the study area predicted as prospective based on the integrated values of *Bel*. The grey and black dots represent classes of integrated values of *Bel* that correspond spatially with certain numbers of training proxy deposit locations (in grey) and certain numbers of testing deposit locations (in black), respectively.

locations of epithermal Au deposits (Fig. 8-19A) if more than 20% of the study area is considered prospective. However, if 5% of the study area is considered prospective, then the latter map delineates 31% of the testing deposit locations, whereas the latter map delineates 23% of the testing deposit locations. Therefore, because mineral prospectivity mapping aims to constrain the sizes of exploration targets in order to increase the chance of mineral deposit discovery, then the map of integrated values of *Bel* based on the training set of 86 coherent proxy locations of epithermal Au deposits (Fig. 8-19A) is better than the map of integrated values of *Bel* based on a training set of 86 randomly-selected proxy locations of epithermal Au deposits (Fig. 8-18A). In addition, the former map has lower values of integrated *Unc* (Fig. 8-19B) than the latter map (Fig. 8-18B). These results illustrate the advantage of, not just proxy but, coherent proxy deposit-type locations in predictive modeling of mineral prospectivity. The results also imply the disadvantage of random selection of training (proxy) deposit-type locations for predictive modeling of mineral prospectivity.

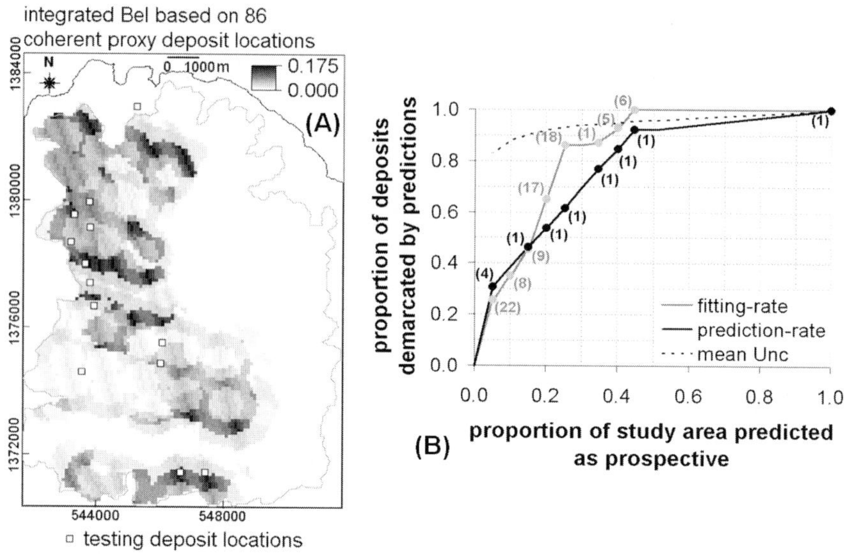

Fig. 8-19. (**A**) Epithermal Au prospectivity map of Aroroy district (Philippines) portrayed as integrated values of *Bel* of spatial evidence layers with respect to a training set of 86 coherent proxy locations of epithermal Au deposits (Fig. 8-8). Polygon outlined in grey is area of stream sediment sample catchment basins (see Fig. 4-11). The testing set of locations of 13 epithermal Au deposits is shown as reference to the prediction-rate. (**B**) Fitting and prediction-rate curves of, respectively, proportions of training proxy deposits (grey dots) and testing deposits (black dots) demarcated by the predictions versus proportion of the study area predicted as prospective based on the integrated values of *Bel*. The grey and black dots represent classes of integrated values of *Bel* that correspond spatially with certain numbers of training proxy deposit locations (in grey) and certain numbers of testing deposit locations (in black), respectively.

The results of the case study demonstrate the usefulness of data-driven evidential belief modeling of mineral prospectivity. Despite of the caveats of data-driven estimation of EBFs of classes of individual spatial evidence layers, Dempster's (1968) rule of combination provides for calibration experiments in integrating predictor maps in order to emulate the inter-play of processes involved in mineralisation (see Chapter 6). Emulating the simultaneous interactions of various processes involved in mineralisation is the main difficultly in predictive modeling of mineral prospectivity. A way to overcome this difficulty is to quantify simultaneously the spatial associations of the predictor variables with the target variables. This is conceivably the reason why there are more multivariate techniques (Table 8-II) than bivariate techniques (Table 8-I) for mineral prospectivity mapping. This is not to say, however, that multivariate techniques are automatically superior to bivariate techniques because many of the latter techniques do not involve inference systems for combining predictor maps of mineral prospectivity. Let us now turn to one of these multivariate techniques – discriminant analysis.

DISCRIMINANT ANALYSIS OF MINERAL PROSPECTIVITY

Discriminant analysis (DA) has a long history of application in exploration geochemistry (e.g., Bull and Mazzucchelli, 1975; Govett et al., 1975; Beauchamp et al., 1980; Howarth, 1983b; Clarke et al., 1989; Chork and Rousseeuw, 1992; Yusta et al., 1998; Singer and Kouda, 2001) and mineral prospectivity mapping (see references in Table 8-II). The common chief objective of the applications of DA to geochemical anomaly and mineral prospectivity mapping is to classify every location in a study area into two mutually exclusive groups – prospective (anomalous or mineralised) and non-prospective (background or barren) – based on a training set of known deposit-type and non-deposit locations and multiple sets of data of discriminating (i.e., predictor/explanatory) variables at these locations.

Exhaustive explanations of DA are not given here, but readers are referred to Davis (2002, pp. 471-479) for a thorough introduction to DA and to Harris and Pan (1999) and Pan and Harris (2000, pp.411-414) for explanations of different methods of DA that can be applied to mineral prospectivity modeling. The treatment of DA here is limited to the basic principles and application of the two-group DA method, which is also called Fisher linear DA (Fisher, 1936) and demonstration of a GIS-based technique for representation of data of predictor/explanatory variables in DA.

Basic principles and assumptions of linear discriminant analysis

In general, the maximum number of discriminant functions that can be derived for the classification of groups of data is either equal to the number of groups minus one or equal to the number of predictor variables, whichever is less. In the two-group DA (hereafter denoted as LDA; L stands for linear), therefore, there is only one discriminant function, which is a linear combination of the predictor variables with the following mathematical form:

$$S_{DL} = b_0 + b_1 X_{1DL} + b_2 X_{2DL} + \ldots + b_p X_{pDL} \tag{8.11}$$

where S_{DL} is the discriminant score for case (location) L in group D, X_{pDL} is the value of predictor variable p (=1,2,…,n) for case (location) L in group D, b_0 is a constant and b_1, b_2 and b_p are function coefficients. A discriminant function is generated from a training set of L cases (locations), for which membership in group D (e.g., deposit or non-deposit) is known. There are two types of function coefficients derived in DA - standardised function coefficients and unstandardised function coefficients (Tabachnick and Fidell, 2007). Note that there is no b_0 among the standardised function coefficients. On the one hand, the standardised function coefficients are used for assessing the relative importance of the predictor variables in classifying the target variable (in this case deposit and non-deposit locations). On the other hand, the unstandardised function coefficients are the ones used in equation (8.11) in order to derive values of S_{DL} for the classification of unknown cases (i.e., unvisited locations) with values for the same sets of predictor variables used to generate the discriminant function. The values of S_{DL} can then be used for mapping geo-objects (e.g., prospective areas) of interest.

In any method of DA, there are five basic assumptions about the data of the predictor variables. Firstly, the total number of cases must be at least five times the number of predictor variables (Tabachnick and Fidell, 2007). The number of cases (locations) for each group D can be equal or unequal, but if they are unequal the number of cases in the smallest (or smaller) group must be greater than the number of predictor variables. Secondly, the data of the predictor variables for the cases of each group represent samples from a multivariate normal distribution. This assumption is difficult to justify in mineral prospectivity mapping especially because the 'deposit-type' cases and, thus, the data of the predictor variables for most of these cases are likely not representative of samples derived from a multivariate normal distribution (see Fig. 8-6). Fortunately, DA is not seriously affected by violations of the normality assumption as long as non-normality is not due to outliers (Davis, 2002; Tabachnick and Fidell, 2007). Thirdly, the variance-covariance matrices of the groups should be equal, although inequality of variances is, like violation of normality, not 'fatal' to DA (Davis, 2002; Tabachnick and Fidell, 2007). Fourthly, the predictor variables are neither completely redundant nor conditionally dependent (i.e., highly correlated) because, if they are, the matrix is said to be ill-conditioned and thus cannot be inverted. Like the normality assumption, the assumption of conditional independence is difficult to justify for data of the predictor variables at deposit-type locations. Finally, none of the cases used to derive the discriminant function are misclassified (i.e., none of the cases from one group belongs to another group).

In the case study (see below), all the aforementioned five basic assumptions of DA, except the third basic assumption, are addressed as follows. With respect to the first basic assumption of DA, the deposit-type locations, which are very few compared to the number of predictor variables (see below), are not used for training but for testing. Instead, proxy deposit-type locations are used for training. Two training sets, each consisting of equal numbers of proxy deposit-type locations and non-deposit locations, are used in LDA. With respect to the second basic assumption of DA, a one training set consisting of coherent proxy deposit-type locations (Fig. 8-8) is used in order to address the problem of non-normality due to outliers. In order to illustrate the utility of coherent proxy deposit-type locations in data-driven modeling of mineral prospectivity, another training set consisting of randomly-selected proxy deposit-type locations is used. With respect to the fourth basic assumption of DA, it is considered that the predictor variables at the coherent proxy deposit-type locations are not completely redundant because they are not completely coherent (see Fig. 8-7). With respect to the fifth basic assumption of DA, non-deposit locations that are highly dissimilar from the coherent proxy deposit-type locations (see Fig. 8-7) are used in the two training sets described above.

There are two statistical tests of significance in DA (Tabachnick and Fidell, 2007). First, an F-test (Wilks' lambda) is applied to test the null hypothesis that two groups under examination have identical multivariate means (i.e., if the discriminant model as a whole is significant). The smaller the value of Wilks' lambda, the more statistically significant is a discriminant model. Second, if a discriminant model as a whole is statistically significant, then the individual predictor variables are assessed with an F-test (Wilks' lambda) to determine which of them contribute significantly to the discriminant model (i.e., to determine which predictor variables have significantly different means between groups). Predictor variables that do not contribute significantly to the discrimination of the groups are excluded in the final discriminant model.

GIS-based spatial evidence representation for discriminant analysis

A scheme of spatial evidence representation of categorical predictor variables is adopted here (Fig. 8-20) for the GIS-based application of LDA to the case study area so that the results can be compared properly with the earlier results of the application of data-driven EBFs in modeling of epithermal Au prospectivity in the case study area. Carranza and Hale (2001b) and Carranza (2002) demonstrated this scheme of spatial evidence representation for logistic regression modeling of mineral prospectivity in certain case study areas, the results of which are comparable to the results of weights-of-evidence modeling (Carranza and Hale, 2000; Carranza, 2002) and data-driven evidential belief modeling (Carranza, 2002; Carranza and Hale, 2003) of mineral prospectivity in the same case study areas.

The scheme of evidential data presentation presented in Fig. 8-20 is applicable in a raster-based GIS. First, the study area is partitioned into unit cells of a suitable size (i.e., 100×100 m, see above) and each unit cell is given a unique ID. Each unit cell represents a location (L), which can be either a deposit-type location (D=1) or a non-deposit location (D=0). The values of D per unit cell are used as the target variable in LDA. Next, the map of an evidential data E (with n classes C_n) is partitioned further into sub-unit cells, in this case 10×10 m (i.e., each unit cell contains 100 sub-unit cells). Then, the numbers of sub-unit cells of individual classes of evidential data (E_{Cn}) per unit cell are determined via a map overlay (or cross) operation. The numbers of sub-unit cells of E_{Cn} are used as predictor variables in LDA. From the table (or database) of derived hypothetical data exemplified in Fig. 8-20, it is clear that D=1 is associated with E_{C1} whilst D=0 is associated with E_{C2}. By application of this scheme of spatial evidence representation to the properly calibrated classes of evidential data used earlier in the data-driven evidential belief modeling of epithermal Au prospectivity in the case study area (Table 8-IV), the application of LDA to the case study can be expected to yield in the same or similar classes of evidential data that are associated spatially with locations of epithermal Au deposits and thus the results can be compared and contrasted with the data-driven estimates of EBFs shown in Fig. 8-15.

Case study application of discriminant analysis

The objective of the case study is to illustrate the utility of coherent proxy deposit-type locations in modeling of mineral prospectivity via the application of a multivariate

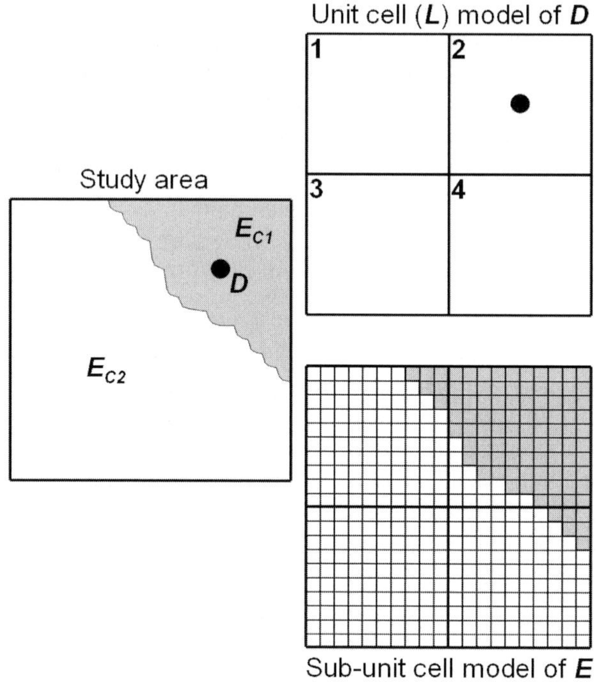

Data representation for DA

	Target variable	Predictor variables	
Unit cell (*L*)	***D***	E_{C1}	E_{C2}
1	0	9	91
2	1	86	14
3	0	0	100
4	0	9	91

Fig. 8-20. A scheme of spatial evidence representation for raster-based GIS application of linear discriminant analysis to mineral prospectivity mapping. Maps of a study area are partitioned into equal-area unit cells (or pixels). Each unit cell is given a unique identifier. If a unit cell contains a deposit-type occurrence (D), then it is given a score of 1; otherwise, it is given a score of 0. A map of spatial evidence (E), with n number of evidential classes (C_n), is partitioned further into sub-unit cells. In each unit cell, the numbers of sub-unit cells of per E_{Cn} are counted. The values of D and the values of E_{Cn} for unit cells representing deposit-type and non-deposit locations are used as the target and predictor variables, respectively, in discriminant analysis.

technique (in this case LDA) to situations of few deposit-type locations. Experiments with two somewhat different training sets (A and B) are performed in order to demonstrate the advantage of using not just proxy but coherent proxy deposit-type locations in modeling of mineral prospectivity. Thus, on the one hand, the training set A consists of 86 randomly-selected (out of 104) proxy deposit-type locations (Fig. 8-8) and 86 non-deposit locations with the lowest predicted mineral occurrence scores (out of 117 non-deposit locations in set 2; Fig. 8-7). On the other hand, the training set B consists of 86 coherent proxy deposit-type locations (Fig. 8-8), which were used earlier in the application of data-driven evidential belief modeling, and the same 86 non-deposit locations in training set A. Because the 86 coherent proxy deposit-type locations in training set B were derived by analysis using the set 1 non-deposit locations (Fig. 8-7), the 86 non-deposit locations in both training sets A and B are drawn randomly from the

117 non-deposit locations of set 2 (Fig. 8-4) in order to avoid bias against the 86 randomly selected proxy deposit-type locations in training set A. The use of training sets with equal numbers of deposit-type locations and non-deposit locations is adopted from (a) the use of equal number of 'zeros' (e.g., non-deposit locations) and 'ones' (e.g., deposit-type locations) in logistic regression analyses when the latter are rare (Breslow and Cain, 1988; Schill et al., 1993; King and Zeng, 2001) and (b) the suggestion of Brown et al. (2000) and Porwal et al (2003a) that a gross imbalance between deposit-type locations and non-deposit locations results in poor recognition of prospective zones via application of artificial neural networks.

Of the properly calibrated classes of individual evidential data or predictor variables (Table 8-IV), class ANOM5 ('no data') is excluded from the application of LDA because it can result in multivariate outliers due to missing geochemical data. The alternative of replacing missing data of a predictor variable with the mean of this variable is also not considered because various parts of geochemical landscapes cannot be appropriately represented by uniform mean uni-element concentrations or mean multi-element scores. Thus, the training set A is left with 79 randomly-selected proxy deposit-type locations and 81 non-deposit locations, whilst the training set B is left with 86 coherent proxy deposit-type locations and 81 non-deposit locations. The small (i.e., 8-9%) difference between the numbers of deposit-type locations in training sets A and B is not remedied because the results are a preliminary indication of the advantage of using coherent rather than just (i.e., randomly-selected) proxy deposit-type locations in modeling of mineral prospectivity. Based on the training sets of deposit-type and non-deposit locations with data for all predictor variables, the predictive models of epithermal Au prospectivity in the case study area derived via the application of LDA are, as in the application of evidential belief modeling (Fig. 8-19), cross-validated against the 13 known locations of epithermal Au deposits.

Table 8-V shows that the discriminant model based on training set B is slightly better (i.e., lower Wilks' lambda) than the discriminant model based on training set A. Both discriminant models based on training sets A and B have common statistically significant predictor variables. This is probably because most of the 79 randomly-selected proxy deposit-type locations in training set A are the same as most the 86 coherent proxy deposit-type locations in training set B. However, except for the standardised function coefficients of the 'FI' predictor variables (i.e., classes of proximity to intersections of NNW- and NW-trending faults/fractures), most of the standardised function coefficients of the predictor variables in the discriminant model based on training set B are, to varying degrees, higher than the standardised function coefficients of corresponding predictor variables in the discriminant model based on training set A. In particular, the standardised function coefficients for the 'NNW' predictor variables (i.e., classes of proximity to NNW-trending faults/fractures) in the discriminant model based on training set B are much higher than the standardised function coefficients of the same predictor variables in the discriminant model based on training set A. However, the standardised function coefficients for the 'FI' predictor

TABLE 8-V

Two models of discriminant functions for predictive mapping of epithermal Au prospectivity, Aroroy district (Philippines) based on training sets each with nearly balanced numbers of proxy deposit-type and non-deposit locations. Values in bold represent predictor variables, per set of spatial evidence, with strong positive spatial associations with the training deposit-type locations.

Discriminant analysis using training set A[1] (Wilks' lambda = 0.192; α=0.0001)			Discriminant analysis using training set B[2] (Wilks' lambda = 0.172; α=0.0001)		
Predictor variables[3]	Function coefficients Standardised	Function coefficients Unstandardised	Predictor variables[3]	Function coefficients Standardised	Function coefficients Unstandardised
NNW1	0.239	0.092	NNW1	0.244	0.088
NNW2	**0.512**	0.017	NNW2	**0.727**	0.024
NNW3	**0.554**	0.021	NNW3	**0.713**	0.026
NNW4	0.326	0.009	NNW4	0.537	0.014
NW1	-0.229	-0.006	NW1	-0.221	-0.006
NW2	**0.094**	0.003	NW2	**0.069**	0.002
NW3	-0.071	-0.002	NW3	-0.069	-0.002
NW4	-0.047	-0.001	NW4	-0.080	-0.002
FI1	**1.497**	0.043	FI1	**1.282**	0.037
FI2	**1.006**	0.028	FI2	**0.939**	0.025
FI3	0.467	0.013	FI3	0.467	0.012
FI4	-0.017	-0.001	FI4	-0.044	-0.001
ANOM1	**0.524**	0.013	ANOM1	**0.517**	0.013
ANOM2	**0.163**	0.004	ANOM2	**0.233**	0.006
ANOM3	0.113	0.003	ANOM3	0.145	0.004
Constant	-	-3.017	Constant	-	-3.323

[1]Consists of 79 randomly-selected proxy deposit-type locations (Fig. 8-8) and 81 non-deposit locations with lowest predicted mineral occurrences (from set 2 non-deposit locations; Figs. 8-4 and 8-7B). [2]Consists of 86 coherent proxy deposit-type locations (Fig. 8-8) and the same 81 non-deposit locations in training data set A. [3]Statistically significant predictor variables in the discriminant models (see 'class code' columns in Table 8-IV for explanations of variable names).

variables in the discriminant model based on training set A are not so much higher than the standardised function coefficients of the same predictor variables in the discriminant model based on training set B. These results mean that, on the one hand, the contributions of the 'FI' predictor variables are more important to the discrimination between the randomly-selected proxy deposit-type locations and non-deposit locations in training set A than to the discrimination between the coherent proxy deposit-type locations and non-deposit locations in training set B. This implies that any of the proxy deposit-type locations chosen randomly has stronger spatial association with intersections of NNW- and NW-trending faults/fractures than to either of these individual sets of faults/fractures. On the other hand, in addition to the contributions of the 'FI' predictor variables, the contributions of the 'NNW' variables are more important to the discrimination between the coherent proxy deposit-type locations and non-deposit locations in training set B than to the discrimination between the random-selected proxy deposit-type locations and non-deposit locations in training set A. This implies that the coherent proxy deposit-type locations have stronger spatial association with NNW-trending faults/fractures compared to randomly-selected proxy deposit-type locations.

In both of the two discriminant models, the contributions of the 'ANOM' predictor variables (i.e., classes of high integrated PC2 and PC3 scores obtained from the catchment basin analysis) are more-or-less the same but are subordinate to the contributions of the 'FI' and 'NNW' predictor variables. In both of the two discriminant models, the contributions of the 'NW' predictor variables (i.e., classes of proximity to NW-trending faults/fractures) are the most inferior. These results suggest that the presence of multi-element geochemical anomalies is a more important predictor of epithermal Au prospectivity in the case study area than proximity to NW-trending faults/fractures.

Both of the discriminant models based on training sets A and B indicate that (a) proximity to intersections of NNW- and NW-trending faults/fractures is a more important control on epithermal Au mineralisation in the case study area than proximity to either NNW- or NW-trending faults/fractures and (b) proximity to NNW-trending faults/fractures is a more important control on epithermal Au mineralisation in the case study area than proximity to NW-trending faults/fractures. These results contrast somewhat with the implications of the results of the analyses of spatial associations in Chapter 6 and the data-driven estimates of EBFs earlier in this chapter. Nevertheless, the multivariate spatial associations depicted by the results shown in Table 8-V are consistent with the knowledge that the presence and/or proximity to dilational jogs or zones of extensions at/near either discontinuities or intersections of faults/fractures are more important controls on hydrothermal mineralisation than faults/fractures alone (Sibson, 1987, 1996, 2000, 2001). These results underscore the advantage of multivariate techniques compared to bivariate techniques in terms of simultaneous analysis and synergistic interpretation of empirical spatial associations between deposit-type locations and indicative geological features.

If the magnitudes of the standardised function coefficients are compared and contrasted with each set of spatial evidence rather than among the classes of spatial evidence, then the two discriminant models (Table 8-V) indicate that epithermal Au deposits in the case study area mostly occur within (a) about 200 m of NNW-trending faults/fractures, (b) about 750 m of NW-trending faults/fractures and (c) about 1 km of intersections of NNW- and NW-trending faults/fractures. These results are consistent with the empirical spatial associations between epithermal Au deposits and indicative geological features as quantified via the distance correlation method rather than as quantified via the distance distribution method (see Chapter 6, Table 6-IX). In addition, the two discriminant models (Table 8-V) indicate that integrated PC2 and PC3 scores (obtained via catchment basins analysis; Chapter 5) greater than 0.25 are associated spatially with most of the known epithermal Au deposits and therefore represent significant anomalies. The overall results of the application of LDA are therefore

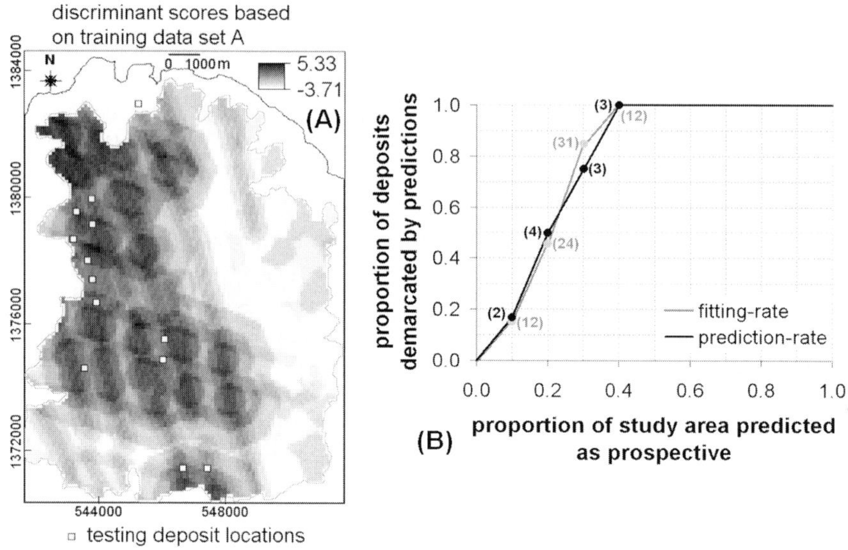

Fig. 8-21. (**A**) Epithermal Au prospectivity map of Aroroy district (Philippines) portrayed as discriminant scores of spatial evidence layers with respect to training set A of 79 randomly-selected proxy deposit-type locations (Fig. 8-8) and 81 non-deposit locations. Polygon outlined in grey is area of stream sediment sample catchment basins (see Fig. 4-11). The testing set of locations of 13 epithermal Au deposits is shown as reference to the prediction-rate. (**B**) Fitting and prediction-rate curves of, respectively, proportions of randomly-selected training proxy deposits (grey dots) and testing deposits (black dots) demarcated by the predictions versus proportion of the study area predicted as prospective based on the discriminant scores. The grey and black dots represent classes of discriminant scores that correspond spatially with certain numbers of training randomly-selected proxy deposit-type locations (in grey) and certain numbers of testing deposit-type locations (in black), respectively.

generally consistent with the data-driven estimates of *Bel* shown in Fig. 8-15. It follows that the scheme of spatial evidence representation for raster-based GIS application of LDA to mineral prospectivity mapping (Fig. 8-20) allows proper comparison with the results of the application of data-driven evidential belief modeling.

The maps of discriminant scores based on training set A (Fig. 8-21A) and training set B (Fig. 8-22A) both show circular patterns and NNW-trending linear patterns of intermediate and high values reflecting the spatial evidence of proximity to intersections of NNW- and NW-trending faults/fractures and proximity to NNW-trending faults/fractures. The discriminant scores are highest mostly where the circular and the NNW-trending linear patterns intersect. The patterns of intermediate and high discriminant scores in both maps (Figs. 8-21A and 8-22A) are consistent with the conceptual model of epithermal Au mineralisation in dilational or extensional settings as depicted in Fig. 6-16.

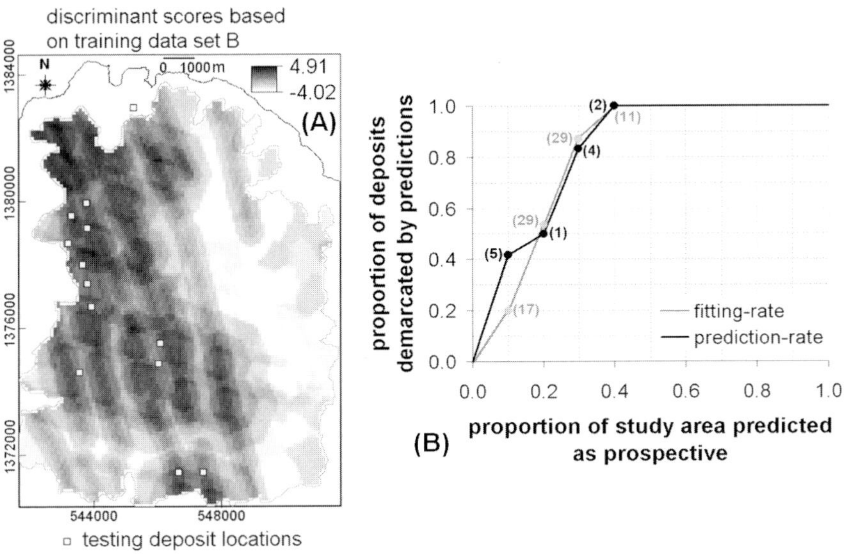

Fig. 8-22. (**A**) Epithermal Au prospectivity map of Aroroy district (Philippines) portrayed as discriminant scores of spatial evidence layers with respect to training set B of 86 coherent proxy deposit-type locations (Fig. 8-8) and 81 non-deposit locations. Polygon outlined in grey is area of stream sediment sample catchment basins (see Fig. 4-11). The testing set of locations of 13 epithermal Au deposits is shown as reference to the prediction-rate. (**B**) Fitting and prediction-rate curves of, respectively, proportions of coherent training proxy deposits (grey dots) and testing deposits (black dots) demarcated by the predictions versus proportion of the study area predicted as prospective based on the discriminant scores. The grey and black dots represent classes of discriminant scores that correspond spatially with certain numbers of training coherent proxy deposit-type locations (in grey) and certain numbers of testing deposit-type locations (in black), respectively.

The fitting-rates of the map of discriminant scores based on training set B (Fig. 8-22B) are better than the fitting-rates of the map of discriminant scores based on training set A (Fig. 8-21B). For example, if 10-30% of the study area is considered prospective, then the former map delineates 20-87% of the training coherent proxy deposit-type locations, whereas the latter map delineates 15-85% of the training randomly-selected proxy deposit-type locations. The prediction-rates of the map of discriminant scores based on training set B (Fig. 8-22B) are better than the prediction-rates of the map of discriminant scores based on training set A (Fig. 8-21B). For example, if 10-30% of the study area is considered prospective, then the former map delineates 42-83% of the training coherent proxy deposit-type locations whereas the latter map delineates 17-75% of the training randomly-selected proxy deposit-type locations. These results demonstrate further the advantage of using coherent (proxy) deposit-type locations in predictive modeling of mineral prospectivity.

Comparing and contrasting the performance of the map of discriminant scores in Fig. 8-22A with the map of integrated *Bel* in Fig. 8-19A, both of which are created using 86 coherent proxy-deposit-type locations, show the following. It is apparent that the former is better than the latter because, if 40% of the case study area is considered prospective, then the map of discriminant scores in Fig. 8-22A delineates correctly 100% of the training coherent proxy deposit-type locations and 100% of the testing deposit-type locations (Fig. 8-22B), whereas the map of integrated *Bel* in Fig. 8-19A delineates correctly 93% of the training coherent proxy deposit-type locations and 85% of the testing deposit-type locations (Fig. 8-19B). However, If 10-20% of the case study area is considered prospective, then the map of integrated *Bel* in Fig. 8-19A delineates correctly 35-65% of the training coherent proxy deposit-type locations and 39-54% of the testing deposit-type locations (Fig. 8-19B), whereas the map of discriminant scores in Fig. 8-22A delineated correctly 20-54% of the training coherent proxy deposit-type locations and 42-50% of the testing deposit-type locations (Fig. 8-22B). If 5% of the study area is considered prospective, then the map of integrated *Bel* in Fig. 8-19A delineates correctly 26% of the training coherent proxy deposit-type locations and 31% of the testing deposit-type locations (Fig. 8-19B), whereas the map of discriminant scores in Fig. 8-22A delineated correctly 10% of the training coherent proxy deposit-type locations and 23% of the testing deposit-type locations (Fig. 8-22B). Therefore, because mineral prospectivity mapping aims to constrain the sizes of exploration targets in order to increase the chance of mineral deposit discovery, the cross-validation results indicate that the map of integrated *Bel* in Fig. 8-19A is a better predictive model of epithermal Au prospectivity in the case study area compared to the map of discriminant scores in Fig. 8-22A.

The poorer model performance of the map of discriminant scores in Fig. 8-22A compared to the map of integrated *Bel* in Fig. 8-19A can probably be ascribed to the use of training data sets with (almost) equal numbers of deposit-type locations and non-deposit locations in the application of LDA. In contrast, note that the data-driven estimates of EBFs are based on all non-deposit locations (see equations (8.8b) and (8.9b)). In addition, Skabar (2005) demonstrated that, in contrast to the findings of Brown et al. (2000) and Porwal et al (2003a), using a training set of known deposit-type and all known non-deposit locations optimises the performance of artificial neural networks in data-driven modeling of prospectivity. Further experiments, explained in the following paragraphs, were performed in order to show that the arguments of Skabar (2005) for the application of artificial neural networks to data-driven modeling of mineral prospectivity are also valid for the application of LDA to data-driven modeling of mineral prospectivity.

The part of the study area with data for all predictor variables consists of 9719 unit cells (each measuring 100×100 m). Thus, training data set A is modified to training data set AA, which now consists of 79 randomly-selected proxy deposit-type locations and 9640 non-deposit locations; whilst training set B is modified to training set BB, which now consists of 86 coherent proxy deposit-type locations and 9633 non-deposit

TABLE 8-VI

Two models of discriminant functions for predictive mapping of epithermal Au prospectivity, Aroroy district (Philippines) based on training sets each with grossly imbalanced numbers of proxy deposit-type and non-deposit pixels. Values in bold represent predictor variables, per set of spatial evidence, with strong positive spatial associations with the training deposit-type locations.

Discriminant analysis using training set AA[1] (Wilks' lambda = 0.988; α=0.0001)			Discriminant analysis using training set BB[2] (Wilks' lambda = 0.983; α=0.0001)		
Predictor variables[3]	Function coefficients		Predictor variables[3]	Function coefficients	
	Standardised	Unstandardised		Standardised	Unstandardised
NNW1	0.048	0.020	NNW1	0.121	0.051
NNW2	**0.399**	0.013	NNW2	**0.504**	0.017
NNW3	**0.257**	0.009	NNW3	**0.357**	0.013
NNW4	-0.002	0.000	NNW4	0.145	0.004
NW1	-0.357	-0.009	NW1	-0.321	-0.008
NW2	-0.198	-0.005	NW2	-0.252	-0.006
NW3	**0.122**	0.003	NW3	**0.092**	0.002
NW4	-0.027	-0.001	NW4	-0.099	-0.003
FI1	**0.676**	0.017	FI1	**0.391**	0.010
FI2	**0.758**	0.020	FI2	**0.661**	0.017
FI3	0.155	0.004	FI3	0.128	0.003
FI4	0.005	0.000	FI4	-0.013	0.000
ANOM1	**0.327**	0.008	ANOM1	**0.313**	0.008
ANOM2	**0.227**	0.006	ANOM2	**0.234**	0.006
ANOM3	0.149	0.004	ANOM3	0.226	0.006
Constant	-	-1.283	Constant		-1.252

[1]Consists of 79 randomly-selected proxy deposit-type locations (Fig. 8-8) and 9640 non-deposit locations. [2]Consists of 86 coherent proxy deposit-type locations (Fig. 8-8) and 9633 non-deposit locations. [3]Statistically significant predictor variables in the discriminant models (see 'class code' columns in Table 8-IV for explanations of variable names).

locations. Table 8-VI shows that the discriminant model based on training set BB is slightly better (i.e., lower Wilks' lambda) than the discriminant model based on training set AA. The discriminant models based on training sets AA and BB have common statistically significant predictor variables. The standardised and unstandardised function coefficients based on training sets AA and BB are mostly lower than the standardised and unstandardised function coefficients based on training sets A and B (Table 8-V). However, the results shown in Tables 8-V and 8-VI show that quantified relative degrees of spatial associations between individual predictor variables and deposit-type locations are similar either when non-deposit locations equal in number to deposit-type locations are used or when all known non-deposit locations are used. That is, the 'FI' predictor variables are the most important, followed by the 'NNW' predictor variables, then by 'ANOM' predictor variables and then by the 'NW' predictor variables. Therefore, using all known non-deposit locations together with (proxy) deposit-type locations for training

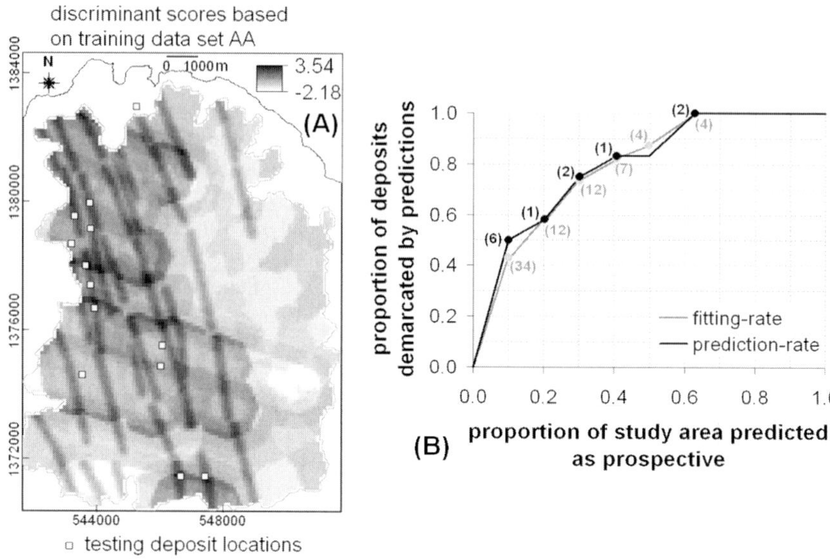

Fig. 8-23. (**A**) Epithermal Au prospectivity map of Aroroy district (Philippines) portrayed as discriminant scores of spatial evidence layers with respect to training set AA of 79 coherent proxy deposit-type locations (Fig. 8-8) and 9640 non-deposit locations. Polygon outlined in grey is area of stream sediment sample catchment basins (see Fig. 4-11). The testing set of locations of 13 epithermal Au deposits is shown as reference to the prediction-rate. (**B**) Fitting and prediction-rate curves of, respectively, proportions of coherent training proxy deposits (grey dots) and testing deposits (black dots) demarcated by the predictions versus proportion of the study area predicted as prospective based on the discriminant scores. The grey and black dots represent classes of discriminant scores that correspond spatially with certain numbers of training coherent proxy deposit-type locations (in grey) and certain numbers of testing deposit-type locations (in black), respectively.

in LDA (a) results in empirical spatial associations between epithermal Au deposits and indicative geological features that are consistent with the results of the application of the distance correlation method (see Chapter 6, Table 6-IX) and with the data-driven estimates of *Bel* shown in Fig. 8-15 and (b) does not undermine the geological significance of the predictor variables with respect to the target variable. It also follows that the scheme of spatial evidence representation for raster-based GIS application of LDA to mineral prospectivity mapping (Fig. 8-20) allows proper comparison with the results of the application of data-driven evidential belief modeling.

The maps of discriminant scores based on training set AA (Fig. 8-23A) and training set BB (Fig. 8-24A) portray mostly NNW-trending linear patterns of intermediate and high values reflecting the spatial evidence of proximity to NNW-trending faults/fractures. However, the discriminant scores are highest mostly where the NNW-trending linear patterns intersect with circular patterns reflecting the spatial evidence of

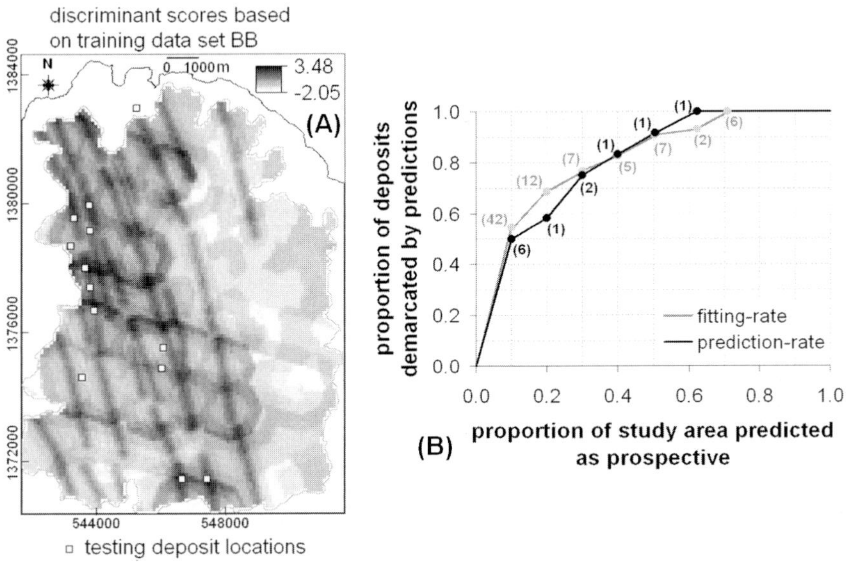

Fig. 8-24. (**A**) Epithermal Au prospectivity map of Aroroy district (Philippines) portrayed as discriminant scores of spatial evidence layers with respect to training set BB of 86 coherent proxy deposit-type locations (Fig. 8-8) and 9633 non-deposit locations. Polygon outlined in grey is area of stream sediment sample catchment basins (see Fig. 4-11). The testing set of locations of 13 epithermal Au deposits is shown as reference to the prediction-rate. (**B**) Fitting and prediction-rate curves of, respectively, proportions of coherent training proxy deposits (grey dots) and testing deposits (black dots) demarcated by the predictions versus proportion of the study area predicted as prospective based on the discriminant scores. The grey and black dots represent classes of discriminant scores that correspond spatially with certain numbers of training coherent proxy deposit-type locations (in grey) and certain numbers of testing deposit-type locations (in black), respectively.

proximity to intersections of NNW- and NW-trending faults/fractures. The patterns of intermediate and high discriminant scores in Figs. 8-23A and 8-24A are somewhat similar to the patterns of intermediate and high discriminant scores in Fig. 8-21A and 8-22A and thus are more-or-less consistent with the conceptual model of epithermal mineralisation in dilational or extensional settings as depicted in Fig. 6-16.

The fitting- and prediction-rates of the maps of discriminant scores based on training set AA (Fig. 8-23A) and based on training set BB (Fig. 8-24A) are similar. However, if 20% of the study area is considered prospective, then the fitting-rates of the map of discriminant scores based on training set BB (Fig. 8-24B) are better than the fitting-rates of the map of discriminant scores based on training set AA (Fig. 8-23B). This means that, in mineral prospectivity mapping, using coherent (proxy) deposit-type locations is better than using randomly-selected (proxy) deposit-type locations. The results also mean that the map of discriminant scores based on training set BB (Fig. 8-24A) is better than the map of discriminant scores based on training set AA (Fig. 8-23A).

Comparing and contrasting the performance of the maps of discriminant scores in Figs. 8-23A and 8-24A with the maps of discriminant scores in Figs. 8-21A and 8-22A indicate the following. If 40% of the case study area is considered prospective, then the maps of discriminant scores in Figs. 8-23A and 8-24A have poorer fitting- and prediction-rates (Figs. 8-23B and 8-24B) than those of the maps of discriminant scores in Figs. 8-21A and 8-22A. If 20% of the case study area is considered prospective, then the maps of discriminant scores in Figs. 8-23A and 8-24A have better fitting- and prediction-rates (Figs. 8-23B and 8-24B) than those of the maps of discriminant scores in Figs. 8-21A and 8-22A. Therefore, because mineral prospectivity mapping aims to constrain the sizes of exploration targets in order to increase the chance of mineral deposit discovery, the cross-validation results show that predictive modeling of mineral prospectivity via application of LDA generally produces better predictive models by using training sets consisting of all known non-deposit locations together with coherent (proxy) deposit-type locations rather than by using training sets consisting of balanced numbers of coherent (proxy) deposit-type locations and non-deposit locations. These findings in the application of LDA are consistent with the findings of Skabar (2005) in the application of artificial neural networks that using a training set of known deposit-type and all known non-deposit locations results in better data-driven models of prospectivity than when using a training set consisting of balanced numbers of deposit-type locations and non-deposit locations.

Comparing and contrasting the performance of the maps of discriminant scores in Figs. 8-23A and 8-24A with the maps of integrated *Bel* in Figs. 8-18A and 8-19A indicate the following. If 40% of the case study area is considered prospective, then the maps of discriminant scores in Figs. 8-23A and 8-24A have poorer fitting- and prediction-rates (Figs. 8-23B and 8-24B) than those of the maps of integrated *Bel* in Figs. 8-18A and 8-19A. If 20% of the case study area is considered prospective, then the maps of discriminant scores in Figs. 8-23A and 8-24A have better fitting- and prediction-rates (Figs. 8-23B and 8-24B) than those of the maps of discriminant scores in Figs. 8-18A and 8-19A. These results illustrate that the application of LDA, using training sets consisting of all known non-deposit locations together with (proxy) deposit-type locations, generally produces better predictive models of mineral prospectivity than the application of data-driven evidential belief modeling.

The more-or-less similar performances of the maps of discriminant scores in Figs. 8-21 to 8-24 and the maps of integrated *Bel* in Figs. 8-16 to 8-19, depending on the composition of training sets, is attributed to using the same sets of spatial evidence and to the application of a scheme of spatial evidence representation (Fig. 8-20) in order to adapt the spatial evidence layers used in data-driven evidential belief modeling in the application of LDA. The scheme of spatial evidence representation for GIS-based application of LDA (Fig. 8-20) was deemed necessary in performing controlled experiments of bivariate and multivariate modeling of mineral prospectivity in the case study area by using the same sets of spatial evidence layers represented in a highly similar fashion (i.e., as numbers of unit cells) in order to demonstrate objectively the utility of coherent (proxy) deposit-type locations. In fact the application of LDA is advantageous compared to the application of data-driven EBFs because the former method, like the other multivariate methods (Table 8-II), can take on raw values (i.e., unclassified data) of continuous fields as predictor variables whilst the latter method, like the weights-of-evidence method (Table 8-I), requires classification of data of continuous fields.

Classification of data of continuous fields, however, is useful in predictive modeling of mineral prospectivity in terms of recognising spurious spatial evidence. For example, Carranza et al. (2008a) used band4/band6 ratios of ASTER (Advanced Spaceborne Thermal Emission Reflection Radiometer) data as a set representing hydrothermal alteration intensity evidence for modeling prospectivity for epithermal Au deposits in the Cabo de Gata area (Spain). They found that empirical spatial associations of epithermal Au deposits with classes of high ASTER band4/band6 ratios are partly spurious because these classes of spatial evidence coincide either with hydrothermally-altered volcanic rocks or with greenhouses for which the Cabo de Gata area is (in)famous. Therefore, proper classification of data and proper calibrations or adjustments of classified (i.e., categorical) spatial evidence data is useful not only in the application of bivariate methods, like data-driven evidential belief modeling, but also in the application of multivariate methods, like LDA, to predictive modeling of mineral prospectivity.

A disadvantage of LDA, like most multivariate methods (Table 8-II), is that the way predictor variables are integrated (see, for example, equation (8.11)) is chiefly mathematical but not necessarily logical representations of the inter-play of geological processes involved in mineral deposit formation. In contrast, in data-driven evidential belief modeling predictor maps are integrated not only mathematically but also logically through the application of an inference network, which reflects inferences about the inter-relationships of processes that control the occurrence of a geo-object (e.g., mineral deposits) and spatial features that indicate the presence of that geo-object. Hybridisation of some mathematical methods by incorporation of inference systems thus potentially results in better predictive models of mineral prospectivity. For example, a hybrid neuro-fuzzy model of mineral prospectivity (Porwal et al., 2004; Porwal, 2006) is better than a neural network model of mineral prospectivity (Porwal et al., 2003a; Porwal, 2006).

Application of data-driven models/methods can be problematic in cases of (a) limited number of occurrences of mineral deposits of the type sought and (b) incomplete sampling or data over locations of known mineral deposit occurrences. In such cases, approaches to mineral prospectivity mapping lead to so-called geographic expert systems (GES), whereby elements of methods usually employed in knowledge-driven methods (e.g., expert-based assignment of evidential weights, inference-based integration of evidence) and elements of methods usually employed in data-driven methods (e.g., Bayesian 'updating' of evidence) are combined in so-called artificial intelligence (AI) techniques (Duda et al., 1978; Campbell et al., 1982; McCammon, 1989, 1994; Katz,

1991; Reddy et al., 1992; Chinn and Ascough, 1997; Chinn, 2006). GES- and AI-based techniques for mineral prospectivity mapping are beyond the scope of this volume. Readers are referred, nonetheless, to Yatabe and Fabbri (1988, 1989) for review of geoscience applications of AI methods and to Pan and Harris (2000, pp. 422-438) for a review of expert systems and AI methods applicable to mineral exploration.

DISCUSSION AND CONCLUSIONS

There are various bivariate and multivariate methods of GIS-based data-driven modeling of mineral prospectivity and these methods are now mostly well-established. This chapter has demonstrated two methods for GIS-based data-driven modeling of mineral prospectivity (one bivariate method, data-driven evidential belief modeling, and one multivariate method, linear discriminant analysis) and two techniques supporting these methods in order to properly create and integrate predictor maps in determining new targets for further exploration of undiscovered occurrences of mineral deposit of the type sought in a study area. These two latter techniques each address the issues of (a) objective selection of a suitable unit cell size for data-driven modeling of mineral prospectivity and (b) selection of coherent deposit-type locations for data-driven modeling of mineral prospectivity. In addition, a brief review of cross-validation strategies in data-driven modeling of mineral prospectivity has been provided here.

Until now, the choice of a suitable unit cell (or pixel) size [denoted as $N(\bullet)$] for raster-based-GIS data-driven modeling of mineral prospectivity has been subjective. Point pattern analysis (Boots and Getis, 1988; Rowlingson and Diggle, 1993) of the range of distances in which there is zero probability of one neighbour deposit-type location situated next to another deposit-type location is useful in deriving a preliminary set of choices for a suitable $N(\bullet)$. Then, using this range of distances, the analysis of the rate of increase in the ratio $[N(D)]$: $[N(T)–N(D)]$ as function of equal-interval change in $N(\bullet)$ is robust, regardless of the number of deposit-type locations and the size of a study area, in determining the most suitable $N(\bullet)$. Although these propositions have not been demonstrated through formal testing, the empirical spatial associations between deposit-type locations and indicative geological features in the case study area as quantified via data-driven estimation of EBFs (this chapter) and as quantified via the distance distribution method and distance correlation method (Chapter 6) are very similar. On the one hand, the pixel size used in the spatial association analyses in Chapter 6 is, in fact, 10 m. This small pixel size was necessary in the analysis in Chapter 6 because we know (or have learned from Chapter 4) that spatial measurements (e.g., distances) are fractals so that an infinitesimally small pixel or unit cell is necessary for accurate spatial measurements. On the other hand, the pixel size used in all the experiments of data-driven modeling of mineral prospectivity presented in this chapter is 100 m, as determined objectively via the proposed techniques. A pixel size of 100 m is a practical spatial representation of the epithermal Au deposit occurrences in the case study area, whilst a pixel size of 10 m is not. Nevertheless, the proposed techniques for objective selection of a suitable unit cell size for spatial representation of deposit-type locations requires further testing in GIS-based data-driven modeling of mineral prospectivity.

The experiments of data-driven modeling of mineral prospectivity presented in this chapter show that using coherent deposit-type (or proxy deposit-type) locations results in better predictive models than using just any or randomly-selected deposit-type (or proxy deposit-type) locations. The results presented here, therefore, show further the usefulness of the two-stage methodology proposed by Carranza et al. (2008b) and explained here for objective selection of coherent deposit-type (or proxy deposit-type) locations. The results of the experiments presented here also show that using not just any but coherent proxy deposit-type locations (i.e., unit cells immediately surrounding a unit cell representing a deposit-type location) is useful in cases where the number of occurrences of mineral deposits of the type sought is considered and/or found insufficient to derive a proper (e.g., statistically significant) data-driven model of mineral prospectivity.

The experiments presented here were tested through a combination of *N–n* and deposit-type classification strategies for cross-validation of predictive models of mineral prospectivity. In practice, one must adopt or adapt a cross-validation strategy depending on factors like (a) number of known occurrences of mineral deposits of the type sought in a study area, (b) relevant attributes of deposit-type locations (e.g., grade, tonnage, mine status, etc.) and (c) spatial (geological) coherence of different parts of a moderately- to well-sampled mineralised landscape under investigation. The cross-validation strategy adopted or adapted must enable comparison of predictive models, derived via either one method or different methods, in order to determine the best predictive model. Finding the 'best' predictive model of mineral prospectivity in the case study area was, however, not an objective of this chapter. The cross-validation strategies adopted here were, nonetheless, necessary in order to demonstrate the utility of coherent (proxy) deposit-type locations in data-driven modeling of mineral prospectivity via at least two different methods – evidential belief modeling and linear discriminant analysis.

The methods for data-driven evidential belief modeling of mineral prospectivity presented here are relatively new, whilst the techniques for linear discriminant analysis for mineral prospectivity mapping were first demonstrated more than three decades ago. The latter techniques are able to handle both quantitative and qualitative predictor variables, whilst the former method is able to handle qualitative predictor variables (although some of these variables are derived via classification of quantitative variables). Because evidential belief modeling and discriminant analysis are very different methods, a scheme of spatial representation of qualitative evidential data (Fig. 8-20) was devised so that (a) the same predictor variables used in evidential belief modeling can be used in linear discriminant analysis and (b) testing of the utility of coherent (proxy) deposit-type locations using the two different methods is objective. However, the scheme of spatial representation of qualitative evidential data presented here is not only useful for the sorts of analyses performed in this volume. In fact, Carranza and Castro (2006) employed this scheme of spatial representation of qualitative evidential data in the application of weights-of-evidence modeling, evidential belief modeling and logistic regression for predictive modeling of volcanic lahar-inundation zones based on the same sets of predictor variables.

In closing, two recommendations can be made for further studies. Firstly, further research is required in the proper representation and evaluation of uncertainties associated with predictive models of geochemical anomalies and/or mineral prospectivity. Although the application of EBFs is useful in representation of evidential uncertainty, further studies are needed in monitoring the propagation of uncertainty from the input maps to the output map(s). This is a challenging task because, for example, most, if not all, geological maps invariably do no contain information about their accuracy. Secondly, although cross-validation strategies allow us to estimate empirically the likelihood of discovery of a new deposit-type location within the predicted prospective zones in a study area and although mineral prospectivity models can be used to estimate undiscovered mineral endowment (e.g., McCammon and Kork, 1992), further research is required to answer the following logical question regarding the efficacy of a mineral prospectivity map: "Where, in the predicted (most) prospective zones, should targets for further exploration of undiscovered deposit-type locations be focused?". This is also a challenging task because, after all, the ultimate goal and thus real validation of mineral prospectivity mapping is finding undiscovered mineral deposits. Thus, because methods for mapping mineral prospectivity are now mostly well-established, scientific progress in developing new techniques for improving the accuracy and utility of mineral prospectivity maps is more desirable than scientific progress in developing new techniques for creating and integrating predictor maps for mineral prospectivity mapping.

REFERENCES

Agterberg, F.P., 1974. Automatic contouring of geological maps to detect target areas for mineral exploration. Mathematical Geology 6(4): 373-395.

Agterberg, F.P., 1988. Application of recent developments of regression analysis in regional mineral resource evaluation. In: C.F. Chung, A.G. Fabbri, R. Sinding-Larsen (Eds.), Quantitative Analysis of Mineral and Energy Resources, D. Reidel Publishing Company, Dordrecht, pp. 1-28.

Agterberg, F.P., 1989. Computer programs for mineral exploration. Science 245(4913): 76-81.

Agterberg, F.P., 1992. Combining indicator patterns in weights of evidence modeling for resource evaluation. Nonrenewable Resources 1(1): 39-50.

Agterberg, F.P., 1993. Calculation of the variance of mean values for blocks in regional resource evaluation studies. Nonrenewable Resources 2(4): 312-324.

Agterberg, F.P., 1994. Fractal, multifractals, and change of support. In: R. Dimitrakopoulus (Ed.), Geostatistics for the Next Century, Kluwer, Dordrecht, pp. 223-234.

Agterberg, F.P., 2001. Multifractal simulation of geochemical map patterns. In: D.F. Merriam, J.C. Davis (Eds.), Geologic Modeling and Simulation: Sedimentary Systems, Kluwer-Plenum Publishers, New York, pp. 327-346.

Agterberg, F.P., 2007. New application of the model of de Wijs in regional geochemistry. Mathematical Geology 39(1): 1-25.

Agterberg, F.P., Bonham-Carter, G.F., 2005. Measuring performance of mineral-potential maps. Natural Resources Research 14(1): 1-17.

Agterberg, F.P., Bonham-Carter, G.F., Cheng, Q., Wright, D.F., 1993a. Weights of evidence modeling and weighted logistic regression in mineral potential mapping. In: J.C. Davis, U.C. Herzfeld (Eds.), Computers in Geology, Oxford University Press, New York, pp. 13-32.

Agterberg, F.P., Bonham-Carter, G.F., Wright, D.F., 1990. Statistical pattern integration for mineral exploration. In: G. Gaál, D.F. Merriam (Eds.), Computer Applications in Resource Estimation, Pergamon Press, Oxford, pp. 1-21.

Agterberg, F.P., Cheng, Q., 2002. Conditional independence test for weights-of-evidence modeling. Natural Resources Research 11(4): 249-255

Agterberg, F.P., Cheng, Q., Wright, D.F., 1993b. Fractal modeling of mineral deposits. In: J. Elbrond, X. Tang (Eds.), Application of Computers and Operations Research in the Mineral Industry, Vol. 1, Proceedings of the 24th APCOM Symposium (Montreal), Canadian Institute of Mining, Metallurgy and Petroleum Engineering, pp. 43-53.

Allison, P.D., 2002. Missing Data. Sage Publications, Thousand Oaks, California.

Alms, R., Balovnev, O., Breunig, M., Cremers, M., Cremers, A.B., Jentzsch, T., Siehl, A., 1998. Space-time modelling of the Lower Rhine Basin supported by an object-oriented database. Physics and Chemistry of the Earth 23(3): 251-260.

An, P., 1992. Spatial Reasoning Techniques and Integration of Geological and Geophysical Information for Resource Exploration. Ph.D. dissertation, The University of Manitoba, Canada. 162 pp.

An, P., Moon, W.M., Bonham-Carter, G.F., 1992. On knowledge-based approach on integrating remote sensing, geophysical and geological information. Proceedings of International Geoscience and Remote Sensing Symposium (IGARSS) 1992, pp. 34-38.

An, P., Moon, W.M., Bonham-Carter, G.F., 1994a. An object-oriented knowledge representation structure for exploration data integration. Nonrenewable Resources 3(2): 132-145.

An, P., Moon, W.M., Bonham-Carter, G.F., 1994b. Uncertainty management in integration of exploration data using the belief function. Nonrenewable Resources 3(1): 60-71.

An, P., Moon, W.M., Rencz, A., 1991. Application of fuzzy set theory for integration of geological, geophysical and remote sensing data. Canadian Journal of Exploration Geophysics, 27(1): 1-11.

Ansoult, M.M., Soille, P.J., Loodts, J.A., 1990. Mathematical morphology: a tool for automated GIS data acquisition from scanned thematic maps. Photogrammetric Engineering and Remote Sensing 56(9): 1263-1271.

Arribas, A., Hedenquist, J.W., Itaya, T., Okada, T., Concepcion, R.A., Garcia, J.S., 1995. Contemporaneous formation of adjacent porphyry and epithermal Cu-Au deposits over 300 ka in northern Luzon, Philippines. Geology 23(4): 227-340.

Aurelio, M.A., Barrier, E., Gaulon, R., Rangin, C., 1997. Deformation stress states along the segment of the Philippine Fault: implications to wrench fault tectonics. Journal of Asian Earth Sciences 15(2-3): 107-119.

Aurelio, M.A., Barrier, E., Rangin, C., Müller, C., 1991. The Philippine Fault in the Late Cenozoic tectonic evolution of the Bondoc-Masbate-N. Leyte area, Central Philippines. Journal of Southeast Asian Earth Sciences 6(3-4): 221-238.

Baldi, P., Bonvalots, S., Briole, P., Coltelli, M., Gwinner, K., Marsella, M., Puglis, G., Remy, D., 2002. Validation and comparison of different techniques for derivation of digital elevation models and volcano monitoring (Vulcano Island, Italy). International Journal of Remote Sensing 23(22): 4783-4800.

Bárdossy, G., Fodor, J., 2005. Assessment of the completeness of mineral exploration the application of fuzzy arithmetic and prior information. Acta Polytechnica Hungarica 2(1): 15-31.

Bartier, P.M., Keller, C.P., 1991. Integrating bedrock geology with stream-sediment geochemistry in a geographic information system (GIS): case study NTS 92H. Geological Fieldwork 1990, Paper 1991-1, British Columbia Geological Survey Branch, pp. 315-321.

Basilevsky, A., 1994. Statistical Factor Analysis and Related Methods, Theory and Application, John Wiley & Sons, New York.

Baybayan, N.Q., Matos, A.M., 1986. Geology of Masbate Island, Central Philippines. Technical Report, Philippine Bureau of Mines and Geosciences, Quezon, City.

Beauchamp, J.J., Begovich, C.L., Kane, V.E., Wolf, D.A., 1980. Application of discriminant analysis and generalized distance measure to uranium exploration. Mathematical Geology 12(6): 539–558.

Bellehumeur, C., Marcotte, D., Jébrak, M., 1994. Multi-element relationships and spatial structures of regional geochemical data from stream sediments, southwestern Quebec, Canada. Journal of Geochemical Exploration 51(1): 11-35.

Bellot, J.-P., 2008. Hydrothermal fluids assisted crustal-scale slip on the Argentat fault zone. Tectonophysics 450(1-4): 21-33.

Berger, B.R., Drew, L.J., 2002. Mineral-deposit models: new developments. In: A.G. Fabbri, Gaál, G., McCammon, R.B. (Eds.), Deposit and Geoenvironmental Models for Resource Exploitation and Environmental Security. NATO Science Series 2, Environmental Security, Vol. 80, Kluwer Academic Publishers, Dordrecht, pp.121-134.

Berman, M., 1977. Distance distributions associated with Poisson processes of geometric figures. Journal of Applied Probability 14(1): 195-199.

Berman, M., 1986. Testing for spatial association between a point processes and another stochastic process. Applied Statistics 35(1): 64-62.

Bierlein, F.P., Murphy, F.C., Weinberg, R.F., Lees, T., 2006. Distribution of orogenic gold deposits in relation to fault zones and gravity gradients: targeting tools applied to the Eastern Goldfields, Yilgarn Craton, Western Australia. Mineralium Deposita 41(2): 107-126.

Bierlein, F.P., Northover, H.J., Groves, D.I., Goldfarb, R.J., Marsh, 2008. Controls on mineralisation in the Sierra Foothills gold province, central California, USA: a GIS-based reconnaissance prospectivity analysis. Australian Journal of Earth Sciences 55(1): 61-78.

Billa, M., Cassard, D., Lips, A.L.W., Bouchot, V., Tourliére, B., Stein, G., Guillou-Frottier, L., 2004. Predicting gold-rich epithermal and porphyry systems in the central Andes with a continental-scale metallogenic GIS. Ore Geology Reviews 25(1-2): 39-67.

Blenkinsop, T.G., Kadzviti, S., 2006. Fluid flow in shear zones: insights from the geometry and evolution of ore bodies at Renco gold mine, Zimbabwe. Geofluids 6(4): 334-345.

Bölviken, B., Stokke, P.R., Feder, J., Jössang, T., 1992. The fractal nature of geochemical landscapes. Journal of Geochemical Exploration 43(2): 91-109.

Bonham-Carter, G.F., 1985. Statistical association of gold occurrences with Landsat-derived lineaments, Timmins-Kirkland Lake area, Ontario. Canadian Journal of Remote Sensing 11(2): 195-211.

Bonham-Carter, G.F., 1991. Integration of geoscientific data using GIS. In: D.J. Maguire, M.F. Goodchild, D.W. Rhind (Eds.), Geographic Information Systems: Principles and Applications, Vol. 2, Longman, London, pp. 171-184.

Bonham-Carter, G.F., 1994. Geographic Information Systems for Geoscientists: Modelling with GIS, Pergamon, Ontario.

Bonham-Carter, G.F., Agterberg, F.P., 1990. Application of a microcomputer-based geographic information systems to mineral potential mapping. In: T. Hanley, D.F. Merriam (Eds.), Microcomputer Applications in Geology, Vol. 2, Pergamon Press, Oxford, pp. 49-74.

Bonham-Carter, G.F., Agterberg, F.P., Wright, D.F., 1988. Integration of geological datasets for gold exploration in Nova Scotia. Photogrammetric Engineering and Remote Sensing 54(11): 1585-1592.

Bonham-Carter, G.F., Agterberg, F.P., Wright, D.F., 1989. Weights of evidence modelling: a new approach to mapping mineral potential. In F.P. Agterberg, G.F. Bonham-Carter (Eds.),

Statistical Applications in the Earth Sciences, Geological Survey of Canada, Paper 89-9, pp. 171-183.

Bonham-Carter, G.F., Chung, C.F., 1983. Integration of mineral resource data for Kasmere Lake area, Northwest Manitoba, with emphasis on uranium. Mathematical Geology 15(1): 25-45.

Bonham-Carter, G., Goodfellow, W.D., 1984. Autocorrelation structure of stream sediment geochemical data: interpretation of zinc and lead anomalies, Nahanni river area, Yukon-Northwest Territories, Canada. In: G. Verly et al. (Eds.), Geostatistics for Natural Resources Characterization, Part 2, D. Reidel, Dordrecht, pp. 817-829.

Bonham-Carter, G.F., Goodfellow, W.D., 1986. Background corrections to stream geochemical data using digitized drainage and geological maps: application to Selwyn Basin, Yukon and Northwest Territories. Journal of Geochemical Exploration 25(1-2): 139-155.

Bonham-Carter, G.F., Rencz, A.N., Harris, J.R., 1985. Spatial relationship of gold occurrences with lineaments derived from Landsat and Seasat imagery, Meguma Group, Nova Scotia. Proceedings of the 4th Thematic Conference on Remote Sensing for Exploration Geology, San Francisco, April 1985, pp. 755-768.

Bonham-Carter, G.F., Rogers, P.J., Ellwood, D.J., 1987. Catchment basin analysis applied to surficial geochemical data, Cobequid Highlands, Nova Scotia. Journal of Geochemical Exploration 29(1-3): 259-278.

Boots, B.N, Getis, A., 1988. Point Pattern Analysis. Sage University Scientific Geography Series no. 8, Sage Publications, Beverly Hills.

Botbol, J.M., Sinding-Larsen, R., McCammon, R.B., Gott, G.B., 1977. Weighted characteristics analysis of spatially dependent mineral deposit data. Mathematical Geology 9(3): 309-311.

Botbol, J.M., Sinding-Larsen, R., McCammon, R.B., Gott, G.B., 1978. A regionalized multivariate approach to target area selection in geochemical exploration. Economic Geology 73(4): 534-546.

Boroushaki, S., Malczewski, J., 2008. Implementing an extension of the analytical hierarchy process using ordered weighted averaging operators with fuzzy quantifiers in ArcGIS. Computers & Geosciences 34(4): 399-410.

Bougrain, L., Gonzalez, M., Bouchot, V., Cassard, D., Lips, A.L.W., Alexandre, F., Stein, G., 2003. Knowledge recovery for continental-scale mineral exploration by neural networks. Natural Resources Research 12(3): 173-181.

Bounessah, M., Atkin, B.P., 2003. An application of exploratory data analysis (EDA) as a robust non-parametric technique for geochemical mapping in a semi-arid climate. Applied Geochemistry 18(8): 1185–1195.

Bradshaw, P.M.D. (Ed.), 1975. Conceptual Models in Exploration Geochemistry: The Canadian Cordillera and Canadian Shield, Journal of Geochemical Exploration 4(1): 1-213.

Brand, N.W., 1999. Element ratio in nickel sulphide exploration: vectoring towards ore environments. Journal of Geochemical Exploration 67(1-3): 145-165.

Breslow, N.E., Cain, K.C., 1988. Logistic regression for two-stage case-control data. Biometrika 75(1): 11-20.

Breunig, M., Cremers, A.B., Götze, H.-J., Schmidt, S., Seidemann, R., Shumilov, S., Siehl, A., 1999. First steps towards an interoperable GIS – an example from Southern Lower Saxony. Physics and Chemistry of the Earth (Part A: Solid Earth and Geology) 24(3): 179-189.

Brown, W.M., Gedeon, T.D., Groves, D.I., 2003. Use of noise to augment training data: a neural network method for mineral-potential mapping in regions of limited known deposit examples. Natural Resources Research 12(2): 141-152.

Brown, W.M., Gedeon, T.D., Groves, D.I., Barnes, R.G., 2000. Artificial neural networks: a new method for mineral prospectivity mapping. Australian Journal of Earth Sciences 47(4): 757-770.

Brown, D.G., Riolo, R., Robinson, D.T., North, M., Rand, W., 2005. Spatial processes and data models: toward integration of agent-based models and GIS. Journal of Geographical Systems 7(1): 25-47.

Bull, A.J., Mazzucchelli, R.H., 1975. Application of discriminant analysis to the geochemical evaluation of gossans. In: I.L. Elliott and W.K. Fletcher (Eds.), Geochemical Exploration 1974, Elsevier, Amsterdam, pp. 219–316.

Burrough, P.A., 1987. Principles of Geographic Information Systems for Land Resource Assessment. Clarendon Press, Oxford.

Burrough, P.A., McDonnell, R.A., 1998. Principles of Geographical Information Systems, Oxford University Press, Oxford.

Butt, C.R.M., Smith, R.E. (Eds.), 1980. Conceptual Models in Exploration Geochemistry: Australia. Journal of Geochemical Exploration 12(2-3): 89-365.

Butt, C.R.M., Zeegers, H. (Eds.), 1992. Regolith Exploration Geochemistry in Tropical and Subtropical Terrains, Handbook of Exploration Geochemistry, Vol. 4, Elsevier, Amsterdam.

Campbell, A.N., Hollister, V.F., Duda, R.O., 1982. Recognition of a hidden mineral deposit by an artificial intelligence program. Science 217(3): 927-929.

Campbell, N.A., 1982. Statistical treatment of geochemical data. In: R.E. Smith (Ed.), Geochemical Exploration in Deeply Weathered Terrain. CSIRO Institute of Energy and Earth Resources, Floreat Park, Western Australia, pp. 141-144.

Carlson, C.A., 1991. Spatial distribution of ore deposits. Geology 19(2): 111-114.

Carr, J.R., 1995. Numerical Analysis for the Geological Sciences. Prentice-Hall, New Jersey.

Carr, J.R., 1997. Statistical self-affinity, fractal dimension, and geological interpretation. Engineering Geology 48(3-4): 269-282.

Carranza, E.J.M., 2002. Geologically-Constrained Mineral Potential Mapping (Examples from the Philippines), Ph.D. Thesis, Delft University of Technology, The Netherlands, ITC (International Institute for Geo-Information Science and Earth Observation) Publication No. 86, Enschede, 480 pp.

Carranza, E.J.M., 2004a. Usefulness of stream order to detect stream sediment geochemical anomalies. Geochemistry: Exploration, Environment, Analysis 4(4): 341-352.

Carranza, E.J.M., 2004b. Weights-of-evidence modelling of mineral potential: a case study using small number of prospects, Abra, Philippines. Natural Resources Research 13(3): pp. 173-187.

Carranza, E.J.M., Castro, A.T., 2006. Predicting lahar-inundation zones: case study in West Mount Pinatubo, Philippines. Natural Hazards 37(3): 331-372.

Carranza, E.J.M., Hale, M., 1997. A catchment basin approach to the analysis of geochemical-geological data from Albay province, Philippines, Journal of Geochemical Exploration 60(2): 157-171.

Carranza, E.J.M., Hale, M., 2000. Geologically constrained probabilistic mapping of gold potential, Baguio district, Philippines. Natural Resources Research 9(3): 237-253.

Carranza, E.J.M., Hale, M., 2001a. Geologically-constrained fuzzy mapping of gold mineralization potential, Baguio district, Philippines. Natural Resources Research 10(2): 125-136.

Carranza, E.J.M., Hale, M., 2001b. Logistic regression for geologically-constrained mapping of gold mineralization potential, Baguio district, Philippines. Exploration and Mining Geology Journal 10(3): 165-175.

Carranza, E.J.M., Hale, M., 2002a. Mineral imaging with Landsat TM data for hydrothermal alteration mapping in heavily-vegetated terrane. International Journal of Remote Sensing 23(22): 4827-4852.

Carranza, E.J.M., Hale, M., 2002b. Spatial association of mineral occurrences and curvi-linear geological features. Mathematical Geology 34(2): 199-217.

Carranza, E.J.M., Hale, M., 2002c. Where are porphyry copper deposits spatially localized? A case study in Benguet province, Philippines. Natural Resources Research 11(1): 45-59.

Carranza, E.J.M., Hale, M., 2002d. Wildcat mapping of gold potential, Baguio district, Philippines. Transactions of the Institution of Mining and Metallurgy (Section B – Applied Earth Science) 111(2): 100-105.

Carranza, E.J.M. and Hale, M., 2003. Evidential belief functions for geologically constrained mapping of gold potential, Baguio district, Philippines. Ore Geology Reviews 22(1-2): 117-132.

Carranza, E.J.M., Hale, M., Faassen, C., 2008b. Selection of coherent deposit-type locations and their application in data-driven mineral prospectivity mapping. Ore Geology Reviews 33(3-4): 536-558.

Carranza, E.J.M., Mangaoang, J.C., Hale, M., 1999. Application of mineral exploration models and GIS to generate mineral potential maps as input for optimum land-use planning in the Philippines. Natural Resources Research 8(2): 165-173.

Carranza, E.J.M., Van Ruitenbeek, F.J.A., Hecker, C.A., Van der Meijde, M., Van der Meer, F.D., 2008a. Knowledge-guided data-driven evidential belief modeling of mineral prospectivity in Cabo de Gata, SE Spain. International Journal of Applied Earth Observation and Geoinformation, doi:10.1016/j.jag.2008.02.008.

Carranza, E.J.M, Wibowo, H., Barritt, S.D., Sumintadireja, P., 2008c. Spatial data analysis and integration for regional-scale geothermal potential mapping, West Java, Indonesia. Geothermics 33(3): 267-299.

Carranza, E.J.M., Woldai, T., Chikambwe, E.M., 2005. Application of data-driven evidential belief functions to prospectivity mapping for aquamarine-bearing pegmatites, Lundazi district, Zambia. Natural Resources Research 14(1): 47-63.

Chambers, J.M., Cleveland, W.S., Kleiner, B., Tukey, P.A., 1983. Graphical Methods for Data Analysis. Wadsworth International Group, Belmont, California, Duxbury Press, Boston.

Cheng, Q., 1994. Multifractal Modeling and Spatial Analysis with GIS: Gold Mineral Potential Estimation in the Mitchell-Sulphurets Area, Northwestern British Columbia, Ph.D. Thesis, University of Ottawa, Ottawa, 268 pp.

Cheng, Q., 1995. The perimeter-area fractal model and its application in geology. Mathematical Geology 27(1): 69–82.

Cheng, Q., 1999a. Multifractal interpolation. In: S.J. Lippard, A. Naess, R. Sinding-Larsen, R. (Eds.), Proceedings of the Firth Annual Conference of the International Association for Mathematical Geology, Trondheim, Norway, vol. 1, pp. 245–250.

Cheng, Q., 1999b. Spatial and scaling modelling for geochemical anomaly separation. Journal of Geochemical Exploration 65(3): 175-194.

Cheng, Q., Agterberg, F.P., 1995. Multifractal modeling and spatial point processes. Mathematical Geology 27(7): 831-845.

Cheng, Q., Agterberg, F.P., 1996. Multifractal modeling and spatial statistics. Mathematical Geology 28(1): 1-16.

Cheng, Q., Agterberg, F.P., 1999. Fuzzy weights of evidence and its application in mineral potential mapping. Natural Resources Research 8(1): 27-35.

Cheng, Q., Agterberg, F.P., Ballantyne, S.B., 1994. The separation of geochemical anomalies from background by fractal methods 54(2): 109-130.

Cheng, Q., Agterberg, F.P., Bonham-Carter, 1996. A spatial analysis method for geochemical anomaly separation. Journal of Geochemical Exploration 56(3): 183-195.

Cheng, Q., Bonham-Carter, G.F., Hall, G.E.M., Bajc, A., 1997. Statistical study of trace elements in the soluble organic and amorphous Fe-Mn phases of surficial sediments, Sudbury Basin. 1. Multivariate and spatial analysis. Journal of Geochemical Exploration 59(1): 27-46.

Cheng, Q., Xu, Y., Grunsky, E., 2000. Integrated spatial and spectrum method for geochemical anomaly separation. Natural Resources Research 9(1): 43-52.

Chica-Olmo, M., Abarca, F., Rigol, J.P., 2002. Development of a decision support system based on remote sensing and GIS techniques for gold-rich area identification in SE Spain. International Journal of Remote Sensing 23(22): 4801-4814.

Chinn, G.T., 2006. Geographic expert system modelling at Noranda. In: J.R. Harris (Ed.), GIS for the Earth Sciences, Geological Association of Canada Special Publication 44, Geological Association of Canada, St. John's, pp. 181-189.

Chinn, G.T., Ascough, G.L., 1997. Mineral potential mapping using an expert system and GIS. In: A.G. Gubins (Ed.), Proceedings of Exploration 97: 4th Decennial International Conference on Mineral Exploration, Toronto, Canada, 14-18 September 1997, pp. 105-114.

Chork, C.Y., Mazzucchelli, R.H., 1989. Spatial filtering of exploration geochemical data using EDA and robust statistics. Journal of Geochemical Exploration 34(3): 221-243.

Chork, C.Y., Rousseeuw, P.J., 1992. Integrating a high-breakdown option into discriminant analysis in exploration geochemistry. Journal of Geochemical Exploration 43(4): 191-203.

Chung, C.F., 1977. An application of discriminant analysis for the evaluation of mineral potential. In: R.V. Ramani (Ed.), Application of Computer Methods in the Mineral Industry, Proceedings of the 14th APCOM Symposium, Society of Mining Engineers of American Institute of Mining, Metallurgical, and Petroleum Engineers, New York, pp. 299–311.

Chung, C.F., 1978. Computer program for the logistic model to estimate the probability of occurrence of discrete events. Geological Survey of Canada, Paper 78-12, 23 pp.

Chung, C.F., 1983. SIMSAG; integrated computer system for use in evaluation of mineral and energy resources. Mathematical Geology 15(1): 47-58.

Chung, C.F., 2003. Use of airborne geophysical surveys for constructing mineral potential maps. In: W.D. Goodfellow, S.R. McCutcheon, S.R., J.M. Peter (Eds.), Massive Sulfide Deposits of the Bathurst Mining Camp, New Brunswick, and Northern Maine. Economic Geology Monograph 11, Society of Economic Geologists, Colorado, pp. 879-891.

Chung, C.F., Agterberg, F.P., 1980. Regression models for estimating mineral resources from geological map data. Mathematical Geology 12(5): 472-488.

Chung, C.F., Agterberg, F.P., 1988. Poisson regression analysis and its application. In: C.F. Chung, A.G. Fabbri, R. Sinding-Larsen (Eds.), Quantitative Analysis of Mineral and Energy Resources, D. Reidel Publishing Company, Dordrecht, pp. 29-36.

Chung, C.F., Fabbri, A.G., 1993. The representation of geoscience information for data integration. Nonrenewable Resources 2(2): 122-139.

Chung, C.F., Fabbri, A.G., 2005. On mineral potential maps and how to make them useful. In: Q. Cheng, G.F. Bonham-Carter (Eds.), GIS and Spatial Analysis, Proceedings of IAMG '05, The Annual Conference of the International Association of Mathematical Geology, 21-26 August 2005, Toronto, Canada, pp. 533-538.

Chung, C.F., Fabbri, A.G., Chi, K.H., 2002. A strategy for sustainable development of nonrenewable resources using spatial prediction models. In: A.G. Fabbri, G. Gáal, R.B. McCammon (Eds.), Geoenvironmental Deposit Models for Resource Exploitation and Environmental Security. Kluwer, Dordrecht, pp. 101–118.

Chung, C.F., Keating, P.B., 2002. Mineral potential evaluation based on airborne geophysical data. Exploration Geophysics 33(1): 28-34.

Chung, C.F., Moon, W.M., 1991. Combination rules of spatial geoscience data for mineral exploration. Geoinformatics 2(2): 159-169.

Clarke, D.B., Reardon, N.C., Chatterjee, A.K., Gregoire, D.C., 1989, Tourmaline composition as a guide to mineral exploration; a reconnaissance study from Nova Scotia using discriminant function analysis. Economic Geology 84(7): 1921-1935.

Clarke, K.C., 1995. Analytical and Computer Cartography, 2nd edn., Prentice Hall, Englewood Cliffs.

Codd, E., 1970. A relational model for large shared data banks. Communications of the Association for Computing Machinery 13(6): 377-387.

Cohen, P.R., 1985. Heuristic Reasoning About Uncertainty: An Artificial Intelligence Approach, Pitman, London.

Colman-Sadd, S.P., Ash, J.S., Nolan, L.W., 1997. GeoLegend: a database system for managing geological map units in a geographic information system. Computers & Geosciences 23(7): 715-724.

Cook, S.J., Fletcher W.K., 1993. Distribution and behaviour of platinum in soils, sediments and waters in the Tulameen ultramafic complex, southern British Columbia, Canada. Journal of Geochemical Exploration 46(3): 279-308.

Cooke, D.R., Bloom, M.S., 1990. Epithermal and subjacent porphyry mineralization, Acupan, Baguio district, Philippines: a fluid-inclusion and paragenetic study. Journal of Geochemical Exploration 35(1-3): 341-362.

Coolbaugh, M.F., Bedell, R.L., 2006. A simplification of weights of evidence using a density function and fuzzy distributions; geothermal systems, Nevada. In: J.R. Harris (Ed.), GIS for the Earth Sciences, Geological Association of Canada Special Publication 44, Geological Association of Canada, St. John's, pp. 115-130.

Coolbaugh, M.F., Taranik, J.V., Raines, G.L., Shevenell, L.A., Sawatzky, D.L., Minor, T.B., Bedell, R.L., 2002. A geothermal GIS for Nevada: defining regional controls and favorable exploration terrains for extensional geothermal systems. Geothermal Resources Council Transactions 26, 485-490.

Cox, D.P., Singer, D.A. (Eds.), 1986. Mineral Deposit Models. U.S. Geological Survey Bulletin 1693, United States Government Printing Office, Washington.

Cox, D.R., Snell, E.J., 1989. Analysis of Binary Data, 2nd edn., Chapman and Hall, London.

Cox, T.F., 1981. Reflexive nearest neighbours. Biometrics 37(2): 367-369.

Date, C.J., 1990. An Introduction to Database Systems, Vol. I, 5th edn., Addison-Wesley, Reading.

Davis, J.C., 2002. Statistics and Data Analysis in Geology, 3rd edn., John Wiley & Sons, New York.

De Araújo, C.C., Macedo, A.B., 2002. Multicriteria geologic data analysis for mineral favorability mapping: application to a metal sulphide mineralized area, Ribeira Valley metallogenic province, Brazil. Natural Resources Research 11(1): 29-43.

Demicco, R., Klir, G. (Eds.), 2004. Fuzzy Logic in Geology, Elsevier, Amsterdam

Dempster, A.P., 1967. Upper and lower probabilities induced by a multivalued mapping. Annals of Mathematical Statistics 38(2): 325-339.

Dempster, A.P., 1968. A generalization of Bayesian inference. Journal of the Royal Statistical Society B30(2): 205-247.

D'Ercole, C., Groves, D.I., Knox-Robinson, C.M., 2000. Using fuzzy logic in a Geographic Information System environment to enhance conceptually based prospectivity analysis of Mississippi Valley-type mineralisation. Australian Journal of Earth Sciences 47(5): 913-927.

Diepenbroek, M., Grobe, H., Reinke, M., Schindler, U., Schlitzer, R., Sieger, R., Wefer, G., 2002. PANGAEA – an information system for environmental sciences. Computers & Geosciences 28(10): 1201-1210.

Diggle, P.J., 1983. Statistical Analysis of Spatial Point Patterns. Academic Press, London.

Duda, R.O., Hart, P.E., Nilsson, N.J., Sutherland, G.L., 1978. Semantic network representations in rule-based inference systems. In: D.A. Waterman, F. Hayes-Roth (Eds.), Pattern-Directed Inference Systems, Academic Press, London, pp. 203-221.

Eastman, J.R., Jin, W., Kyem, A.K., Toledano, J., 1995. Raster procedures for multi-criteria/multi objective decisions. Photogrammetric Engineering and Remote Sensing 61(5): 539-547.

Eddy, B.G., Bonham-Carter, G.F., Jefferson, C.W., 2006. Mineral potential analyzed and mapped at multiple scales – a modified fuzzy logic method using digital geology. In: J.R. Harris (Ed.), GIS for the Earth Sciences, Geological Association of Canada Special Publication 44, Geological Association of Canada, St. John's, pp. 143-162.

Evans, M.E., 1993. Ore Geology and Industrial Minerals; an Introduction, Blackwell Scientific Publications, Oxford.

Fabbri, A.G., Chung, C.J., 2008. On blind tests and spatial prediction models. In: G.F. Bonham-Carter (Ed.), Forty Years of Progress in Geomathematics, Springer, (in press).

Feder, J., 1988. Fractals. Plenum, New York.

Fisher, R.A., 1936. The use of multiple measurements in taxonomic problems. Annals of Eugenics 7(2): 179-188.

Fletcher, W.K., Muda, J., 1999. Influence of selective logging and sedimentological processes on geochemistry of tropical rainforest streams. Journal of Geochemical Exploration 67(1-3): 211-222.

Ford, A., Blenkinsop, T.G., 2008. Combining fractal analysis of mineral deposit clustering with weights of evidence to evaluate patterns of mineralization: application to copper deposits of the Mount Isa Inlier, NW Queensland, Australia. Ore Geology Reviews 33(3-4): 435-450.

Fordyce, F.M., Green, P.M., Simpson, P.R., 1993. Simulation of regional geochemical survey maps at variable sample density. Journal of Geochemical Exploration 49(1-2): 161-175.

Fry, N., 1979. Random point distributions and strain measurement in rocks. Tectonophysics 60(1-2): 89-105.

Garbrecht, J., Martz, L.W., 1997. Automated channel ordering and node indexing for raster channel networks. Computers & Geosciences 23(9): 961-966.

Gardoll, S.J., Groves, D.I., Knox-Robinson, C.M., Yun, G.Y., Elliott, N., 2000. Developing the tools for geological shape analysis, with regional- to local-scale examples from the Kalgoorlie Terrane of Western Australia. Australia Journal of Earth Sciences 47(5): 943-953.

Garcia, J.A., Fdez-Valdivia, J., 1994. Boundary simplification in cartography preserving the characteristics of the shape features. Computers & Geosciences 20(3): 349-368.

Garrett, R.G., 1988. IDEAS: an interactive computer graphics tool to assist the exploration geochemist. In: Current Research, Part F. Geological Survey of Canada, Paper 88-1F, pp. 1-13.

Garrett, R.G., Banville, R.M., Adcock, S.W., 1990. Regional geochemical data compilation and map preparation, Labrador, Canada. Journal of Geochemical Exploration 39(1-2): 91-166.

Garrett, R.G., Lalor, G.C., 2005. The Fe/Na ratio, a framework for modelling trace element distributions in Jamaican soil. Geochemistry: Exploration, Environment, Analysis 5(2): 147-157.

Garrett, R.G., Kane, V.E., Zeigler, R.K., 1980. The management and analysis of regional geochemical data. Journal of Geochemical Exploration 13(2-3): 115-152.

George, H., Bonham-Carter, G.F., 1989. Spatial modeling of geological data for gold exploration, Star Lake area, Saskatchewan. In: F.P. Agterberg, G.F. Bonham-Carter (Eds.), Statistical Application in the Earth Sciences, Geological Survey of Canada Paper 89-9, pp. 157-169.

Gonçalves, M.A., 2001. Characterization of geochemical distributions using multifractal models. Mathematical Geology 33(1): 41-61.

Gonçalves, M.A., Mateus, A., Oliveira, V., 2001. Geochemical anomaly separation by multifractal modelling. Journal of Geochemical Exploration 72(2): 91-114.

Good, I.J., 1983. The philosophy of exploratory data analysis. Philosophy of Science 50(2): 283-295.

Goodman, L.P., 1954. Kolmogorov-Smirnov test for psychological research, Psychological Bulletin 51(2): 160-168.

Goovaerts, P., 1992. Factorial kriging analysis: a useful tool for exploring the structure of multivariate spatial soil information. European Journal of Soil Science 43(4), 597-619.

Govett, G.J.S., 1983. Rock Geochemistry in Mineral Exploration, Handbook of Exploration Geochemistry, Vol. 3, Elsevier, Amsterdam.

Govett, G.J.S., Goodfellow, W.D., Chapman, A., Chork, C.Y., 1975. Exploration geochemistry distribution of elements and recognition of anomalies. Mathematical Geology 7(5-6): 415-446.

Groves, D.I., Goldfarb, R.J., Knox-Robinson, C.M., Ojala, J., Gardoll, S., Yun, G.Y., Holyland, P., 2000. Late-kinematic timing of orogenic gold deposits and significance for computer-based exploration techniques with emphasis on the Yilgarn Block, Western Australia. Ore Geology Reviews. Ore Geology Reviews 17(1-2): 1-38.

Grunsky, E.C., 2006. The evaluation of geochemical survey data: data analysis and statistical methods using geographic information systems. In: J.R. Harris (Ed.), GIS for the Earth Sciences, Geological Association of Canada Special Publication 44, Geological Association of Canada, St. John's, pp. 229-283.

Grunsky, E.C., Agterberg, F.P., 1992. Spatial relationships of multivariate data. Mathematical Geology 24(6): 731-758.

Grunsky, E.C., Kilby, W.E. and Massey, N.W.D., 1994. Mineral resource assessment in British Columbia, Natural Resources Research, vol. 3, no. 4, pp. 271-283.

Grunsky, E.C., Smee, B.W., 1999. The differentiation of soil types and mineralization from multi-element geochemistry using multivariate methods and digital topography. Journal of Geochemical Exploration 67(1-3): 287-299.

Haklay, M., 2007. Comparing map calculus and map algebra in dynamic GIS. In: J. Drummond, R. Billen, E. João, D. Forrest (Eds.), Dynamic and Mobile GIS, Investigating Changes in Space and Time, CRC Press, Boca Raton, pp. 89-104.

Hale, M. (Ed.), 2000. Geochemical Remote Sensing of the Sub-surface, Handbook of Exploration Geochemistry, Vol. 7, Elsevier, Amsterdam.

Hale, M., Plant, J. (Eds.), 1994. Drainage Geochemistry, Handbook of Exploration Geochemistry, Vol. 6, Elsevier, Amsterdam.

Hampel, F., Ronchetti, E., Rousseeuw, P., Stahel, W., 1986. Robust Statistics: The Approach Based on Influence Functions, Wiley, New York.

Hanna, S.S, Fry, N., 1979. A comparison of methods of strain determination in rocks from southwest Dyfed (Pembrokeshire) and adjacent areas. Journal of Structural Geology 1(2): 155-162.

Harbaugh, J.W., Bonham-Carter, G.F., 1970. Computer Simulation in Geology, Wiley-Interscience, New York.

Harris, D.P., 1984. Mineral Resources Appraisal – Mineral Endowment, Resources, and Potential Supply – Concept, Methods, and Cases, Oxford University Press, New York.

Harris, D.P., Pan, G., 1991. Consistent geological areas for epithermal gold-silver deposits in the Walker Lake quadrangle of Nevada and California delineated by quantitative methods. Economic Geology 86(8): 142-165.

Harris, D.P., Pan, G., 1999. Mineral favorability mapping: a comparison of artificial neural networks, logistic regression and discriminant analysis. Natural Resources Research 8(2): 93-109.

Harris, D.P., Zurcher, L., Stanley, M., Marlow, J., Pan, G., 2003. A comparative analysis of favourability mappings by weights of evidence, probabilistic neural networks, discriminant analysis, and logistic regression. Natural Resources Research 12(4): 241-255.

Harris, J.R., Sanborn-Barrie, M., 2006. Mineral potential mapping: examples from the Red Lake Greenstone Belt, Northwest Ontario. In: J.R. Harris (Ed.), GIS for the Earth Sciences,

Geological Association of Canada Special Publication 44, Geological Association of Canada, St. John's, pp. 1-21.

Harris, J.R., Sanborn-Barrie, M., Panagapko, D.A., Skulski, T., Parker, J.R., 2006. Gold prospectivity maps of the Red Lake greenstone belt: application of GIS technology. Canadian Journal of Earth Sciences 43(7): 865-893.

Harris, J.R., Wilkinson, L., Bernier, M., 2001a. Analysis of geochemical data for mineral exploration using a GIS – a case study from the Swayze greenstone belt, northern Ontario, Canada. In: McClenaghan, M.B., Borrowsky, P.T., Hall, G.E.M., Cook, S.J. (Eds.), Drift Exploration in Glaciated Terrain, Geological Society, London, Special Publications 2001, 185: 165-200.

Harris, J.R., Wilkinson, L., Grunsky, E., Heather, K., Ayer, J., 1999. Techniques for analysis and visualization of lithogeochemical data with applications to the Swayze greenstone belt, Ontario. Journal of Geochemical Exploration 67(1-3): 301-334.

Harris, J.R., Wilkinson, L., Heather, K., Fumerton, S., Bernier, M.A., Ayer, J., Dahn, R., 2001b. Application of GIS processing techniques for producing mineral prospectivity maps – a case study: mesothermal Au in the Swayze Greenstone Belt, Ontario, Canada. Natural Resources Research 10(2): 91-124.

Haslett, J., Bradley, R., Craig, P., Unwin, A., Wills, G., 1991. Dynamic graphics for exploration spatial data with application to locating global and local anomalies. The American Statistician 45(3): 234-242.

Hawkes, H.E., 1976. The downstream dilution of stream sediment anomalies. Journal of Geochemical Exploration 6(1-2): 345-358.

Hedenquist, J.W., 1986. Geothermal systems in the Taupo Volcanic Zone: their characteristics and relation to volcanism and mineralization. In: I.E.M. Smith (Ed.), Late Cenozoic Volcanism in New Zealand, Royal Society of New Zealand Bulletin 23, pp. 134-168

Hedenquist, J.W., Arribas, A., Reynolds, T.J., 1998. Evolution of intrusion-centered hydrothermal system: Far Southeastern – Lepanto porphyry and epithermal Cu-Au deposits, Philippines. Economic Geology 93(4): 373-404.

Hedenquist, J.W., Henley, R.W., 1985. The importance of CO_2 on freezing point measurements of fluid inclusions; evidence from active geothermal systems and implications for epithermal ore deposition. Economic Geology 80(5): 1379-1406.

Hellendoorn, H., Thomas, C., 1993. Defuzzification in fuzzy controllers. Journal of Intelligent Fuzzy Systems 1(2): 109-123.

Hengl, T., 2006. Finding the right pixel size. Computers & Geosciences 32(9): 1283-1298.

Henley, R.W., 1990. Epithermal gold deposits in volcanic terranes. In: R.P. Foster (Ed.), Gold Metallogeny and Exploration, Blackie, Glasgow, pp. 133-164.

Henley, R.W., Adams, D.P.M., 1992. Strike-slip fault reactivation as a control on epithermal vein-style gold mineralization. Geology 20(5): 443-446.

Hill, D.P., 1977. A model for earthquake swarms. Journal of Geophysical Research 82(B8): 347-352.

Hirano, A., Welch, R., Lang, H., 2003. Mapping from ASTER stereo image data: DEM validation and accuracy assessment. ISPRS Journal of Photogrammetry & Remote Sensing 57(5-6): 356-370.

Hoaglin, D.C., Mosteller, F., Tukey, J.W., 2000. Understanding Robust and Exploratory Data Analysis, 2nd edn., John Wiley & Sons, New York.

Hodge, R., Evans, M., Marshall, J., Quigley, J., Walls, L., 2001. Eliciting engineering knowledge about reliability during design – lessons learnt from implementation. Quality and Reliability Engineering International 17(3): 169-179.

Hodkiewicz, P.F., Weinberg, R.F., Gardoll, S.J., Groves, D.I., 2005. Complexity gradients in the Yilgarn Craton: fundamental controls on crustal-scale fluid flow and formation of world-class orogenic-gold deposits. Australian Journal of Earth Sciences 52(6): 831-841.

Holroyd, F., Bell, S.B.M., 1992. Raster GIS: models for raster encoding. Computers & Geosciences 18(4): 419-426.

Hoover, D.B., Heran, W.D., Hill, P.L. (Eds.), 1992. The Geophysical Expression of Selected Mineral Deposit Models, U.S. Geological Survey Open-File Report 92-557, United States Department of the Interior, Denver.

Hosmer, D.W., Lemeshow, S., 2000. Applied Logistic Regression, 2nd edn., John Wiley & Sons, New York.

Hosseinali, F., Alesheikh, A.A., 2008. Weighting spatial information in GIS for copper mining exploration. American Journal of Applied Sciences 5(9): 1187-1198.

Howarth, R.J., 1983a. Mapping. In: R.J. Howarth (Ed.). Statistics and Data Analysis in Geochemical Prospecting, Handbook of Exploration Geochemistry, Vol. 2, Elsevier, Amsterdam, pp. 111-205.

Howarth, R.J. (Ed.), 1983b. Statistics and Data Analysis in Geochemical Prospecting, Handbook of Exploration Geochemistry, Vol. 2, Elsevier, Amsterdam.

Howarth, R.J., 1984. Statistical approach in geochemical prospecting: a survey of recent developments. Journal of Geochemical Exploration 21(1): 41-61.

Howarth, R.J., Sinding-Larsen, R., 1983. Multivariate analyses. In: R.J. Howarth (Ed.). Statistics and Data Analysis in Geochemical Prospecting, Handbook of Exploration Geochemistry, Vol. 2, Elsevier, Amsterdam, pp. 207-289.

Howarth, R.J., Turner, M.S.J., 1987. Statistical graphics in geochemical journals. Mathematical Geology 19(1):1-24.

Hronsky, J.M.A., Groves, D.I., 2008. Science of targeting: definition, strategies, targeting and performance measurement. Australian Journal of Earth Sciences 55(1): 3-12.

Huber, P. 1981. Robust Statistics, Wiley, New York.

Isaaks, E.H., Srivastava, R.M., 1989. An Introduction to Applied Geostatistics, Oxford University Press, New York.

Jaeger, J.C., Cook, N.G., 1976. Fundamentals of Rock Mechanics, John Wiley, New York.

Jenson, S.K., Domingue, J.O., 1988. Extracting topographic structure from digital elevation data for geographic information system analysis. Photogrammetric Engineering and Remote Sensing 54(11): 1593-1600.

JICA-MMAJ, 1986. Report on Mineral Exploration, Mineral Deposits and Tectonics of Two Contrasting Geologic Environments in the Republic of the Philippines, Phase II (Masbate and Leyte Areas), Japan International Cooperating Agency and Metal Mining Agency of Japan, Tokyo.

Jimenez-Espinosa, Chica-Olmo, M., 1993. Application of geostatistics to identify gold-rich areas in the Finisterre-Fervenza region, NW Spain. Applied Geochemistry 14(1): 133-145.

Jones, R., 2002. Algorithms for using a DEM for mapping catchment areas of stream sediment samples. Computers & Geosciences 28(9): 1051-1060.

Jordan, C., Zhang, C., Higgins, A., 2007. Using GIS and statistics to study influences of geology on probability of features of surface soil geochemistry in Northern Ireland. Journal of Geochemical Exploration 93(3): 135-152.

Joseph, I., Bhaumik, B.K., 1997. Improved estimation of the Box-Cox transform parameter and its application to hydrogeochemical data. Mathematical Geology 29(8): 963-976.

Journel, A., Huijbregts, C.J., 1978. Mining Geostatistics, Academic Press, London.

Karger, M., Sandomirsky, S., 2001. Multidimensional statistical technique for detection of low contrast geochemical anomalies. Journal of Geochemical Exploration 72(1): 47-58.

Katz, S.S., 1991. Emulating the Prospector expert system with a raster GIS. Computers & Geosciences 17(7): 1033-1050.

Kauranne, L.K. (Ed.), 1975. Conceptual Models in Exploration Geochemistry: Norden 1975. Journal of Geochemical Exploration 5(3): 173-420.

Kauranne, L.K., Salminen, R., Eriksson, K. (Eds.), 1992. Regolith Exploration Geochemistry in Arctic and Temperate Terrains, Handbook of Exploration Geochemistry, Vol. 5, Elsevier, Amsterdam.

Kaye, B.H., 1989. A Random Walk Through Fractal Dimensions, VCH Publishers, New York.

Kearey, P., Brooks, M., Hill, I., 2002. An Introduction to Geophysical Exploration, 3rd edn., Blackwell Scientific Publications, Oxford.

Kim, H., Swain, P.H., 1989. Multisource data analysis in remote sensing and geographic information systems based on Shafer's theory of evidence. Proceedings of 1989 International Geoscience and Remote Sensing Symposium (IGARSS), pp. 829-832.

King, G., Zeng, L., 2001. Logistic regression in rare events data. Political Analysis 9(2): 137-163.

Klinkenberg, B., 1994. A review of methods used to determine the fractal dimension of linear features. Mathematical Geology 26(1): 23-46.

Knox-Robinson, C.M., 2000. Vectorial fuzzy logic: a novel technique for enhanced mineral prospectivity mapping with reference to the orogenic gold mineralisation potential of the Kalgoorlie Terrane, Western Australia. Australian Journal of Earth Sciences 47(5): 929-942.

Kotz, S., Johnson, N.L., 1985. Encyclopedia of Statistical Sciences, Volume 3, John Wiley & Sons, New York.

Kramar, U., 1995. Application of limited fuzzy clusters to anomaly recognition in complex geological environments. Journal of Geochemical Exploration 55(1-3): 81-92.

Kreuzer, O.P., Blenkinsop, T.G., Morrison, R.J., Peters, S.G., 2007. Ore controls in the Charters Towers goldfield, NE Australia: constraints from geological, geological and numerical analyses. Ore Geology Reviews 32(1-4): 37-80.

Kürzl, H., 1988. Exploratory data analysis: recent advances for the interpretation of geochemical data. Journal of Geochemical Exploration 30(1-3): 309-322.

Laing, W.P., 2004. Tension vein arrays in progressive strain: complex but predictable architecture, and major host of ore deposits. Journal of Structural Geology 26(6-7): 1303-1315.

Laxton, J.L., Becken, K., 1996. The design and implementation of a spatial database for the production of geological maps. Computers & Geosciences 22(7): 723-733.

Lee, T., Richards, J.A., Swain, P.H., 1987. Probabilistic and evidential approaches for multisource data analysis. IEEE Transactions on Geoscience and Remote Sensing GE-25(3): 283-293.

Le Pichon, X., Francheteau, J., Bonnin, J., 1973. Plate Tectonics. Developments in Geotectonics 6, Elsevier, Amsterdam.

Levinson, A.A., 1974. Introduction to Exploration Geochemistry, Applied Publishing Ltd., Calgary.

L'Heureux, I, Katsev, S., 2006. Oscillatory zoning in a (Ba,Sr)SO_4 solid solution: macroscopic and cellular automata models. Chemical Geology 225(1-3): 230-243.

Li, C., Ma, T., Shi, J., 2003. Application of a fractal method relating concentrations and distances for separation of geochemical anomalies from background. Journal of Geochemical Exploration 77(2-3): 167-175.

Likkason, O.K., Shemang, E.M., Suh, C.E., 1997. The application of evidential belief functions in the integration of regional geochemical and geological data over the Ife-Ilesha goldfield, Nigeria. Journal of African Earth Sciences 25(3): 491-501.

Lindgren, W., 1933. Mineral Deposits, 4th edn., McGraw-Hill, New York.

Lombard, M., De Bruin, D., Elsenbroek, J.H., 1999. High density regional geochemical mapping of soils and stream sediments in South Africa. Journal of Geochemical Exploration 66(1-2): 145-149.

Lopez-Blanco, J., Villers-Ruiz, L., 1995. Delineating boundaries of environmental units for land management using a geomorphological approach and GIS: a study in Baja California, Mexico. Remote Sensing of Environment 53(2): 109-117.

Lovering, T.G., McCarthy, J.H. (Eds.), 1978. Conceptual Models in Exploration Geochemistry: The Basin and Range Province of the Western United States and Northern Mexico. Journal of Geochemical Exploration 9(2-3): 113-276.

Ludington, S., Folger, H., Kotlyar, B., Mossotti, V.G., Coombs, M.J., Hildenbrand, T.G., 2006. Regional surficial geochemistry of the Northern Great Basin. Economic Geology 101(1): 33-57.

Luo, J., 1990. Statistical mineral prediction without defining a training area. Mathematical Geology 22(3): 253-260.

Luo, X., Dimitrakopoulos, R., 2003. Data-driven fuzzy analysis in quantitative mineral resource assessment. Computers & Geosciences 29(1): 3-13.

Maguire, D.J., Goodchild, M.F., Rhind, D.W. (Eds.), 1991. Geographical Information Systems, Volume 1 – Principles, Volume 2 – Applications, Longman Scientific & Technical, London.

Maling, D.H., 1992. Coordinate Systems and Map Projections, 2nd edn., Pergamon Press, Oxford.

Mandelbrot, B.B., 1967. How long is the coast of Britain: statistical self-similarity and fractional dimension. Science 156(3775): 636-638.

Mandelbrot, B.B., 1975. Stochastic models for the Earth's relief, the shape and fractional dimension of the coastlines, and the number-area rule for islands. Proceedings of the National Academy of Science, USA, 72(10): 3825-2828.

Mandelbrot, B.B., 1982. The Fractal Geometry of Nature, Freeman, New York.

Mandelbrot, B.B., 1983. The Fractal Geometry of Nature (updated and augmented edition), Freeman, New York.

Mandelbrot, B.B., 1985. Self-affine fractals and fractal dimension. Physica Scripta 32(4): 257-260.

Mandl, G., 1988. Mechanics of Tectonic Faulting, Elsevier, Amsterdam.

Martz, L.W., Garbrecht, J., 1993. Automated recognition of drainage network and watershed data from digital elevation models. Water Resources Bulletin, American Water Resources Association 29(6): 901-908.

Massey N.W.D., MacIntyre D.G., Desjardins P.J., Cooney R.T., 2005. Digital Geology Map of British Columbia: Whole Province. B.C. Ministry of Energy and Mines, Geofile 2005-1, scale 1:250,000.

Mather, P.M., 1987. Computer Processing of Remotely-Sensed Images, John Wiley and Sons, Chichester.

Mattiske, T., 1983. Data storage and retrieval. In: Howarth, R.J. (Ed.), 1983. Statistics and Data Analysis in Geochemical Prospecting, Handbook of Exploration Geochemistry, Volume 2. Elsevier, Amsterdam, pp. 3-38.

McCammon, R.B., 1989. Prospector III – Towards a map-based expert system for regional mineral resource assessment. In: F.P. Agterberg, G.F. Bonham-Carter (Eds.), Statistical Applications in the Earth Sciences, Geological Survey of Canada, Paper 89-9, pp. 295-404.

McCammon, R.B., 1994. Prospector II: towards a knowledge base for mineral deposits. Mathematical Geology 26(8): 917-936.

McCammon, R.B., Botbol, J.M., Sinding-Larsen, R., Bowen, R.W., 1983. Characteristics analysis – 1981: final program and a possible discovery. Mathematical Geology 15(1): 59-83.

McCammon, R.B., Botbol, J.M., Sinding-Larsen, R., Bowen, R.W., 1984. The New CHARaracteristic ANalysis (NCHARAN) Program. U.S. Geological Survey Bulletin 1621, 27 pp.

McCammon, R.B., Kork, J.O., 1992. One-level prediction – a numerical method for estimating undiscovered metal endowment. Nonrenewable Resources 1(2): 139-147.

McGrath, S.P., Loveland, P.J., 1992. The Soil Geochemical Atlas of England and Wales, Blackie Academic, London.

McMaster, R.B., Shea, K.S., 1992. Generalization in Digital Cartography, Association of American Geographers, Washington D.C.

Menard, S., 2001. Applied Logistic Regression Analysis. Sage University Papers Series on Quantitative Applications in the Social Sciences, Vol. 07-106. Sage, Thousand Oaks, California.

Miesch, A.T., 1977. Log-transformation in geochemistry. Mathematical Geology 9(2): 191-194.

Mihalasky, M., 1999. Mineral Potential Modelling of Gold and Silver Mineralization in the Nevada Great Basin – a GIS-based Analysis Using Weights of Evidence. Ph.D. Thesis, University of Ottawa, Canada, 360 pp.

Mihalasky, M.J., Bonham-Carter, G.F., 2001. Lithodiversity and its spatial association with metallic mineral sites, Great Basin of Nevada. Natural Resources Research 10(3): 209-226.

Mineter, M.J., 1998. Raster-to-vector conversion. In: R. Healy, S. Dowers, B. Gittings, M.J. Mineter (Eds.), Parallel Algorithms for GIS, Taylor and Francis, London, pp. 253-264.

Mineter, M.J., 2003. A software framework to create vector-topology in parallel GIS operations. International Journal of Geographical Information Science 17(3): 203-222.

Mitchell, A.H.G., Balce, G.R., 1990. Geological features of some epithermal gold systems, Philippines. In: J.W. Hedenquist, N.C. White, G. Siddeley (Eds.), Epithermal Gold Mineralization of the Circum-Pacific: Geology, Geochemistry, Origin and Exploration I, Journal of Geochemical Exploration 36(1-3): 241-296.

Mitchell, A.H.G., Garson, M.S., 1981. Mineral Deposits and Global Tectonic Settings, Academic Press, London.

Mitchell, A.H.G., Leach, T.M. 1991. Epithermal Gold in the Philippines: Island Arc Metallogenesis, Geothermal Systems and Geology. Academic Press, London.

Moon, C.J., 1999. Towards a quantitative model of downstream dilution of point source geochemical anomalies. Journal of Geochemical Exploration 65(2): 111-132.

Moon, W.M., 1990. Integration of geophysical and geological data using evidential belief function. IEEE Transactions on Geoscience and Remote Sensing 28(4): 711-720.

Moon, W.M., 1993. On mathematical representation and integration of multiple geoscience data sets. Canadian Journal of Remote Sensing 19(1): 663-667.

Moon, W.M., Chung, C.F., An, P., 1991. Representation and integration of geological, geophysical and remote sensing data. Geoinformatics 2(2): 177-188.

Moreira, F.R.S., Almeida-Filho, R., Câmara, G., 2003. Spatial analysis techniques applied to mineral prospecting: an evaluation in the Poços de Caldas Plateau. Revista Brasileira de Geociências 33(2-Suppl.): 183-190.

Muntean, J.L., Einaudi, M.T., 2001. Porphyry-epithermal transition: Maricunga belt, northern Chile. Economic Geology 96(4): 743-772.

Nelson, E.P., 2006. Drill-hole design for dilational ore shoot targets in fault-fill veins. Economic Geology 101(5): 1079-1085.

Nykänen, V., 2008. Radial basis functional link nets used as a prospectivity mapping tool for orogenic gold deposits within the Central Lapland Greenstone Belt, Northern Fennoscandian Shield. Natural Resources Research 17(1): 29-48.

Nykänen, V., Groves, D.I., Ojala, V.J., Eilu, P., Gardoll, S.J., 2008a. Reconnaissance scale conceptual fuzzy-logic prospectivity modelling for iron oxide copper – gold deposits in the northern Fennoscandian Shield, Finland. Australian Journal of Earth Sciences 55(1): 25-38.

Nykänen, V., Groves, D.I., Ojala, V.J., Eilu, P., Gardoll, S.J., 2008b. Combined conceptual/empirical prospectivity mapping for orogenic gold in the northern Fennoscandian Shield, Finland. Australian Journal of Earth Sciences 55(1): 39-59.

Nykänen, V., Raines, G.L., 2006. Quantitative analysis of scale of aeromagnetic data raises questions about geologic-map scale. Natural Resources Research 15(4): 213-222.

Oxborrow, E, P., 1989. Databases and Database Systems: Concepts and Issues, 2nd edn., Chartwell-Bratt Ltd., Sweden.

Pan, G.C., 1989. Concepts and Methods of Multivariate Information Synthesis for Mineral Resources Estimation. Ph.D. Thesis, University of Arizona, Tucson, Arizona, U.S.A., 302 pp.

Pan, G.C., 1993a. Canonical favourability model for data integration and mineral potential mapping. Computers & Geosciences 19(8): 1077-1100.

Pan, G.C., 1993b. Indicator favorability theory for mineral potential mapping. Nonrenewable Resources 2(4): 292-311.

Pan, G.C., 1993c. Regionalized favorability theory for information synthesis in mineral exploration. Mathematical Geology 25(5): 603-631.

Pan, G.C., Harris, D.P., 1990. Quantitative analysis of anomalous sources and geochemical signatures in the Walker Lake quadrangle of Nevada and California. Journal of Geochemical Exploration 38(3): 299-321.

Pan, G.C., Harris, D.P., 1992a. Decomposed and weighted characteristic analysis for the quantitative estimation of mineral resources. Mathematical Geology 24(7): 807-823.

Pan, G.C., Harris, D.P., 1992b. Estimating a favourability equation for the integration of geodata and selection of mineral exploration targets. Mathematical Geology 24(2): 177-202.

Pan, G.C., Harris, D.P., 2000. Information Synthesis for Mineral Exploration, Oxford University Press, Inc., New York.

Pan, G., Harris, D.P., Heiner, T., 1992. Fundamental issues in quantitative estimation of mineral resources. Natural Resources Research 1(4): 281-292.

Pan, G.C., Portefield, B., 1995. Large-scale mineral potential estimation for blind precious metal ore deposits. Nonrenewable Resources 4(2): 187-207.

Panahi, A., Cheng, Q., 2004. Multifractality as a measure of spatial distribution of geochemical patterns. Mathematical Geology 36(7): 827-846.

Panahi, A., Cheng, Q., Bonham-Carter, G.F., 2004. Modelling lake sediment and geochemical distribution using principal component, indicator kriging and multifractal power-spectrum analysis: a case study from Gowganda, Ontario. Geochemistry: Exploration, Environment, Analysis 4(1): 59-70.

Parasnis, D.S., 1997, Principles of Applied Geophysics, 5th edn., Chapman and Hall, London.

Pardo-Igúzquiza, E., Chica-Olmo, M., 2005. Interpolation and mapping of probabilities for geochemical variables exhibiting spatial intermittency. Applied Geochemistry 20(1): 157-168.

Park, N.-W., Chi, K.-H., Kwon, D.-B., 2007. Accounting for spatial patterns of multiple geological data sets in geological thematic mapping using GIS-based spatial analysis. Environmental Geology 51(7): 1147-1155.

Peh, Z., Miko, S., Mileusnic, M., 2006. Areal versus linear evaluation of relationship between drainage basin lithology and geochemistry of stream and overbank sediments in low-order mountainous drainage basin. Environmental Geology 49(8): 1102-1115.

Pendharkar, P.C., 2003. Characterization of aggregate fuzzy membership functions using Saaty's eigenvalue approach. Computers & Operations Research 30(2): 199-212.

Peter, C., Stuart, N., 1999.Modelling river floodplain inundation in space and time. In: B. Gittings (Ed.), Integrating Information Infrastructures with GI Technology, Innovations in GIS 6, CRC Press, London, pp. 255-267.

Pirajno, F., 1992. Hydrothermal Mineral Deposits, Principles and Fundamental Concepts for the Exploration Geologists, Springer-Verlag, Berlin.

Plimer, I.R., Elliott, S.M., 1979. The use of Rb/Sr ratios as a guide to mineralization. Journal of Geochemical Exploration 12(1): 21-34.

Polikarpochkin, V.V., 1971. The quantitative estimation of ore-bearing areas from sample data of the drainage system. In: R.W. Boyle (Ed.), Transactions 3rd International Geochemical

Exploration Symposium, Special Volume 11, Canadian Institute of Mining and Metallurgy, pp. 585-586.

Ponce, D.A., Glen, J.M.G., 2002. Relationship of epithermal gold deposits to large-scale fractures in Northern, Nevada. Economic Geology 97(1): 3-9.

Porwal, A., 2006. Mineral Potential Mapping with Mathematical Geological Models, Ph.D. Thesis, University of Utrecht, The Netherlands, ITC (International Institute for Geo-Information Science and Earth Observation) Publication No. 130, Enschede, 289 pp.

Porwal, A., Carranza, E.J.M., 2008. Classifiers for modelling of mineral potential. In: O. Pourret, P. Naïm, B. Marcot (Eds.), Bayesian Networks: A Practical Guide to Applications. John Wiley & Sons, Chichester, pp. 149-171.

Porwal, A., Carranza, E.J.M., Hale, M., 2001. Extended weights-of-evidence modelling for predictive mapping of base metal deposit potential in Aravalli province, western India. Exploration and Mining Geology 10(4): 273-287.

Porwal, A., Carranza, E.J.M., Hale, M., 2003a. Artificial neural networks for mineral potential mapping. Natural Resources Research 12(3): 155-171.

Porwal, A., Carranza, E.J.M., Hale, M., 2003b. Knowledge-driven and data-driven fuzzy models for predictive mineral potential mapping. Natural Resources Research 12(1): 1-25.

Porwal, A., Carranza, E.J.M., Hale, M., 2004. A hybrid neuro-fuzzy model for mineral potential mapping. Mathematical Geology 36(7): 803-826.

Porwal, A., Carranza, E.J.M., Hale, M., 2006a. A hybrid fuzzy weights-of-evidence model for mineral potential mapping. Natural Resources Research 15(1): 1-14.

Porwal, A., Carranza, E.J.M., Hale, M., 2006b. Bayesian network classifiers for mineral potential mapping. Computers & Geosciences 32(1): 1-16.

Porwal, A., Carranza, E.J.M., Hale, M., 2006c. Tectonostratigraphy and base-metal mineralization controls, Aravalli province (western India): new interpretations from geophysical data analysis. Ore Geology Reviews 29(3-4): 287-306.

Porwal, A., Sides, E.J., 2000. A predictive model for base metal exploration in a GIS environment. International Archives of Photogrammetry and Remote Sensing XXXIII(B7): 1178-1184.

Pouch, G.W., 1997. An interactive program for computer-aided map design, display and query: EM_{AP}KGS2. Computers & Geosciences 23(3): 259-266.

Prelat, A.E., 1977. Discriminant analysis as a method of predicting mineral occurrence potentials in central Norway. Mathematical Geology 9(4): 343-367.

Prol-Ledesma, R.M., 2000. Evaluation of the reconnaissance results in geothermal exploration using GIS. Geothermics 29(1): 83-103.

Raines, G.L., 1999. Evaluation of weights of evidence to predict epithermal gold deposits in the Great Basin of the western United States. Natural Resources Research 8(4): 257-276.

Raines, G.L., Connors, K.A., Chorlton, L.B., 2007. Porphyry copper deposit tract definition – a global analysis comparing geologic map scales. Natural Resources Research 16(2): 191-198.

Ranjbar, H., Honarmand, M., 2004. Integration and analysis of airborne geophysical and ETM+ data for exploration of porphyry type deposits in the Central Iranian Volcanic Belt using fuzzy classification. International Journal of Remote Sensing 25(21): 4729-4741.

Rantitsch, G., 2000. Application of fuzzy clusters to quantify lithological background concentrations in stream sediment geochemistry. Journal of Geochemical Exploration 71(1): 73-82.

Rantitsch, G., 2001. The fractal properties of geochemical landscapes as an indicator of weathering and transport processes within the Eastern Alps. Journal of Geochemical Exploration 73(1): 27-42.

Raper, J., 1989. A 3-dimensional geoscientific mapping and modelling system: a conceptual design. In: J. Raper (Ed.), Three Dimensional Applications in Geographical Information Systems, Taylor and Francis, London, pp. 11-20.

Reddy, R.K.T., Bonham-Carter, G.F., Galley, A.G., 1992. Developing a geographic expert system for regional mapping of volcanogenic massive sulphide (VMS) deposit potential. Nonrenewable Resources 1(2): 112-124.

Reimann, C., 2005. Geochemical mapping: technique or art? Geochemistry: Exploration, Environment, Analysis 5(4): 359-370.

Reimann, C., Filzmoser, P., 1999. Normal and lognormal data distribution in geochemistry: death of a myth. Consequences for the statistical treatment of geochemical and environmental data. Environmental Geology 39(9): 1001-1014.

Reimann, C., Filzmoser, P., Garrett, R.G., 2002. Factor analysis applied to regional geochemical data: problems and possibilities. Applied Geochemistry 17(3): 185-206.

Reimann, C., Filzmoser, P., Garrett, R.G., 2005. Background and threshold: critical comparison of methods of determination. Science of the Total Environment 346(1-3): 1-16.

Reimann, C., Garrett, R.G., 2005. Geochemical background – concept and reality. Science of the total Environment 350 (1-3): 12-27.

Reis, A.P., Sousa, A.J., Cardoso Fonseca, E., 2003. Application of geostatistical methods in gold geochemical anomalies identification (Montemor-O-Novo, Portugal). Journal of Geochemical Exploration 77(1): 45-63.

Richards, J.P., Tosdal, R.M. (Eds.), 2001. Structural Controls on Ore Genesis, Reviews in Economic Geology 14, Society of Economic Geologists, El Paso.

Rigol-Sanchez, J.P., Chica-Olmo, M., Abarca-Hernandez, F., 2003. Artificial neural networks as a tool for mineral potential mapping with GIS. International Journal of Remote Sensing 24(5): 1151-1156.

Robb, L., 2004. Introduction to Ore-forming Processes, Blackwell, Oxford.

Roberts, R.G., Sheahan, P., Cherry, M.E. (Eds.), 1988. Ore Deposit Models, Geoscience Canada Reprint Series 3, Geological Association of Canada, Newfoundland.

Robinove, C.J., 1989. Principles of logic and the use of digital geographic information systems. In: W.J. Ripple (Ed.), Fundamentals of Geographic Information Systems, a Compendium, American Society for Photogrammetry and Remote Sensing, pp.112-124.

Robinson, V.B., 2003. A perspective on the fundamentals of fuzzy sets and their use in geographic information systems. Transactions in GIS 7(1): 3-30.

Rock, N.M.S., 1988a. Numerical Geology. Springer-Verlag, Berlin.

Rock, N.M.S., 1988b. Summary statistics in geochemistry: a study of the performance of robust estimates. Mathematical Geology 20(3): 243-275.

Rogge, D.M., Halden, N.M., Beaumont-Smith, C., 2006. Application of data integration for shear-hosted Au potential modelling: Lynn Lake Greenstone Belt, Northwestern Manitoba, Canada. In: J.R. Harris (Ed.), GIS for the Earth Sciences, Geological Association of Canada Special Publication 44, Geological Association of Canada, St. John's, pp. 191-210.

Rose, A.W., Dahlberg, E.C., Keith, M.L., 1970. Multiple regression technique for adjusting background values in stream sediment geochemistry. Economic Geology 65(2): 156-165.

Rose, A.W., Hawkes, H.E., Webb, J.S., 1979. Geochemistry in Mineral Exploration, 2nd edn., Academic Press, London.

Rowlingson, B., Diggle, P., 1993. Splancs: spatial point pattern analysis code in S-plus. Computers & Geosciences 19(5): 627–655.

Roy, R., Cassard, D., Cobbold, P.R., Rossello, E.A., Billa, M., Bailly, L., Lips, A.L.W., 2006. Predictive mapping for copper-gold magmatic-hydrothermal systems in NW Argentina: use of a regional-scale GIS, application of an expert-guided data-driven approach, and comparison with results from a continental-scale GIS. Ore Geology Reviews 29(3-4): 260-286.

Rubin, D.B., 1996. Multiple imputation after 18+ years. Journal of the American Statistical Association 91(434): 473-489.

Saaty, T.L., 1977. A scaling method for priorities in hierarchical structures. Journal of Mathematical Psychology 15(3): 234-281.

Saaty, T.L., 1980. The Analytical Hierarchy Process: Planning, Priority Setting, Resource Allocation. McGraw-Hill, New York.

Saaty, T.L., 1994. Fundamentals of Decision Making and Priority Theory with the Analytical Hierarchy Process. RWS Publications, Pittsburgh.

Sabins, F.F., 1999. Remote sensing for mineral exploration. Ore Geology Reviews 14(3): 157-183.

Sahoo, N.R., Pandalai, H.S., 1999. Integration of sparse geologic information in gold targeting using logistic regression analysis in the Hutti-Maski schist belt, Raichur, Karnataka, India – a case study. Natural Resources Research 8(2): 233-250.

Sahu, B.K., 1982. Stochastic modelling of mineral deposits. Mineralium Deposita 17(1): 99-105.

Sahu, B.K., Raiker, P.S., 1985. Univariate and multivariate modeling of the chemical composition of iron ores in Northern Goa, India. Mathematical Geology 17(3): 317-325.

Sawkins, F.J., 1989. Metal Deposits in Relation to Plate Tectonics, 2nd edn., Springer-Verlag, Berlin.

Schill, W., Jockel, K.-H., Drescher, K., Timm, J., 1993. Logistic analysis in case-control studies under validation sampling. Biometrika 80(2): 339-352.

Schmatz, D.S., Engi, M., Lieberman, J.E., 1995. ParaDIS: a relational database for the consistent documentation and analysis of metamorphic mineral assemblage. Computers & Geosciences 21(9): 1021-1041.

Schuenemeyer, J.H., 2002. A framework for expert judgment to assess oil and gas resources. Natural Resources Research 11(2): 97-107.

Segall, P., Pollard, D.D., 1980. Mechanics of discontinuous faults. Journal of Geophysical Research 85(B8): 4337-4350.

Seoane, J.C.S., De Barros Silva, A., 1999. Gold-anomalous catchment basins: a GIS prioritization model considering drainage sinuosity. Journal of Geochemical Exploration 67(1-3): 335-344.

Shafer, G., 1976. A Mathematical Theory of Evidence. Princeton University Press, Princeton, N.J.

Shen, W., Cohen, D.R., 2005. Fractally invariant distributions and an application in geochemical exploration. Mathematical Geology 37(8): 895-913.

Sibson, R.H., 1987. Earthquake rupturing as mineralizing agent in hydrothermal systems. Geology 15(8): 701–704.

Sibson, R.H., 1996. Structural permeability of fluid-driven fault-fracture meshes. Journal of Structural Geology 18(8): 1031–1042.

Sibson, R.H., 2000. Tectonic controls on maximum sustainable overpressure: fluid redistribution from stress transitions. Journal of Geochemical Exploration 69-70: 471-475.

Sibson, R.H., 2001. Seismogenic framework for hydrothermal transport and ore deposition. Society of Economic Geologists Reviews 14, 25–50.

Siegel, S., 1956. Non-Parametric Statistics for the Behavioral Sciences, McGraw-Hill, New York.

Sillitoe, R.H., 1993. Epithermal models: genetic types, geometric controls and shallow features. Geological Association of Canada, Special Volume 40: 403-417.

Sillitoe, R.H., Bonham, H.F., 1984. Volcanic landforms and ore deposits. Economic Geology 79(6): 1286-1298

Sim, B.L., Agterberg, F.P., Beaudry, C., 1999. Determining the cutoff between background and relative base metal smelter contamination levels using multifractal methods. Computers & Geosciences 25(9): 1023-1041.

Simpson, C.J., Huntington, J.F., Leishman, J, Green, A.A., 1980. A study of the Pine Creek Geosyncline using integrated Landsat and aeromagnetic data. In: J. Ferguson, A. Goleby (Eds.), Uranium in the Pine Creek Geosyncline, International Atomic Energy Agency, Vienna, pp. 141-155.

Sinclair, A.J., 1974. Selection of threshold values in geochemical data using probability graphs. Journal of Geochemical Exploration 3(2): 129-149.

Sinclair, A.J., 1983. Univariate analysis. In: R.J. Howarth (Ed.). Statistics and Data Analysis in Geochemical Prospecting, Handbook of Exploration Geochemistry, Vol. 2, Elsevier, Amsterdam, pp. 59-81.

Singer, B., 2000. Temporal evolution of arc magmatism and hydrothermal activity including epithermal gold veins, Borovitsa Caldera, southern Bulgaria. Economic Geology 95(5): 1155-1164.

Singer, D.A., 1993. Basic concepts in three-part quantitative assessments of undiscovered mineral resources. Nonrenewable Resources 2(2): 69-81.

Singer, D.A., Kouda, R., 1996. Application of a feedforward neural network in the search for Kuruko deposits in the Hokuroku district, Japan. Mathematical Geology 28(8): 1017-1023.

Singer, D.A., Kouda, R., 1997. Use of neural network to integrate geoscience information in the classification of mineral deposits and occurrences. In: A.G. Gubins (Ed.), Proceedings of Exploration 97: 4th Decennial International Conference on Mineral Exploration, pp. 127-134.

Singer, D.A., Kouda, R., 1999. A comparison of the weights-of-evidence method and probabilistic neural networks. Natural Resources Research 8(4): 287-298.

Singer, D.A., Kouda, R., 2001. Some simple guides to finding useful information in exploration geochemical data. Natural Resources Research 10(2) 137-147.

Skabar, A.A., 2005. Mapping mineralization probabilities using multilayer perceptrons. Natural Resources Research 14(2): 109-123.

Skabar, A., 2007a. Mineral potential mapping using Bayesian learning for multilayer perceptrons. Mathematical Geology 39(5): 439-451.

Skabar, A., 2007b. Modeling the spatial distribution of mineral deposits using neural networks. Natural Resource Modeling 20(3): 435-450.

Sloan, T., 1998. Vector-to-raster conversion. In: R. Healy, S. Dowers, B. Gittings, M.J. Mineter (Eds.), Parallel Algorithms for GIS, Taylor and Francis, London, pp. 233-252.

Smee, B.W., 1998. A new theory to explain the formation of soil geochemical responses overly deeply covered gold mineralization in arid environments. Journal of Geochemical Exploration 61(1-3): 149-172.

Smith, R.E., 1987. Some conceptual models for geochemistry in areas of preglacial deep weathering. Journal of Geochemical Exploration 28(1-3): 337-352.

Smith, R.E., Campbell, N.A., Perdix, J.L., 1982. Identification of some Western Australia Cu-Zn and Pb-Zn gossans by multi-element geochemistry. In: R.E. Smith (Ed.), Geochemical Exploration in Deeply Weathered Terrain. CSIRO Institute of Energy and Earth Resources, Floreat Park, Western Australia, pp. 75-90.

Snyder, J.P., 1993. Flattening the Earth: Two Thousand Years of Map Projections, University of Chicago Press, Chicago.

Spatz, D.M., 1997. Remote sensing characteristics of sediment- and volcanic-hosted precious metal system: imagery selection for exploration and development. International Journal of Remote Sensing 18(7): 1413-1438.

Stanley, C.R., 2006. Numerical transformations of geochemical data: 1. Maximizing geochemical contrast to facilitate information extraction and improve data presentation. Geochemistry: Exploration, Environment, Analysis 6(1): 69-78.

Stanley, H.E., Meakin, P., 1988. Multifractal phenomena in physics and chemistry. Nature 335(6189): 405-409.

Stein, A., Riley, J., Halberg, N., 2001. Issues of scale for environmental factors. Agriculture, Ecosystems & Environment 87(2): 215-232.

Stensgaard, B.M., Chung, C.J., Rasmussen, T.M., Stendal, H., 2006. Assessment of mineral potential using cross-validation techniques and statistical analysis: a case study from the Paleoproterozoic of West Greenland. Economic Geology 101(7): 1297-1413.

Stern, N.B., 1995. Computing in the Information Age, 2nd edn., Wiley and Sons, New York.

Strahler, A.N., 1957. Quantitative analysis of watershed geomorphology. Transactions American Geophysical Union 38(6): 913-920.

Stubley, M.P., 2004. Spatial distribution of kimberlite in the Slave craton, Canada: a geometrical approach. Lithos 77(1-4): 683-693.

Tabachnick, B.G., Fidell, L.S., 2007. Using Multivariate Statistics, 5th edn., Allyn & Bacon, Boston.

Tangestani, M.H., Moore, F., 2002. The use of Dempster-Shafer model and GIS in integration of geoscientific data for porphyry copper potential mapping, north of Shahr-e-Babak, Iran. International Journal of Applied Earth Observation and Geoinformation 4(1): 65-74.

Tangestani, M.H., Moore, F., 2003. Mapping porphyry copper potential with a fuzzy model, northern Shahr-e-Babak, Iran. Australian Journal of Earth Sciences 50(3): 311-317.

Tardy, Y., Bustillo, V., Boeglin, J.-L., 2004. Geochemistry applied to the watershed survey: hydrograph separation, erosion and soil dynamics. A case study: the basin of the Niger River, Africa. Applied Geochemistry 19(4): 469-518.

Telford, W.M., Geldart, L.P., Sheriff, R.E., 1990. Applied Geophysics, 2nd edn., Cambridge University Press, Cambridge.

Tennant, C.B., White, M.L., 1959. Study of the distribution of some geochemical data. Economic Geology 54(7): 1281-1290.

Thiart, C., De Wit, M., 2000. Linking spatial statistics to GIS: exploring potential gold and tin models of Africa. South African Journal of Geology 103(3-4): 215-230.

Thole, U., Zimmermann, H.-J., Zysno, P., 1979. On the suitability of minimum and product operators for intersection of fuzzy sets. Fuzzy Sets and Systems 2(3): 167-180.

Titley, S. R., Beane, R. E., 1981, Porphyry copper deposits – Part I: Geological settings, petrology and tectogenesis: Economic Geology 75th Anniversary Volume (Special Issue), 214–235.

Triantaphyllou, E., 1990. An evaluation of the Eigenvalue approach for determining the membership values in fuzzy sets. Fuzzy Sets and Systems 35(3): 295-301.

Trochimczyk, J., Chayes, F., 1978. Some properties of principal components. Mathematical Geology 10(1): 43-52.

Tukey, J.W., 1977. Exploratory Data Analysis, Addison-Wesley, Reading.

Tukey, P.A., Tukey, J.W., 1981. Summarization, smoothing, supplemental views. In: V. Barnett (Ed.), Interpreting Multivariate Data. John Wiley & Sons, Chichester, pp. 245-275.

Turcotte, D.L., 1997. Fractals and Chaos in Geology and Geophysics, 2nd edn., Cambridge University Press, Cambridge.

UNDP, 1987. Geochemical nature of epithermal gold mineralization and related anomalies in the Philippines. Technical Report No. 4, DP/UN/PHI-85-001/6, United Nations Development Programme, New York.

Varnes, D.J., 1974. The logic of geological maps, with reference to their interpretation and use for engineering purposes. United States Geological Survey Professional Paper 837, 48 pp.

Vearncombe, J., Vearncombe, S., 1999. The spatial distribution of mineralization: applications of Fry analysis. Economic Geology 94(4): 475-486.

Vearncombe, S., Vearncombe, J.R., 2002. Tectonic controls on kimberlite location, southern Africa. Journal of Structural Geology 24(10): 1619-1625.

Velleman, P.F., Hoaglin, D.C., 1981. Applications, Basics and Computing of Exploratory Data Analysis, Duxbury Press, Boston.

Venkataraman, G., Babu Madhavan, B., Ratha, D.S., Antony, J.P., Goyal, R.S., Banglani, S., Sinha Roy, S., 2000. Spatial modeling for base-metal mineral exploration through integration of geological data sets. Natural Resources Research 9(1): 27-42.

Vieux, B.E., 2004. Distributed Hydrologic Modeling Using GIS. Water Science and Technology Library, Volume 48. Springer, Dordrecht.

Vistelius, A.B., 1960. The skew frequency distributions and the fundamental law of the geochemical processes. Journal of Geology 68(1):1-22.

Vriend, S.P., Van Gaans, P.F.M., Middelburg, J., De Nijs, A., 1988. The application of fuzzy *c*-means and non-linear mapping to geochemical datasets: examples from Portugal. Applied Geochemistry 3(2): 213-224.

Wackernagel, H., 1995. Multivariate Geostatistics, Springer-Verlag, Berlin.

Wackernagel, H., Butenuth, C., 1989. Caractérisation d'anomalies géochemiques par la gépstatistique multivariable. Journal of Geochemical Exploration 32(1-3): 437-444.

Walley, P., 1987. Belief function representations of statistical evidence. The Annals of Statistics, 15(4): 1439-1465.

Watson, D.F., 1992. Contouring: A Guide to the Analysis and Display of Spatial Data, Pergamon Press, Oxford.

Wei, S., Pengda, Z., 2002. Multidimensional self-affine distribution with application in geochemistry. Mathematical Geology 34(2): 109-123.

Weiberg, R.F., Hodkiewicz, P.F., Groves, D.I., 2004. What controls gold distribution in Archean terranes? Geology 32(7): 545-548.

White, F.M., 1991. Viscous Fluid Flow, 2nd edn., McGraw-Hill, New York.

White, N.C., Hedenquist, J.W., 1990. Epithermal environments and styles of mineralization; variations and their causes, and guidelines for exploration. In: J.W. Hedenquist, N.C. White, G. Siddeley (Eds.), Epithermal Gold Mineralization of the Circum-Pacific: Geology, Geochemistry, Origin and Exploration II, Journal of Geochemical Exploration 36(1-3): 454-474.

Whitten, D.G.A., Brooks, J.R.V., 1972. Dictionary of Geology, Penguin Books Ltd., London.

Wielemaker, W.G., de Bruin, S., Epema, G.F., Veldkamp, A., 2003. Significance and application of the multi-hierarchical land system in soil mapping. Catena 43(1): 15-34.

Winter, S., 1998. Bridging vector and raster representation in GIS. In: R. Laurini, K. Makki, N. Pissinou (Eds.), Proceedings of the 6th Symposium on Advances in Geographic Information Systems, ACM Press, Washington D.C., pp. 57-62.

Winter, S., Frank, A.U., 2000. Topology in raster and vector representation. GeoInformatica 4(1): 35-65.

Woldai, T., Pistocchi, A., Master, S., 2006. Validation and sensitivity analysis of mineral potential model using favourability functions. Applied GIS 2(1): 2.1-2.19. DOI:10.2104/ag060002.

Wright, D.F., Bonham-Carter, G.F., 1996. VHMS favourability mapping with GIS-based integration models, Chisel Lake – Anderson Lake area. In: G.F. Bonham-Carter, A.G. Galley, G.E.M. Hall (Eds.), EXTECH I: A Multidisciplinary Approach to Massive Sulphide Research in the Rusty Lake – Snow Lake Greenstone Belts, Manitoba. Geological Survey of Canada Bulletin 426, pp. 339-376, 387-401.

Xie, S., Bao, Z., 2004. Fractal and multifractal properties of geochemical fields. Mathematical Geology 36(7): 847-864.

Xu, Y., Cheng, Q., 2001. A fractal filtering technique for processing regional geochemical maps for mineral exploration. Geochemistry: Exploration, Environment, Analysis 1(2): 147-156.

Xu, T., Moore, I., Gallant, J., 1993. Fractals, fractal dimensions and landscapes: a review. Geomorphology 8(4): 245-262.

Yatabe, S.M., Fabbri, A.G., 1988. Artificial intelligence in the geosciences: a review. In: J.J. Royer (Ed.), Geomathematical and Statistical Analysis Applied to Time Dependent Data. Sciences de la Terre, Serie Informatique Geologique 27(1): 37-68.

Yatabe, S.M., Fabbri, A.G., 1989. Putting AI to work in geoscience. Episodes 12(1): 10-17.

Yu, B., Xie, X., 1985. Fuzzy cluster analysis in geochemical exploration. Journal of Geochemical Exploration 23(3): 281-291.

Yumul, G.P., Dimalanta, C.B., Maglambayan, V.B., Tamayo, R.A., 2003. Mineralization controls in island arc settings: insights from Philippine metallic deposits. Gondwana Research 6(4): 767-776.

Yusta, I., Velasco, F., Herrero, J.-M., 1998. Anomaly threshold estimation and data normalization using EDA statistics: application to lithogeochemical exploration in Lower Cretaceous Zn±Pb carbonate-hosted deposits, Northern Spain. Applied Geochemistry 13(4): 421-439.

Zadeh, L.A., 1965. Fuzzy sets. IEEE Information and Control 8(3): 338-353.

Zadeh, L.A., 1973. Outline of a new approach to the analysis of complex systems and decision processes. IEEE Transactions on Systems, Man and Cybernetics 3(1): 28-44.

Zadeh, L.A., 1983. The role of fuzzy logic in the management of uncertainty. Fuzzy Sets and Systems 11(3): 199-227.

Zadeh, L.A., 1986. A simple view of the Dempster-Shafer theory of evidence and its implication for the rule of combination. AI Magazine, vol. 7, no. 2, pp. 85-90.

Zimmerman, H.-J., 1991. Fuzzy Set Theory – and Its Applications, 2nd edn., Kluwer Academic Publisher, Dordrecht.

Zimmerman, H.-J., Zysno, P., 1980. Latent connectives in human decision making. Fuzzy Sets and Systems 4(1): 37-51.

ONLINE SOURCES

American Geological Institute, 2007. Glossary of Geology. http://www.agiweb.org/pubs/glossary/

BCGS, 2007. MINFILE Mineral Inventory. British Columbia Geological Survey (BCGS). http://www.empr.gov.bc.ca/Mining/Geolsurv/

Geoscience Data Repository, 2006a. Canadian Aeromagnetic Data Base. Geological Survey of Canada, Earth Sciences Sector, Natural Resources Canada, Government of Canada. http://gdr.nrcan.gc.ca/index_e.php.

Geoscience Data Repository, 2006b. Canadian Gravity Data Base. Geological Survey of Canada, Earth Sciences Sector, Natural Resources Canada, Government of Canada. http://gdr.nrcan.gc.ca/index_e.php.

Paul, C., Mason, W.M., McCaffrey, D., Fox, S.A., 2003. What should we do about missing data? A case study using logistic regression with missing data on a single covariate. Paper CCPR-028-03, California Center for Population Research, On-line Working Paper Series. http://repositories.cdlib.org/ccpr/olwp/CCPR-028-03. Date last accessed 10 July 2007.

Oxford University Press, 2007. Oxford English Dictionary Online. http://dictionary.oed.com/

Wikimedia Foundation, 2007. Wiktionary. http://en.wiktionary.org/wiki/Wiktionary:Main_Page

AUTHOR INDEX*

Abarca, *317*, *330*
Abarca-Hernandez, *330*
Adams, 158, *322*
Adcock, *320*
Agterberg, VII, 7, 11, 17, 92, 152, 154, 190, 271, 272, *311*, *313*, *317*, *318*, *320*, *321*, *326*, *332*
Alesheikh, 200, *323*
Alexandre, *314*
Allison, 265, *311*
Almeida-Filho, *327*
Alms, 32, *311*
An, 218, 227, 229, 230, 231, 274, 275, 285, *312*, *327*
Ansoult, 30, *312*
Antony, *334*
Arribas, 169, *312*, *322*
Ascough, 308, *317*
Ash, *318*
Atkin, 53, 57, *314*
Aurelio, 159, *312*
Ayer, *322*

Babu Madhavan, *334*
Bailly, *331*
Bajc, *317*
Balce, 64, 159, 168, *327*
Baldi, 125, *312*
Ballantyne, *317*
Balovnev, *311*
Banglani, *334*
Banville, *320*
Bao, 92, *335*
Bárdossy, 217, *312*
Barnes, *315*
Barrier, *312*
Barritt, *316*
Bartier, 39, *312*
Basilevsky, 79, *312*
Baybayan, 159, *312*
Beane, 169, *334*
Beauchamp, 293, *312*
Beaudry, *332*
Beaumont-Smith, *331*
Becken, 32, *325*
Bedell, *318*, *319*
Begovich, *312*
Bell, 27, *323*
Bellehumeur, 11, 85, *313*
Bellot, 185, *313*
Berger, 13, 169, 246, *313*
Berman, 162, 172, *313*
Bernier, *322*
Bhaumik, 51, 66, *324*
Bierlein, 162, 186, *313*
Billa, 169, 208, *313*, *331*
Blenkinsop, 152, 185, *313*, *320*, *324*
Bloom, 169, *318*
Boeglin, *334*
Bölviken, 7, 86, *313*
Bonham, 158, *332*

* Page numbers in italics refer to the list of references and online sources, pp. 311-337.

Bonham-Carter, VII, 7, 11, 14, 15, 17, 18, 33, 38, 44, 116, 117, 118, 119, 121, 123, 124, 132, 134, 137, 144, 162, 165, 173, 190, 196, 199, 218, 229, 271, 272, 274, 280, *311*, *312*, *313*, *314*, *317*, *318*, *319*, *320*, *321*, *326*, *328*, *330*, *335*
Bonnin, *325*
Bonvalots, *312*
Boots, 149, 150, 151, 152, 156, 254, 259, 260, 261, 308, *314*
Boroushaki, 201, *314*
Botbol, *314*, *326*
Bouchot, *313*, *314*
Bougrain, *314*
Bounessah, 53, 57, *314*
Bowen, *326*
Bradley, *322*
Bradshaw, 6, *314*
Brand, 10, *314*
Breslow, 261, 297, *314*
Breunig, 32, *311*, *314*
Briole, *312*
Brooks, 236, *324*, *335*
Brown, 20, 297, 302, *315*
Bull, 293, *315*
Burrough, VII, 39, 40, *315*
Bustillo, *334*
Butenuth, 85, *335*
Butt, 6, 9, *315*

Cain, 261, 297, *314*
Câmara, *327*
Campbell, 53, 307, *315*, *333*
Cardoso Fonseca, *330*
Carlson, 14, 17, 152, 153, 154, 155, *315*
Carr, 86, *315*
Carranza, 11, 14, 15, 17, 64, 116, 117, 121, 123, 128, 137, 141, 144, 155, 162, 169, 173, 175, 182, 199, 210, 217, 221, 229, 236, 239, 241, 245, 258, 259, 260, 261, 269, 271, 272, 274, 276, 284, 295, 307, 309, *315*, *316*, *329*
Cassard, *313*, *314*, *331*
Castro, 309, *315*
Chambers, 53, *316*
Chapman, *320*, *328*
Chatterjee, *318*
Cheng, 11, 85, 87, 89, 90, 91, 92, 94, 98, 107, 113, 114, 136, 144, 152, *311*, *316*, *317*, *318*, *328*, *335*
Cherry, *330*
Chi, *318*, *328*
Chica-Olmo, 85, *317*, *324*, *328*, *330*
Chikambwe, *316*
Chinn, 308, *317*
Chork, 53, 293, *317*, *320*
Chorlton, *329*
Chung, 17, 190, 227, 229, 250, 271, 274, 276, *311*, *314*, *317*, *318*, *319*, *327*, *333*
Clarke, 38, 103, 293, *318*
Cleveland, *316*
Cobbold, *331*
Codd, 32, *318*
Cohen, 92, 100, 114, 227, *318*, *332*
Colman-Sadd, 32, *318*
Coltelli, *312*
Concepcion, *312*
Connors, *329*
Cook, 53, 159, *318*, *322*, *323*
Cooke, 169, *318*
Coolbaugh, 186, *318*, *319*
Coombs, *325*
Cooney, *326*
Cox, 13, 150, 246, 265, *319*, *324*
Craig, *322*
Cremers, *311*, *314*

Dahlberg, *331*
Dahn, *322*
Date, *319*
Davis, 42, 78, 123, 176, 293, 294, *311*, *319*
De Araújo, 200, 208, *319*
De Barros Silva, 11, 37, 142, 144, *331*
De Bruin, *325*
De Nijs, *334*
De Wit, 196, 199, *334*

Demicco, 210, *319*
Dempster, 20, 227, 228, 229, 232, 273, 280, 285, 292, *319*, *333*, *336*
Desjardins, *326*
Diepenbroek, 32, *319*
Diggle, 148, 254, 259, 260, 261, 308, *319*, *331*
Dimalanta, *336*
Dimitrakopoulos, 251, *325*
Domingue, *323*
Drescher, *331*
Drew, 13, 169, 246, *313*
Duda, 307, *315*, *319*

Eastman, 200, *319*
Eddy, 210, *319*
Eilu, *327*
Einaudi, 169, *327*
Elliott, *315*, *320*, *328*
Ellwood, *314*
Elsenbroek, *325*
Engi, *331*
Epema, *335*
Eriksson, *324*
Evans, 6, *319*, *323*

Faassen, *316*
Fabbri, 17, 190, 227, 229, 250, 271, 274, 276, 308, *311*, *313*, *318*, *319*, *335*
Fdez-Valdivia, 38, *320*
Feder, 92, 152, 154, *313*, *319*
Fidell, 293, 294, *333*
Filzmoser, 51, *330*
Fisher, 293, *319*
Fletcher, 53, 142, *315*, *318*, *320*
Folger, *325*
Ford, 152, *320*
Fordyce, 38, *320*
Fox, *337*
Francheteau, *325*
Frank, 20, *335*
Fry, 148, 150, 152, 155, 156, 158, 161, 166, 170, *320*, *321*, *334*
Fumerton, *322*

Gallant, *335*
Galley, *330*, *335*
Garbrecht, 126, 127, *320*, *326*
Garcia, 38, *312*, *320*
Gardoll, 272, *320*, *321*, *323*, *327*
Garrett, 10, 11, 38, 53, 66, *320*, *330*
Garson, 13, 158, 246, *327*
Gaulon, *312*
Gedeon, *315*
Geldart, *334*
George, 123, 124, 137, *320*
Getis, 149, 150, 151, 152, 156, 254, 259, 260, 261, 308, *314*
Glen, 162, *329*
Goldfarb, *313*, *321*
Gonçalves, 92, *320*
Gonzalez, *314*
Good, 53, 201, *320*
Goodchild, *325*
Goodfellow, 11, 116, 117, 121, 134, 144, *314*, *317*, *320*
Goodman, 163, *320*
Goovaerts, 85, *320*
Gott, *314*
Götze, *314*
Govett, 9, 10, 293, *320*
Goyal, *334*
Green, *320*, *332*
Gregoire, *318*
Grobe, *319*
Groves, 12, 186, *313*, *315*, *319*, *320*, *321*, *323*, *327*, *335*
Grunsky, 8, 11, 12, 53, *317*, *321*, *322*
Guillou-Frottier, *313*
Gwinner, *312*

Haklay, 44, *321*
Halberg, *333*
Halden, *331*
Hale, 9, 11, 14, 15, 17, 116, 117, 121, 123, 137, 141, 144, 162, 169, 173, 182, 210, 221, 236, 239, 245, 260, 272, 276, 295, *315*, *316*, *321*, *329*
Hall, *315*, *317*, *318*, *322*, *328*, *335*
Hampel, 53, *321*
Hanna, 155, *321*
Harbaugh, 7, *321*
Harris, VII, 7, 11, 20, 116, 196, 208, 210, 273, 293, 308, *314*, *321*, *322*, *328*
Hart, *319*
Haslett, 20, *322*
Hawkes, 11, 115, 121, 142, 144, *322*, *331*
Heather, *322*
Hecker, *316*
Hedenquist, 158, 159, 169, *312*, *322*, *327*, *335*
Heiner, *328*
Hellendoorn, 224, *322*
Hengl, 256, 259, *322*
Henley, 158, 159, *322*
Heran, *323*
Herrero, *336*
Higgins, *324*
Hildenbrand, *325*
Hill, 160, *322*, *323*, *324*, *325*, *331*, *332*, *335*
Hirano, 125, *322*
Hoaglin, 53, 56, *323*, *334*
Hodge, 189, *323*
Hodkiewicz, 152, *323*, *335*
Hollister, *315*
Holroyd, 27, *323*
Holyland, *321*
Honarmand, 210, *329*
Hoover, *323*
Hosmer, 265, *323*
Hosseinali, 200, *323*
Howarth, 9, 10, 20, 53, 78, 123, 293, *323*, *326*, *332*
Hronsky, 12, 186, *323*
Huber, 53, *323*
Huijbregts, 87, *324*
Huntington, *332*

Itaya, *312*

Jaeger, 159, *323*
Jébrak, *313*
Jefferson, *319*
Jenson, *323*
Jentzsch, *311*
JICA-MMAJ, 64, 159, *323*
Jimenez-Espinosa, 85, *324*
Jin, *319*
Jockel, *331*
Johnson, 56, *324*
Jones, 127, *324*
Jordan, 11, *324*
Joseph, 51, 66, *324*
Jössang, *313*
Journel, 87, *324*

Kadzviti, 185, *313*
Kane, *312*, *320*
Karger, 11, *324*
Katsev, 7, *325*
Katz, 307, *324*
Kauranne, 6, 9, *324*
Kaye, 91, *324*
Kearey, 15, *324*
Keating, *318*
Keith, *331*
Keller, 39, *312*
Kilby, *321*
Kim, 227, *324*
King, 261, 297, *324*
Kleiner, *316*
Klinkenberg, 86, *324*
Klir, 210, *319*
Knox-Robinson, 210, *319*, *320*, *321*, *324*
Kork, 310, *326*
Kotlyar, *325*
Kotz, 56, *324*

Kouda, 8, 293, *332*
Kramar, 10, *324*
Kreuzer, 155, *324*
Kürzl, 53, 59, *324*
Kwon, *328*
Kyem, *319*

Laing, 185, *324*
Lalor, 10, 11, *320*
Lang, *322*
Laxton, 32, *325*
Le Pichon, 158, *325*
Leach, 64, 159, 168, 186, *327*
Lee, 227, *325*
Lees, *313*
Leishman, *332*
Lemeshow, 265, *323*
Levinson, 9, 10, 103, *325*
Li, 114, 173, 175, 176, 180, *325*
Lieberman, *331*
Likkason, 229, 274, *325*
Lindgren, 6, *325*
Lips, *313*, *314*, *331*
Lombard, 11, *325*
Loodts, *312*
Lopez-Blanco, 32, *325*
Loveland, 51, *326*
Lovering, 6, *325*
Ludington, 98, *325*
Luo, 240, 251, *325*

Ma, *325*
Macedo, 200, 208, *319*
MacIntyre, *326*
Maglambayan, *336*
Maguire, VII, *325*
Malczewski, 201, *314*
Maling, 29, 37, *325*
Mandelbrot, 7, 24, 86, 87, 89, 152, 153, *325*, *326*
Mandl, 159, *326*
Mangaoang, *316*
Marcotte, *313*
Marsella, *312*
Marsh, *313*
Marshall, *323*
Martz, 126, 127, *320*, *326*
Mason, *337*
Massey, 259, *321*, *326*
Master, *335*
Mateus, *320*
Mather, 37, *326*
Matos, 159, *312*
Mattiske, 32, *326*
Mazzucchelli, 53, 293, *315*, *317*
McCaffrey, *337*
McCammon, 307, 310, *313*, *314*, *318*, *326*
McCarthy, 6, *325*
McDonnell, 39, 40, *315*
McGrath, 51, *326*
McMaster, 38, *326*
Middelburg, *334*
Miesch, 51, 66, *326*
Mihalasky, 44, *326*
Miko, *328*
Mileusnic, *328*
Mineter, 20, 38, *326*, *327*, *333*
Minor, *319*
Mitchell, 13, 64, 158, 159, 168, 186, 246, *327*
Moon, 11, 115, 127, 144, 229, 274, *312*, *318*, *327*
Moore, 210, 229, 274, *333*, *335*
Moreira, 200, *327*
Morrison, *324*
Mossotti, *325*
Mosteller, *323*
Muda, 142, *320*
Müller, *312*
Muntean, 169, *327*
Murphy, *313*

Nelson, 185, *327*
Nilsson, *319*
Nolan, *318*
North, *315*

Northover, *313*
Nykänen, 210, 256, *327*

Ojala, *321*, *327*
Okada, *312*
Oliveira, *320*
Oxborrow, 32, *327*

Pan, VII, 8, 116, 293, 308, *321*, *327*, *328*
Panagapko, *322*
Pandalai, *331*
Parasnis, 15, *328*
Pardo-Igúzquiza, 85, *328*
Park, 162, *315*, *328*, *333*
Parker, *322*
Paul, 265, *337*
Peh, 116, *328*
Pendharkar, 217, *328*
Pengda, 92, 152, 154, *335*
Peter, 44, *317*, *328*
Peters, *324*
Pirajno, 6, 13, 246, *328*
Pistocchi, *335*
Plant, 9, *321*
Plimer, 10, *328*
Polikarpochkin, 11, 115, 127, *328*
Pollard, 161, *331*
Ponce, 162, *329*
Portefield, *328*
Porwal, 186, 210, 235, 297, 302, 307, *329*
Pouch, 32, *329*
Prelat, *329*
Prol-Ledesma, *329*
Puglis, *312*

Quigley, *323*

Raines, 256, 284, *319*, *327*, *329*
Rand, *315*
Rangin, *312*
Ranjbar, 210, *329*
Rantitsch, 7, 10, 92, *330*
Raper, 24, *330*
Rasmussen, *333*
Ratha, *334*
Reardon, *318*
Reddy, 308, *330*
Reimann, VII, 21, 51, 53, 57, 59, 60, 66, 72, 73, 78, *330*
Reinke, *319*
Reis, 85, *330*
Remy, *312*
Rencz, *312*, *314*
Reynolds, *322*
Rhind, *325*
Richards, 6, *325*, *330*
Rigol, *317*, *330*
Rigol-Sanchez, *330*
Riley, *333*
Riolo, *315*
Robb, 6, 13, 158, 246, *330*
Roberts, 13, 246, *330*
Robinove, 196, *330*
Robinson, 212, *315*, *330*
Rock, 53, 265, *320*, *323*, *330*
Rogers, *314*
Rogge, 210, 229, *331*
Ronchetti, *321*
Rose, 9, 10, 11, 103, 115, 121, *331*
Rossello, *331*
Rousseeuw, 293, *317*, *321*
Rowlingson, 254, 259, 260, 308, *331*
Roy, 285, *331*
Rubin, 265, *331*

Saaty, 199, 202, 217, *328*, *331*
Sabins, 15, *331*
Sahoo, *331*
Sahu, 7, *331*
Salminen, *324*
Sanborn-Barrie, 210, 273, *321*, *322*
Sandomirsky, 11, *324*
Sawatzky, *319*
Sawkins, 13, 158, 246, *331*
Schill, 261, 297, *331*
Schindler, *319*

Schlitzer, *319*
Schmatz, 32, *331*
Schmidt, *314*
Schuenemeyer, 189, *331*
Segall, 161, *331*
Seidemann, *314*
Seoane, 11, 37, 142, 144, *331*
Shafer, 227, 273, 280, *324*, *331*, *333*, *336*
Sheahan, *330*
Shemang, *325*
Shen, 92, 100, 114, *332*
Sheriff, *334*
Shevenell, *319*
Shi, *325*
Shumilov, *314*
Sibson, 160, 185, 299, *332*
Sides, 210, *329*
Siegel, 163, *332*
Sieger, *319*
Siehl, *311*, *314*
Sillitoe, 158, *332*
Sim, 92, *332*
Simpson, 162, *320*, *332*
Sinclair, 10, 57, 68, 129, *332*
Sinding-Larsen, 10, 78, 123, *311*, *314*, *317*, *318*, *323*, *326*
Singer, 8, 13, 159, 246, 293, *319*, *332*
Sinha Roy, *334*
Skabar, 271, 302, *332*, *333*
Skulski, *322*
Sloan, 38, *333*
Smee, 12, *321*, *333*
Smith, 6, 53, *315*, *322*, *333*
Snell, 265, *319*
Snyder, 29, *333*
Soille, *312*
Sousa, *330*
Spatz, 15, *333*
Srivastava, *323*
Stahel, *321*
Stanley, 92, *321*, *333*
Stein, 256, *313*, *314*, *333*
Stendal, *333*
Stensgaard, 17, 260, 269, *333*
Stern, 31, *333*
Stokke, *313*
Strahler, 128, *333*
Stuart, 44, *328*
Stubley, 155, *333*
Suh, *325*
Sumintadireja, *316*
Sutherland, *319*
Swain, 227, *324*, *325*

Tabachnick, 293, 294, *333*
Tamayo, *336*
Tangestani, 210, 229, 274, *333*
Taranik, *319*
Tardy, 9, *334*
Telford, 15, *334*
Tennant, 57, *334*
Thiart, 196, 199, *334*
Thole, 218, *334*
Thomas, 224, *322*
Timm, *331*
Titley, 169, *334*
Toledano, *319*
Tosdal, 6, *330*
Tourliére, *313*
Triantaphyllou, 217, *334*
Trochimczyk, 79, *334*
Tukey, 52, 53, 56, 57, 59, *316*, *323*, *334*
Turcotte, 86, *334*
Turner, 53, *323*

UNDP, 159, *334*
Unwin, *322*

Van der Meer, *316*
Van der Meijde, *316*
Van Gaans, *334*
Van Ruitenbeek, *316*
Varnes, 196, *334*
Vearncombe, 14, 17, 155, *334*
Velasco, *336*
Veldkamp, *335*

Velleman, 53, *334*
Venkataraman, *334*
Vieux, 44, *334*
Villers-Ruiz, 32, *325*
Vistelius, 51, *334*
Vriend, 10, *334*

Wackernagel, 85, 87, *335*
Walley, 227, 285, *335*
Walls, *323*
Watson, 39, *335*
Webb, *331*
Wefer, *319*
Wei, 92, 152, 154, *335*
Weiberg, 152, *335*
Weinberg, *313*, *323*
Welch, *322*
White, 57, 158, 161, *327*, *334*, *335*
Whitten, 236, *335*
Wibowo, *316*
Wielemaker, 32, *335*
Wilkinson, *322*
Wills, *322*
Winter, 20, *335*
Woldai, 273, *316*, *335*
Wolf, *312*
Wright, 229, 274, *311*, *313*, *335*

Xie, 10, 92, *335*, *336*
Xu, 86, 114, *317*, *335*

Yatabe, 308, *335*
Yu, 10, 53, 57, 58, 59, 293, *336*
Yumul, 159, *336*
Yun, *320*, *321*
Yusta, 53, 57, 58, 59, 293, *336*

Zadeh, 10, 210, 218, 227, *336*
Zeegers, 9, *315*
Zeigler, *320*
Zeng, 261, 297, *324*
Zhang, *324*
Zimmerman, 212, 218, 220, *336*
Zimmermann, *334*
Zurcher, *321*
Zysno, 220, *334*, *336*

SUBJECT INDEX

Aggregation functions, 192
Alkalic porphyry Cu-Au, 254
Analytical hierarchy process, 199
    pairwise comparison of criteria, 200
Anomaly
    anthropogenic, 3, 11
    downstream dilution, 115
    geochemical, 3
    geogenic, 3, 11
    geological, 3
    geophysical, 3
    permissive regions, 3
    prospective areas, 3
    significant, 3, 9, 24
Artificial intelligence, 307
Artificial neural networks, 251
Attribute
    classification, 284
    map, 18
    spatial, 28
    table, 18
    temporal, 28
    thematic, 28

Boolean
    evidential map, 196
    logic, 196
    operators, 34, 196
    search query, 42
Boxplot, 53
    5-number summary statistics, 56
    hinge width, 55
    inter-quartile range, 55
    lower hinge, 55
    lower inner fence, 55
    lower outer fence, 55
    lower whisker, 56
    threshold, 57
    upper hinge, 55
    upper inner fence, 56
    upper outer fence, 56
    upper whisker, 56
Brownfields, 249
Buffer analysis, 162

Cartographic representation, 4
Catchment basin, 11
    areal proportions of rock units, 116, 117, 127
    drainage sinuosity, 11, 37
Cluster analysis, 10, 78, 123
    fuzzy, 10
Conditional independence, 294
Conditional probability, 171, 276, 277
Consistency index, 203
Consistency ratio, 203
Continuous fields, 98, 191, 217, 279, 284, 307
Coordinate system, 29
    ellipsoid, 29
    projection, 29
    spheroid, 29
Correlation
    Pearson, 10, 173
    Spearman, 123, 137

Data analysis, 18
    confirmatory, 53
    exploratory, 53
Data manipulation, 15

Data transformation, 15
- area-to-area, 38, 41
- area-to-point, 38
- gridding methods, 40
- line-to-area, 38, 41
- line-to-surface, 38
- non-interpolative, 38
- normalisation, 51
- point-to-area, 38, 41
- point-to-surface, 38, 41
- raster-to-vector, 38
- resampling, 37
- spatial interpolation, 39
- standardisation, 58
- surface-to-area, 38
- triangulation methods, 40
- vector-to-raster, 38

Database, 15, 20
- hierarchical, 32
- management, 32
- network, 32
- object-oriented, 32
- relational, 32

Digital elevation model, 42
Discrete fields, 191
Discrete geo-objects, 191, 217
Discriminant analysis, 10, 78, 293
Discriminant function coefficients, 293

Eigenvalue, 202
Eigenvector, 201
Element ratios, 10
Epithermal Au, 64
Euclidean distance, 162
Euclidean geometry, 86
Evidence
- binary, 194
- multi-class, 205

Evidential belief theory, 227, 273
- belief, 227
- disbelief, 227
- plausibility, 227
- uncertainty, 227

Factor analysis, 78, 123
Fault/fracture
- dilational jog, 185, 200, 299
- extension, 185
- shear, 185

Fault/fracture density, 37
Fractal, 7, 24, 86
- box-counting method, 88
- dimension, 11, 86
- geometry, 7, 86
- Hausdorff-Besicovitch dimension, 86, 152
- monofractal, 92
- multifractal, 92
- number-area relation, 86
- perimeter-area relation, 87
- power-law relation, 89, 92, 152, 153
- self-affine, 86
- self-similar, 86

Frequency distribution, 9, 11
- cumulative, 165
- relative, 163

Fuzzy logic, 210
Fuzzy membership, 211
Fuzzy membership functions, 212
- linear, 212
- logistic, 237
- non-linear, 213, 214

Fuzzy set theory, 10, 210

Geochemical
- anomaly, 9
- background, 9
- dispersion, 6
- landscape, 7, 9, 11, 24
- residuals, 5
- threshold, 9

Geographic expert systems, 307
Geometric correction, 37
Geo-objects, 5, 23, 224
- definition-limited, 24, 253
- discretisation of, 28
- geometry, 24

linear, 24, 27
point, 24, 27
polygonal, 24, 27
quantisation of, 28
sampling-limited, 24, 253
Georeference, 27
Geothermal, 155
Goodness-of-fit, 8, 17, 117, 265
Greenfields, 189

Histogram, 10, 54
map, 36
table, 192
Hydrothermal alteration, 3, 15, 64, 307
Hydrothermal mineralisation, 160, 237

Inference network, 20, 196
Boolean, 196
calibration, 204, 235
combining belief functions, 232
fuzzy, 210, 222
Intersect operation, 45
cross, 45, 127, 165
spatial join, 45
zonal statistics, 165
Inverse distance, 237

Jack-knife cross-validation, 271
Join operation
table, 62, 127

Kernel, 42, 99
Knowledge elicitation, 189
Knowledge-guided data-driven technique, 236, 284
Kriging, 99

Logical functions, 20, 191
Logistic regression, 10, 18, 251, 261, 265
Log-normal distribution, 51

Map
binary, 88, 90
Boolean, 199
multi-class, 189, 191
Map overlay operations, 18, 44, 127, 176, 191, 263, 295
unique conditions map, 46
unique polygons map, 46
Mapping, 3, 5
Matrix
consistent, 202
correlation, 78, 79, 123, 137
covariance, 78, 79, 123
pairwise comparison, 200
singularity, 117
Median absolute deviation, 57
Mineral deposit model, 13, 186
Mineral exploration
area selection, 3
reserve definition, 3, 7
resource estimation, 7
resource evaluation, 3
target generation, 3, 6, 7, 12, 17
Mineral favourability. *See* Mineral prospectivity
Mineral potential. *See* Mineral prospectivity
Mineral prospectivity, 12
conceptual model, 12, 147
evidential features, 14
predictive model parameters, 14
predictor maps, 14
recognition criteria, 14, 148, 184, 187, 190, 249
Mineralisation controls, 190
heat-source, 15, 237
host, 160
structural, 15, 148, 237
Model, 4
calibration, 17, 204, 235, 279
conceptual, 12
cross-validation, 16, 191
exploration, 3, 186, *316*
fitting-rate, 17, 270
prediction-rate, 17, 191, 270
predictive, 4, 190

prescriptive, 12, 190
Modeling
bivariate, 76, 250
data-driven, 8, 249
deduction, 5
deterministic, 7
dynamic, 8
empirical, 7
four-dimensional, 9
heuristic, 8
induction, 5
knowledge-driven, 8, 189
mathematical, 8
mechanistic, 7
multivariate, 59, 78, 123, 240, 250
one-dimensional, 9
predictive, 4
probabilistic, 8
qualitative empirical, 8
quantitative empirical, 8
static, 8
statistical, 8
stochastic, 7
symbolic, 8
theoretical, 7
three-dimensional, 9
two-dimensional, 9
Multivariate signature, 83, 268

Nearest neighbour, 40
reflexive, 150
Nearest neighbour distance, 149
order, 149
Neighbourhood operations
buffering, 42
dilation, 41
filtering, 42
search functions, 42
Node, 24
Normal distribution, 51

Outliers, 56

Pathfinder, 9, 72, 107, 122
Pixel, 27
grid cell, 253
unit cell, 253
Point pattern
clustered, 148
complete spatial randomness, 148
fractal, 152
Fry plots, 155
measures of arrangement, 150
measures of dispersion, 149, 254
Poisson, 149, 153, 162
regular, 148
Porphyry Cu, 169
Principal components analysis, 10, 78, 123, 136
Probability distribution, 7, 92, 227
Probability plot, 10, 57
Proximity analysis, 27, 162
Pythagorean theorem, 149

Q-Q plot, 57

Regression analysis, 10, 78, 117, 123
Robust statistic, 53
Rose diagram, 36

Sampling
prospective, 254
retrospective, 254
Sampling density, 7, 64, 99, 118, 253
Scatterplot, 76
one-dimensional, 53
Scavenging, 10, 51, 75, 77, 78, 79, 104, 137
Spatial association
bivariate, 250
multivariate, 250
negative, 163, 175
positive, 163, 175, 239
Spatial data model, 23
Delaunay triangulation, 26
Dirichlet polygon, 26
raster, 27

Thiessen polygon, 26
topology, 25
triangulated irregular networks (TIN), 26
vector, 24
Voronoi polygon, 26
Spatial independence, 51, 163
Spatial query, 33
atomic condition, 34
composite condition, 34
Spatial resolution, 7, 27, 87, 125, 253
Standardised data, 58, 79
Stanley, 92, *321*, *333*
Statistical test
chi-square, 163
Hosmer-Lemeshow, 265
Kolmogorov-Smirnov, 163
t-test, 176
Wald, 265
Stream order, 11, 28, 127, *315*

Topological dimension, 86
Topology, 20, 25
surface, 26

Upper confidence band, 163

Variable
binary, 29
categorical, 29
continuous, 20, 28, 35
dependent, 117
discrete, 28
explanatory, 4, 249
independent, 117
interval, 28
nominal, 28
ordinal, 28
predictor, 4, 7, 8, 10, 18, 20, 249
qualitative, 28
quantitative, 28, 35
ratio, 28
target, 4, 7, 8, 18, 20
ternary, 29
Vector
spaghetti model, 24
topological model, 25

Weighted moving average, 41, 85, 98
Weights-of-evidence, 250, 280, 295, 307,

Zone of influence, 39, 85, 98, 116